Exploring the World through

SOCIAL STATISTICS

Patricia F. Case

Barbara Thomas Coventry

The University of Toledo

Kendall Hunt
publishing company

Cover image © Shutterstock, Inc.

Kendall Hunt
publishing company

www.kendallhunt.com
Send all inquiries to:
4050 Westmark Drive
Dubuque, IA 52004-1840

Contents

Chapter One

Introduction

Why Statistics?

As a social scientist in training, you may wonder why you have to study statistics! The nature of sociology is to study social behaviors. Translating these behaviors into numbers may seem nonsensical at the moment. Statistics, however, are an important tool to allow us to understand group behaviors. They give us the ability to distinguish between perceptions of behaviors and behavioral practices.

We are all social scientists, each in our own ways. We collect data as we go along in our lives. The data that we collect through our daily endeavors supports the world views that we hold. For example, a woman who has grown up with abuse and marries an abusive spouse may believe that all men are abusive. In her experiences, she has data to support this assumption. Likewise, a person who has had only negative encounters with members of another racial or ethnic group may develop a belief system that supports racism against this group, believing that all members of the group are the same. Our experiences fuel our interpretations of the world. The problem is that our personal experiences are limited.

The cultural views that many people have about the average alcoholic is a good example of how statistics can help us to gain a clearer understanding of reality. Culturally, we tend to view the alcoholic as poor, perhaps homeless, dirty and disheveled, unemployed, and uneducated. We may assume that the average alcoholic is nonwhite as well. When we picture this individual, we may think of a man on the street drinking alcohol from a bottle enclosed in a brown paper bag. When researchers started collecting and using statistics to analyze data on drinking behaviors, they discovered that this commonly held belief about alcoholics is incorrect. The wealthy drink more than the poor. The educated drink more than the uneducated. White men drink more than any other group (Klatsky et al. 1977)[1]. These early studies of drinking behaviors led social scientists to an understanding of functional alcoholism, a concept that was unthought-of before. It is possible to be addicted to alcohol and to hold down a job, support a family, pay a mortgage, and be a productive member of society. Using statistics to understand patterns of behavior allows us to dispel myths and stereotypes.

This book will provide you with a guide to understanding statistics and how they can be used to make sense of the social world. Remember, we are not mathematicians. Rather, we use mathematics as a tool to help understand the social patterns that contribute to our experiences of the world we live in. This course is designed to introduce you to the application of statistical analysis to social data. You will learn the difference between outcome data and attitudinal data. You will learn the various methods for sorting, describing, and exploring data, as well as hypothesis testing. You will learn how to distinguish the types of data that are used and how to select the appropriate statistical method to analyze the final product. You will also learn how to use the statistical package for the social sciences (SPSS), the computer program that is favored by many social scientists for analyzing data. Each chapter in this book will provide exercises for you to use to learn the

---

[1] Klatsky, Arthur L., Gary D. Friedman, Abraham B. Siegelaub And Marie J. Gérard (1977) Alcohol consumption among white, black, or oriental men and women: Kaiser-Permanente Multiphasic Health Examination Data. American Journal of Epidemiology, 105(4): 311–323.

techniques required for class. These exercises include math, SPSS skills, and interpretation questions. You will learn how we gather, analyze, and interpret data in the social sciences. You will work with real data that have been collected using a national telephone survey and are used by professional social scientists. In other words, the data provided in this book to be used with the SPSS exercises during this course is the same data that you will use if you become a professional sociologist postgraduation. You will also learn the basics of SPSS, which is the standard for sociology. At the end of this course, you should feel comfortable using not only math, but also, computer resources to manipulate and analyze data.

On Your Way to Analyzing Data

The first step in any research project is to formulate your hypothesis, or your research question. A **research question** is exploratory. We use a question when we don't know enough about the situation that we are interested in studying to make a prediction of a relationship. For example, we might ask: Does income influence how satisfied we are with our lives? Or: Does salary influence how satisfied we are with our jobs? We use a question when we aren't sure enough about the outcomes to make an educated guess about what the data will show. Typically, we use a question when we are planning a **qualitative study.** Qualitative studies are studies that use small samples and focus on the quality of the experiences of our subjects. We use interviews or focus groups to collect these data. We may use some statistical methods to analyze qualitative data, but the analysis is limited to **descriptive statistics.** (Descriptive statistics will be discussed later in this chapter.)

We use a **hypothesis** when we believe that we know enough about the data to make a prediction of the relationships between the variables. A hypothesis must be measurable and testable. We typically predict a causal relationship. In other words, we say that we believe that X causes Y. For example, we might hypothesize that salary and job satisfaction are related. Our hypothesis might be: As salary increases, job satisfaction is likely to increase. A hypothesis predicts a relationship between the variables (salary and job satisfaction) and it predicts a relationship between these variables. This example of a hypothesis predicts a **positive relationship** between the variables. A positive relationship means that the variables move in the same direction. Thus, in our example hypothesis, as salary increases, job satisfaction increases. Another example of a hypothesis that predicts a positive relationship is: As education increases, income is also likely to increase.

We also predict **negative relationships** between variables. This means that the variables move in opposite directions. An example of a hypothesis that predicts a negative relationship is: As people get older, their criminal activity is likely to decrease. This means that the hypothesis predicts that as age goes up, criminality goes down.

Hypotheses are used in **quantitative studies.** Quantitative studies use large samples to test hypotheses that explain the characteristics of a group. Typically, quantitative researchers use surveys to collect data. These surveys can be conducted face-to-face, on the telephone, or via a mailed survey. More recently, researchers have begun to use electronic delivery of surveys though e-mail and online software systems such as Survey-Monkey. Once the data are collected, statistics allow us to test our hypothesis. Was our prediction correct? In the best-case scenario, our statistics always confirm our hypothesis. In reality, we usually sigh in relief when this happens.

Developing a hypothesis always seems problematic to people that are new to research. How do you develop a hypothesis to study? Most researchers are driven by their own interests. Are you interested in social classes and education? If so, you might hypothesize that: As household income decreases, children are less likely to go to college. If your area of interest is families, you might hypothesize that: Parental education is a positive predictor of child education. There are only a few general rules for pursuing your research interests. You should pick a topic that is interesting to you and that you care about. If a topic is uninteresting, you will have difficulty finishing the project. However, the topic should not be something that you care too much about. When you care too much about the outcome, it may be difficult to remain objective. You may overlook important measures that would prove your prediction wrong.

Understanding Data

Variables, Constants, and Constructs

The next step to understanding data analysis is to understand how we collect and store data. The word *variable* was introduced above. A **variable,** put simply, is a measure that varies. An example of a variable might be age. Your age will vary across your lifespan. Yesterday you were younger than you are today and tomorrow you will be older still. Age changes for the individual. Another example of a variable is income. Income varies with the individual and within the population. Most of the data that we collect are collected as variables.

Constructs are developed when we believe that several of our variables are related or that we need several variables to understand the relationships between the variables. For example, think about a study that examines the quality of life of college undergraduates. Our hypothesis might be: As advancement in college (*i.e.,* freshman to senior) increases, satisfaction with living arrangements is likely to increase. However, is how happy undergraduates are with their living arrangements too broad of a concept? There are many factors that influence satisfaction with living arrangements. Does a person live at home? Does he or she have roommates? If so, does the person get along with the roommates? Is the building well maintained? Is the building safe? There are many other factors. We may decide to collapse these various indicators of satisfaction into one measure, which would be a construct. Statistically, there are various methods for creating constructs that we will not go into now. Constructs can be useful for creating one measure that gives an overall assessment of experiences or attitudes when we have several related variables in our data set.

Independent versus Dependent Variables

When we develop a hypothesis, we have at least one independent variable and one dependent variable. The **dependent variable** is the variable that we expect to change, depending on the values of the other variable(s) in our analysis. Take the hypothesis: As education increases, income is also likely to increase. In this hypothesis, we are predicting that income is dependent upon education. Income is the dependent variable. The dependent variable in statistics is referred to as Y.

The **independent variable** is the variable that we believe will cause the change in the dependent variable. In the example hypothesis, the independent variable is education. While it is true that income also effects education, in this hypothesis, we are simply trying to determine whether having a higher education will affect a person's income. We assume that education does not depend upon income. Further, we assume that the level of education that a person has is responsible, at least in part, for the level of income that he or she earns. The independent variable in statistics is referred to as X. When there is more than one independent variable, they are referred to as $X_1, X_2, X_3 \ldots X_i$. (The subscript 'i' after the X means that we continue to number the independent variables until they are all numbered.)

Sometimes the dependent variable is easy to select in a research study. If someone is studying the effect of gender on education, for example, we can easily tell that the researcher believes that education is dependent upon gender, or, in other words, that education is the dependent variable. Being male or female does not depend on the level of education that a person has. In the above example, "As education increases, income is also likely to increase," it may be more difficult for a novice researcher to distinguish between the two. The easiest test for the dependent variable is to ask: Which variable depends on the other to change?

Units of Analysis

Another thing that we need to understand about our data is the **unit of analysis** that we will use to analyze our data. The unit of analysis is what makes up the sample. In the above hypothesis, the individual person is our unit of analysis. We are interested in knowing whether or not the education that a person receives affects the amount of income that he or she earns. When we look at the education within a state, we might choose to use school districts as our unit of analysis. This means that, rather than looking at the education that a person obtains, we might look at average scores on standardized tests or dropout rates for school districts.

In this case, we are less interested in individual outcomes, so the school district becomes our unit of analysis. Another example is that the World Health Organization (WHO) tracks the incidence and prevalence of disease by country. In this case, the country is the unit of analysis.

We select our unit of analysis based on the relationship that we are interested in testing. This book uses data from the General Social Survey (GSS) of 2008. This data set is derived from a national survey that is conducted by telephone. In the GSS, the individual person is the unit of analysis.

Types of Data

There are several different types of variables that we use in data analysis: nominal, ordinal, interval, and ratio. SPSS records interval and ratio data according to a scale. (See the examples from SPSS below.)

Nominal data refers to variables or constants that are complete and mutually exclusive, but the numbers are only a means of counting responses. For example, regarding sex, we assign a 0 to a female and a 1 to a male. These categories are mutually exclusive (no one person can be in both categories), but we could just as easily assign males a 0 and females a 1. The numbers in a nominal variable are placeholders. We use them to allow us to analyze the data in a mathematical equation. The numbers have no true numeric value. We could ask: Do you own a cat or a dog? and assign a 1 for cats and a 2 for dogs. However, this doesn't mean that dogs are twice the value of cats.

Ordinal data are data that can be rank ordered, and using the $>$/$<$ signs makes sense. For example, we can indicate ranking in a variable, such as driving experience. We can say that someone who does not have a license is less-experienced at driving than someone who does have a license. We can also say that someone who has had a license for more than one year is more experienced that someone who has had a license for a year or less. If we assign 1 to a person who does not have a license, 2 to a person who has had a license for one year or less, and 3 to a person who has had a license for more than a year, we can't say that a person who does not have a license is 1/3 as experienced as someone who has had a license for more than one year [1 (no license) divided by 3 (license for more than one year)]. In ordinal data, the numbers have more meaning than in nominal data because we are rank ordering from least to greatest, but, like nominal data, the relationship between the numbers in the variable is not equal to the numeric value of the numbers themselves.

Social scientists prefer to work with interval or ratio data in statistics. In **interval data,** the numbered data are consistent, and the distance between numbers is measurable. The easiest way to understand interval data is to think of a number line. We know that -1 is the same distance from 0 as $+1$. In other words, we can measure the intervals between the numbers and know that they are consistent and that they have true meaning. How do we translate this to social science data that measure behavior? We might ask: On a scale of 1 to 7, in which 1 means unhappy and 7 means very happy, how happy are you? If we ask John and Jane how satisfied they are with their lives, Jane might say very unhappy (1) and John might say somewhat unhappy (2). However, if we interview them independently, we might find that their levels of satisfaction with elements of their lives overall are the same or similar. In contrast, in interval data, we know that everyone understands that 2 is twice 1. Thus, while we don't know what factors influence one person to answer 1 and another person to answer 2, we can be confident that they understand the numerical values that they are given.

Ratio data is similar to interval data. In ratio data, 0 exists and actually has a real value, which means the absence of the value. If we look at an interval variable, such as temperature, in comparison, we see that there is a 0 value but that it doesn't translate into the absence of temperature. With ratio data, 0 means no value. Some examples of ratio variables are grades, age, income, and so on. The distance between the numbers has a true meaning. If John is 21 and Jane is 21, we know that they are both exactly one year older than Tom, who is 20. If Jane makes $40,000 a year and John makes $20,000 a year, then we know that Jane makes double what John makes. See **Figure 1.1** below.

Finally, we have **discrete** and **continuous** variables. Discrete variables are variables whose assigned values are whole numbers. For example, sex is a discrete variable because there can only be a 0 or 1 assigned to the variable. Grade-point average (GPA) would be a continuous variable because the value can fall anywhere within the range of 0 to 4. A GPA of 2.5 can exist, but not a sex of 0.5.

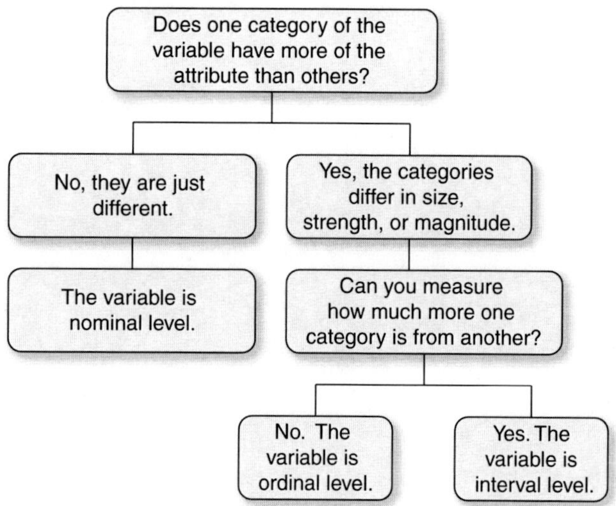

Figure 1.1 Determining the Level of Measurement

GSS and Attitudinal versus Outcome Measures

As you work with the GSS in this course, it is important to note the difference between attitudinal and outcome variables. **Attitudinal variables** measure only attitudes. One variable in the GSS asks whether the respondent believes that whites have the right to segregated neighborhoods (RACSEG). This is an attitudinal variable. It does not ask: Do you live in a segregated neighborhood? Thus, you could use this variable to hypothesize that the more education someone has, the less likely he or she is to believe that whites should be allowed to segregate their neighborhoods. However, you cannot use this variable to test the hypothesis that the less-educated a person is, the more likely it is that he or she will live in a segregated neighborhood.

Another measure in the GSS asks: "How integrated was your high school? (RACHISCH). This is an **outcome variable.** When people respond to this question, they are actually telling you about their high school experiences and not their attitudes about integrated schooling. You could use this variable to test the hypothesis: There is likely to be a strong relationship between family income and integration of high schools. We would expect that the poor and the wealthy would both attend fairly segregated high schools. You could not use this variable to test the hypothesis: As education increases, people are more likely to believe that high schools should be integrated.

The GSS is a secondary data set. This means that its data were collected by others for a different purpose. When you do your own data collection, you can plan your variables around the hypothesis that you want to test, which means that you will know the meaning of each variable. When you use a secondary data set, it is important to familiarize yourself with the variables before you begin to build your hypothesis.

SPSS and Variable Information

You will use SPSS to analyze the data in the GSS. If your professor has been doing this for a while, you will likely hear horror stories of SPSS "back in the day." He or she will cringe while telling stories of main frames and key-punch cards. Fortunately, you are being exposed to SPSS in the age of software such as Windows, in which everything is menu driven. When you open your data file in SPSS, you will see the following screen, or a variation of it depending on the version of SPSS that you are running.

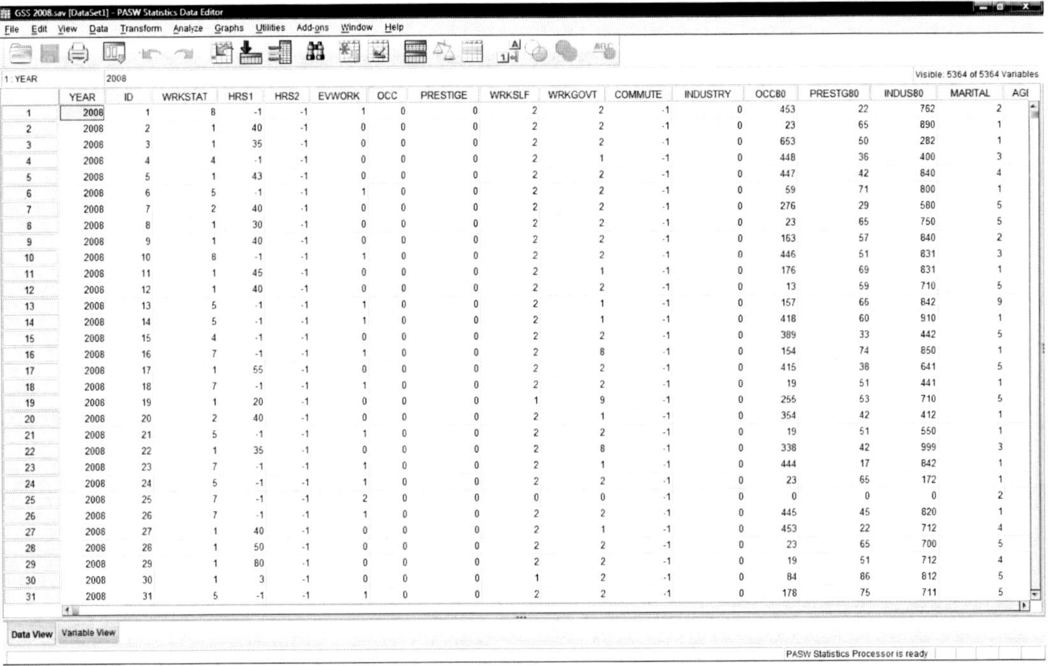

GSS 2008.sav [DataSet1] - PASW Statistics Data Editor

File Edit View Data Transform Analyze Graphs Utilities Add-ons Window Help

1 : YEAR   2008   Visible: 5364 of 5364 Variables

	YEAR	ID	WRKSTAT	HRS1	HRS2	EVWORK	OCC	PRESTIGE	WRKSLF	WRKGOVT	COMMUTE	INDUSTRY	OCC80	PRESTG80	INDUS80	MARITAL	AGE
1	2008	1	8	-1	-1	1	0	0	2	2	-1	0	453	22	762	2	
2	2008	2	1	40	-1	0	0	0	2	2	-1	0	23	65	890	1	
3	2008	3	1	35	-1	0	0	0	2	2	-1	0	653	50	282	1	
4	2008	4	4	-1	-1	0	0	0	2	1	-1	0	448	36	400	3	
5	2008	5	1	43	-1	0	0	0	2	2	-1	0	447	42	840	4	
6	2008	6	5	-1	-1	1	0	0	2	2	-1	0	59	71	800	1	
7	2008	7	2	40	-1	0	0	0	2	2	-1	0	276	29	580	5	
8	2008	8	1	30	-1	0	0	0	2	2	-1	0	23	65	750	5	
9	2008	9	1	40	-1	0	0	0	2	2	-1	0	163	57	840	2	
10	2008	10	8	-1	-1	1	0	0	2	2	-1	0	446	51	831	3	
11	2008	11	1	45	-1	0	0	0	2	1	-1	0	176	69	831	1	
12	2008	12	1	40	-1	0	0	0	2	2	-1	0	13	59	710	5	
13	2008	13	5	-1	-1	1	0	0	2	1	-1	0	157	66	842	9	
14	2008	14	5	-1	-1	1	0	0	2	1	-1	0	418	60	910	1	
15	2008	15	4	-1	-1	0	0	0	2	2	-1	0	389	33	442	5	
16	2008	16	7	-1	-1	1	0	0	2	8	-1	0	154	74	850	1	
17	2008	17	1	55	-1	0	0	0	2	2	-1	0	415	38	641	5	
18	2008	18	7	-1	-1	1	0	0	2	2	-1	0	19	51	441	1	
19	2008	19	1	20	-1	0	0	0	1	9	-1	0	255	53	710	5	
20	2008	20	2	40	-1	0	0	0	2	1	-1	0	354	42	412	1	
21	2008	21	5	-1	-1	1	0	0	2	2	-1	0	19	51	550	1	
22	2008	22	1	35	-1	0	0	0	2	8	-1	0	338	42	999	3	
23	2008	23	7	-1	-1	1	0	0	2	1	-1	0	444	17	842	1	
24	2008	24	5	-1	-1	1	0	0	2	2	-1	0	23	65	172	1	
25	2008	25	7	-1	-1	2	0	0	0	0	-1	0	0	0	0	2	
26	2008	26	7	-1	-1	1	0	0	2	2	-1	0	445	45	820	1	
27	2008	27	1	40	-1	0	0	0	2	1	-1	0	453	22	712	4	
28	2008	28	1	50	-1	0	0	0	2	2	-1	0	23	65	700	5	
29	2008	29	1	80	-1	0	0	0	2	2	-1	0	19	51	712	4	
30	2008	30	1	3	-1	0	0	0	1	2	-1	0	84	86	812	5	
31	2008	31	5	-1	-1	1	0	0	2	2	-1	0	178	75	711	5	

Data View   Variable View

PASW Statistics Processor is ready

This view in SPSS is the data view. Here, we see the actual data as they are stored in the data file. Look first at the bottom of the screen. You will see that Data View is highlighted in yellow in the lower left-hand corner of the screen. This means that you have the data view open. Now look at the variables. Variables are listed in columns, and respondents or cases are listed in rows. This means that each column is one variable and each row is the answers that one person gave on the survey, or one case. The first column is Year. This is the year that the data were collected. Since this book only provides examples from 2008, you will notice that all of the respondents have 2008 listed for this variable. The second column is ID. This is a number that the researchers assigned to each person as he or she answered the survey. You will notice that each row has a different number. This is because each row represents a unique person. Finally, look at the third variable, WRKSTAT. You will notice that the second person who answered this question answered 1. How do you know what 1 means?

When you click on Variable View, you can see the information for the variable.

GSS 2008.sav [DataSet1] - PASW Statistics Data Editor

File Edit View Data Transform Analyze Graphs Utilities Add-ons Window Help

	Name	Type	Width	Decimals	Label	Values	Missing	Columns	Align	Measure	Role
1	YEAR	Numeric	4	0	GSS YEAR FO	None	None	6	Right	Scale	Input
2	ID	Numeric	4	0	RESPONDNT I	None	None	6	Right	Scale	Input
3	WRKSTAT	Numeric	1	0	LABOR FRCE	{0, IAP}...	0, 9	9	Right	Nominal	Input
4	HRS1	Numeric	2	0	NUMBER OF H	{-1, IAP}...	-1, 99	6	Right	Scale	Input
5	HRS2	Numeric	2	0	NUMBER OF H	{-1, IAP}...	-1, 99	6	Right	Scale	Input
6	EVWORK	Numeric	1	0	EVER WORK	{0, IAP}...	0, 9	8	Right	Nominal	Input
7	OCC	Numeric	3	0	RS CENSUS O	{0, DK, NA...	0	5	Right	Scale	Input
8	PRESTIGE	Numeric	2	0	RS OCCUPATI	{0, DK, NA I	0	10	Right	Scale	Input
9	WRKSLF	Numeric	1	0	R SELF-EMP	{0, IAP}...	0, 9	8	Right	Nominal	Input
10	WRKGOVT	Numeric	1	0	GOVT OR PRI	{0, IAP}...	0, 9	9	Right	Nominal	Input
11	COMMUTE	Numeric	2	0	TRAVEL TIME	{-1, IAP}...	-1, 99	9	Right	Scale	Input
12	INDUSTRY	Numeric	3	0	RS INDUSTRY	{0, DK, NA	0	10	Right	Scale	Input
13	OCC80	Numeric	3	0	RS CENSUS O	{0, NAP}...	0, 999	7	Right	Scale	Input
14	PRESTG80	Numeric	2	0	RS OCCUPATI	{0, DK, NA I	0	10	Right	Scale	Input
15	INDUS80	Numeric	3	0	RS INDUSTRY	{0, IAP}...	0, 999	9	Right	Scale	Input
16	MARITAL	Numeric	1	0	MARITAL STAT	{1, MARRIE	9	9	Right	Nominal	Input
17	AGEWED	Numeric	2	0	AGE WHEN FI	{0, IAP}...	0, 99	8	Right	Scale	Input
18	DIVORCE	Numeric	1	0	EVER BEEN D	{0, IAP}...	0, 9	9	Right	Nominal	Input
19	WIDOWED	Numeric	1	0	EVER BEEN	{0, IAP}...	0, 9	9	Right	Nominal	Input
20	SPWRKSTA	Numeric	1	0	SPOUSE LAB	{0, IAP}...	0, 9	10	Right	Nominal	Input
21	SPHRS1	Numeric	2	0	NUMBER OF H	{-1, IAP}...	-1, 99	8	Right	Scale	Input
22	SPHRS2	Numeric	2	0	NO. OF HRS S	{-1, IAP}...	-1, 99	8	Right	Scale	Input
23	SPEVWORK	Numeric	1	0	SPOUSE EVE	{0, IAP}...	0, 9	10	Right	Nominal	Input
24	SPOCC	Numeric	3	0	SPOUSE CEN	{0, DK, NA...	0	7	Right	Scale	Input
25	SPPRES	Numeric	2	0	SPOUSES OC	{0, DK, NA I	0	8	Right	Scale	Input
26	SPWRKSLF	Numeric	1	0	SPOUSE SEL	{0, IAP}...	0, 9	10	Right	Nominal	Input
27	SPIND	Numeric	3	0	SPOUSES IND	{0, DK, NA	0	7	Right	Scale	Input
28	SPOCC80	Numeric	3	0	SPOUSE CEN	{0, NAP}...	0, 999	9	Right	Scale	Input
29	SPPRES80	Numeric	2	0	SPOUSES OC	{0, DK, NA I	0	10	Right	Scale	Input
30	SPINDB0	Numeric	3	0	SPOUSES IND	{0, IAP}...	0, 999	9	Right	Scale	Input
31	PAOCC16	Numeric	3	0	FATHERS CEN	{0, DK, NA...	990 - 999, 0	9	Right	Scale	Input
32	PAPRES16	Numeric	2	0	FATHERS OC	{0, DK, NA I	0	10	Right	Scale	Input

Data View   Variable View

PASW Statistics Processor is ready

Now you will notice that Variable View is highlighted in the lower left-hand corner. Each row provides information on one variable. The variables are listed in the same order that they appear in Data View. If you look at the third row, you will see WRKSTAT. The first column gives the variable name. The second column gives the type of variable that was used to store the data. Whenever we assign numbers, this will be numerical. We have the option of using a date variable for dates, but as you can see here, numericals can be used for the year as well. When we want to type words in the variable, we use String as the variable type. The third column tells how many digits or characters there are in the variable. Year, for example is four digits because the year recorded was 2008, which has four digits. The fourth column tells when there are decimals allowed. Since these are discrete variables, they have a 0 for decimals. If we asked respondents about their GPAs, we would choose 3 for Width and 2 for Decimals so that we could record a GPA as, for example, 2.51 or 3.89. The fifth column is the variable label. The variable label is usually an abbreviation that the researcher can remember. For WRKSTAT, the label is Labor Force Status. We can drag this column to make it wider so that we can read the entire label. We will see that WRKSTAT measures a person's work status: fulltime, part-time, and so on. The next column that we need to concern ourselves with is the Values column. Notice that there is text in this column and a little box with "…" on it. When you click on that box, the value label window opens.

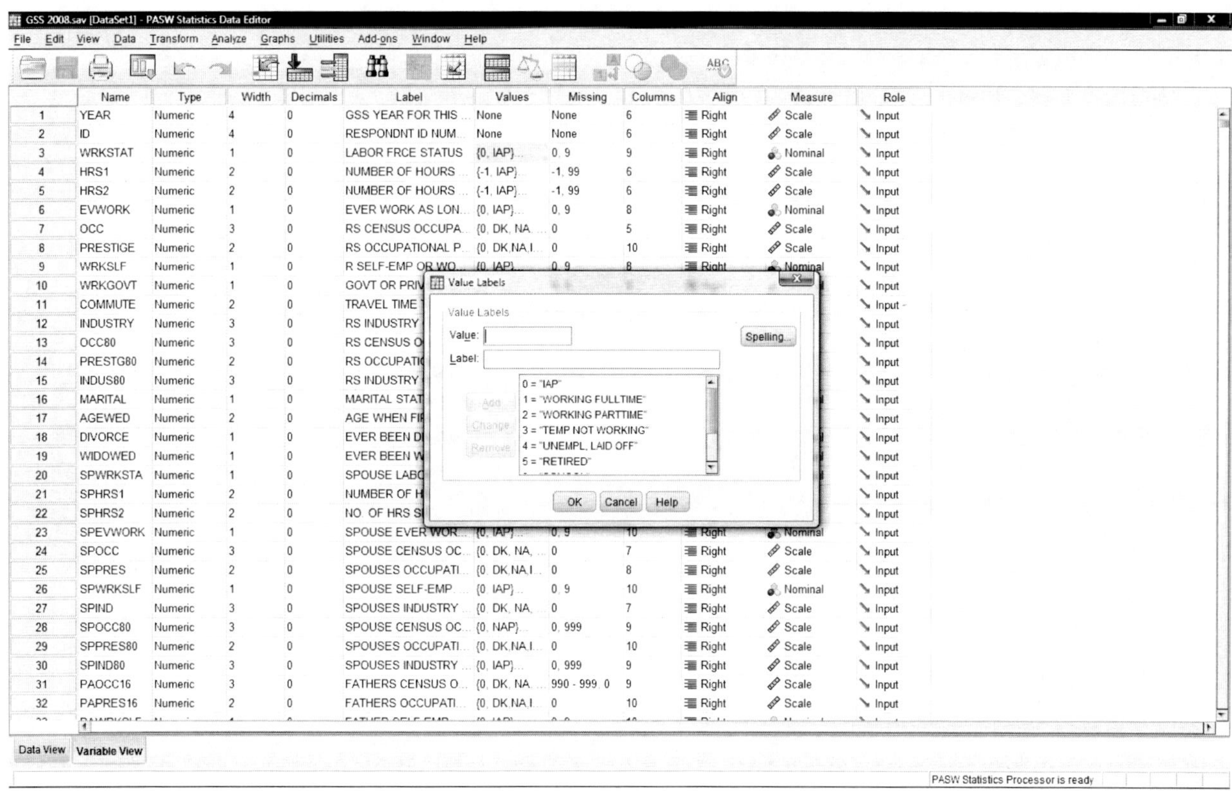

	Name	Type	Width	Decimals	Label	Values	Missing	Columns	Align	Measure	Role
1	YEAR	Numeric	4	0	GSS YEAR FOR THIS …	None	None	6	Right	Scale	Input
2	ID	Numeric	4	0	RESPONDNT ID NUM…	None	None	6	Right	Scale	Input
3	WRKSTAT	Numeric	1	0	LABOR FRCE STATUS	{0, IAP}…	0, 9	9	Right	Nominal	Input
4	HRS1	Numeric	2	0	NUMBER OF HOURS …	{-1, IAP}…	-1, 99	6	Right	Scale	Input
5	HRS2	Numeric	2	0	NUMBER OF HOURS …	{-1, IAP}…	-1, 99	6	Right	Scale	Input
6	EVWORK	Numeric	1	0	EVER WORK AS LON…	{0, IAP}…	0, 9	8	Right	Nominal	Input
7	OCC	Numeric	3	0	RS CENSUS OCCUPA…	{0, DK, NA,…	0	5	Right	Scale	Input
8	PRESTIGE	Numeric	2	0	RS OCCUPATIONAL P…	{0, DK,NA,I…	0	10	Right	Scale	Input
9	WRKSLF	Numeric	1	0	R SELF-EMP OR WO…	{0, IAP}…	0, 9	8	Right	Nominal	Input
10	WRKGOVT	Numeric	1	0	GOVT OR PRIV…						Input
11	COMMUTE	Numeric	2	0	TRAVEL TIME T…						Input
12	INDUSTRY	Numeric	3	0	RS INDUSTRY						Input
13	OCC80	Numeric	3	0	RS CENSUS O…						Input
14	PRESTG80	Numeric	2	0	RS OCCUPATI…						Input
15	INDUS80	Numeric	3	0	RS INDUSTRY						Input
16	MARITAL	Numeric	1	0	MARITAL STAT…						Input
17	AGEWED	Numeric	2	0	AGE WHEN FIR…						Input
18	DIVORCE	Numeric	1	0	EVER BEEN DI…						Input
19	WIDOWED	Numeric	1	0	EVER BEEN W…						Input
20	SPWRKSTA	Numeric	1	0	SPOUSE LABO…						Input
21	SPHRS1	Numeric	2	0	NUMBER OF H…						Input
22	SPHRS2	Numeric	2	0	NO. OF HRS S…						Input
23	SPEVWORK	Numeric	1	0	SPOUSE EVER WOR…	{0, IAP}…	0, 9	10	Right	Nominal	Input
24	SPOCC	Numeric	3	0	SPOUSE CENSUS OC…	{0, DK, NA,…	0	7	Right	Scale	Input
25	SPPRES	Numeric	2	0	SPOUSES OCCUPATI…	{0, DK,NA,I…	0	8	Right	Scale	Input
26	SPWRKSLF	Numeric	1	0	SPOUSE SELF-EMP…	{0, IAP}…	0, 9	10	Right	Nominal	Input
27	SPIND	Numeric	3	0	SPOUSES INDUSTRY …	{0, DK, NA,…	0	7	Right	Scale	Input
28	SPOCC80	Numeric	3	0	SPOUSE CENSUS OC…	{0, NAP}…	0, 999	9	Right	Scale	Input
29	SPPRES80	Numeric	2	0	SPOUSES OCCUPATI…	{0, DK,NA,I…	0	10	Right	Scale	Input
30	SPIND80	Numeric	3	0	SPOUSES INDUSTRY …	{0, IAP}…	0, 999	9	Right	Scale	Input
31	PAOCC16	Numeric	3	0	FATHERS CENSUS O…	{0, DK, NA,…	990 - 999, 0	9	Right	Scale	Input
32	PAPRES16	Numeric	2	0	FATHERS OCCUPATI…	{0, DK,NA,I…	0	10	Right	Scale	Input

The value labels inform us that 1 means working fulltime. We now know that the second person in this data set works fulltime.

The last column that we will concern ourselves with is the one labeled Measure, which tells us what type of variable this is. As we can see, WRKSTAT is a nominal measure. This is because the numbers have no real numeric meaning. For example, 1 means employed fulltime and 5 means retired, but, of course, someone who is retired is not 5 times more employed than someone who works fulltime.

Interpretation

As mentioned earlier, we are not mathematicians. Rather, we use statistics to help us make sense of the social world. This means that while the analysis of the data is important, it is essential that we are able to interpret these data from the perspective of social experience. What does this mean? Consider the example given above on research concerning income, education, and alcoholism. Our statistics tell us that more-educated white males with high educational levels are most likely to drink. But why is that the case? To interpret these findings, we have to examine the real-life experiences that contribute to these relationships. Certainly, people with more money to spend can afford to buy more alcohol. Thus, this is likely one interpretation of the income relationship. Also, the higher the income and the higher the educational level, the more likely a person is to have autonomy or a high level of authority in his or her job. Blue-collar workers don't have "three- martini" lunch meetings, for example. Also, if a low-wage worker is unable to go to work because of a hangover or being under the influence, that person may be likely to lose his or her job. A white-collar worker, who tends to be more educated, may have the ability to call in sick. Since whites are more likely to fall into the higher income and educational range than nonwhites are, we are able to explain the race relationship. What about gender? Historically, women have been less likely to drink because they have been more likely to be sanctioned for drinking. Men who drink too much have been viewed as "sowing wild oats," while women who drink have been seen as having loose morals. Also, women are more likely to be victimized, and drinking increases vulnerability.

This discussion is an example of how we interpret our findings. The statistics will tell us if a relationship exists and if that relationship is significant. However, we have to take those findings and place our understanding within social experiences for the statistics to mean anything. Whenever you get statistical results, you must explain the results from the perspective of the sociological interpretation.

Since we draw conclusions about social situations, we must remember a few important rules. First, the conclusions are only as good as the data that we have collected. If we have a problem in our data, this will result in a problem in our conclusions. Second, we need to make sure that we objectively and ethically interpret our results. It is possible to use data to find support for a value that we personally hold to be true. When we do this, we are misusing the process.

Conclusion

Data analysis can be fun once you get comfortable with the procedures. This chapter has provided you with a discussion of the basic steps to beginning a research project and an explanation of variable types and variations in measurements. The next chapters will introduce you to statistical methods.

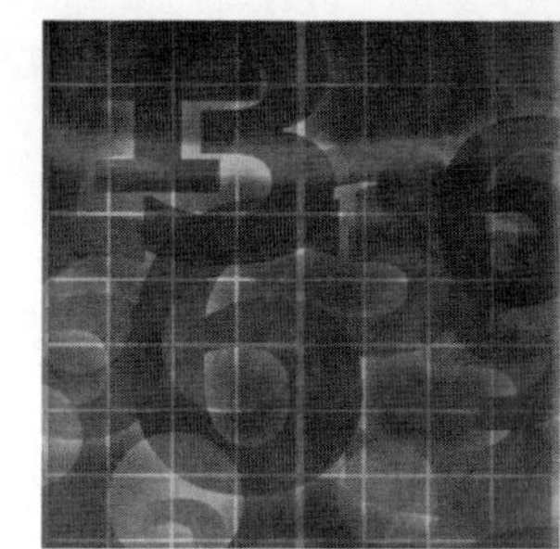

Exercises and Notes

For questions 1–4, match the following measures to the correct level of measurement:

a. nominal   **b.** ordinal   **c.** interval   **d.** ratio

1. What type of pet do you own? If more than one, indicate the type of pet you have owned the longest.

   None   Cat   Dog   Reptile   Other

   1. ______________ (level of measurement)

2. ______________ How old are you?

   2. ______________ (level of measurement)

3. On a scale of 1 to 7, where 1 means not at all and 7 means very likely, how likely are you to go to graduate school?

   3. ______________ (level of measurement)

4. Which of the following is a hypothesis?

   As education increases, liberal political attitudes are likely to increase.

   More education results in more liberal attitudes.

   Does education influence political attitudes?

For questions 5 and 6, identify the dependent and independent variables in the following hypothesis.

Gender is likely to influence the choice of a major for incoming college students.

5. Dependent variable ____________________

6. Independent variable ____________________

For questions 7–9, match the measures to the following:

a. variable   **b.** construct

7. quality of life

8. income

9. A researcher asks, "Do you own a gun?" This is an example of an ______________ variable, because it is measuring an actual event.

10. A researcher asks, "Do you favor stronger gun control laws?" This is an example of an ____________ variable because it measures a belief.

11. A researcher finds that men are more likely to die or be injured in high-risk sports than women are. Interpret these findings from a sociological perspective.

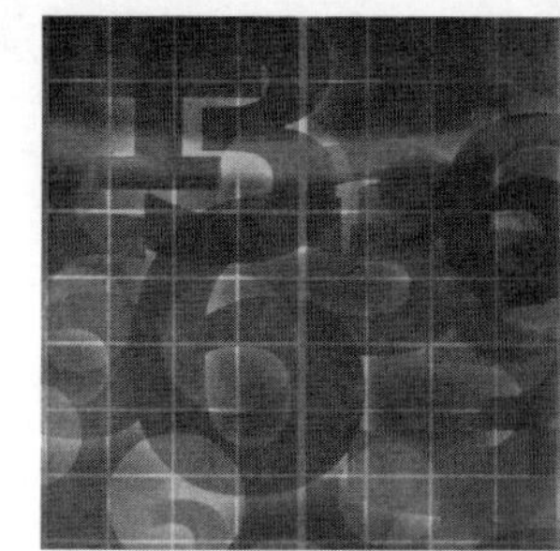

Learning Check

For questions 1–4, match the following measures to the correct level of measurement:

> **a.** nominal   **b.** ordinal   **c.** interval   **d.** ratio

1. How many years of education do you have?

   ______________ (level of measurement)

2. How satisfied are you with your job?

   Very dissatisfied   Somewhat dissatisfied   Somewhat satisfied   Very satisfied

   ______________ (level of measurement)

3. What color hair do you have?

   Blond   Brown   Black   Red   Other

   ______________ (level of measurement)

4. On a scale of 1 to 7, where 1 means not as happy and 7 means much happier, how happy are you today when compared to one year ago?

   ______________ (level of measurement)

5. Which of the following is a hypothesis?

   People with more money get college educations.

   As income increases, education is likely to also increase.

   Are people with more money more likely to go to college?

For questions 6 and 7, identify the dependent and independent variables in the following hypothesis.

The probability of recidivism is likely to increase as poverty increases.

6. Dependent variable ____________________

7. Independent variable ____________________

For questions 8–9, match the measures to the following:

> **a.** variable   **b.** construct

8. Socioeconomic status

9. Political party affiliation

10. A researcher asks, "Do you believe that women should be allowed to have an abortion for any reason?" This is an example of an ______________ variable because it is measuring an actual event.

11. A researcher asks, "Have you ever had an abortion?" This is an example of an ___________ variable because it measures a belief.

12. A researcher finds that people who engage in criminal activities when they are young usually age out of the behavior by the time they are in their late 20s. Interpret these findings from a sociological perspective.

Chapter Two

Organizing Data

After researchers collect quantitative data, the information must be organized for analysis and presentation. One of the most basic ways to organize data is frequency distribution. A **frequency distribution** shows how many cases occur for each outcome of a variable. An **outcome** of a variable is a response category or the value of a variable. For example, the outcomes for the variable of the class rank for an undergraduate student are freshman, sophomore, junior, and senior. The outcomes of a variable should be **mutually exclusive**—each case can be assigned to only one category—and **exhaustive**—there is at least one outcome or category for each case. The outcomes of the variable of rank are mutually exclusive because a student cannot be both a freshman and a sophomore; a student either has enough hours to be a sophomore or not. While at first glance, these outcomes for class rank for undergraduate students appears to be exhaustive. However, there are some students who are undergraduates with a degree. Therefore, the category "other" is often included to be sure that a variable is exhaustive.

Let's look at an example with data to see how a researcher might code and enter data. Suppose we are interested in looking at the sex, ranks, majors, credit hours being taken, and test scores of a group of students. Our first case is a female, senior, criminology major who is taking 12 credit hours and scored 75 on the test. The next case is a male, senior, sociology major who is taking 14 credit hours and scored 84 on the test. The third case is a female senior, nursing major who is taking 15 credit hours and scored a 92 on the test. To organize and prepare the data for statistical analysis, researchers typically enter the data into a computer program, such as SPSS. Two of the variables, test score and credit hours being taken, are interval/ratio measures. Since computer programs such as SPSS typically process numerical data, we need to assign numeric values to the outcomes of the nominal and ordinal variables.

	Sex	Rank	Major	Hours	Score
1	2	4	4	12	75
2	1	4	1	14	84
3	2	4	3	15	92
4	2	4	1	16	84
5	2	3	2	15	70
6	2	3	1	12	80
7	1	4	2	18	86
8	2	3	2	16	56
9	1	4	1	18	94
10	2	2	2	15	71
11	1	3	4	13	71
12	1	4	1	14	83
13	2	4	3	15	92
14	2	4	1	16	86
15	1	3	3	15	73
16	2	3	1	12	80
17	1	4	2	18	71
18	2	3	2	16	63
19	1	4	1	18	97
20	2	2	2	15	67

Therefore, in the data set, you see a 1 under the variable of sex for every man in the sample and a 2 for every woman. Rank and major also have numbers representing the outcomes, with 1 = freshman, 2 = sophomore, 3 = junior, and 4 = senior for rank; and 1 = sociology, 2 = social Work, 3 = nursing, and 4 = criminology for major. Interval/ratio measures are already numeric values, and therefore they need no more additional coding. For example, case 1 scored a 75 on the test and is registered for 12 credit hours. While this may seem like a lot of work for a few variables and 10 cases, a researcher can code and organize data sets with hundreds of variables and thousands—even millions—of cases.

Simple Frequency Distributions

Given that we have just a handful of cases in this data set, we can create a frequency distribution without the aid of a computer so that we can understand what the computer does. We first want to create a frequency distribution for class rank. We will give the frequency distribution the title Students' Class Rank. Next, we need to identify the variable's categories. For class rank, the outcomes are sophomore, junior, and senior. The frequencies represent the number of times each outcome appears in the data. **Table 2.1** shows the frequency distribution for class rank.

Table 2.1 *Students' Class Rank*

Class Rank	f
Sophomore	2
Junior	7
Senior	11
	N = 20

Researchers often find it useful to compare one distribution with another. Suppose we want to compare the class rank of students for this school with the class rank of students at another school. More specifically, if we want to know if seniors comprise a higher composition of the local university versus an out-of-state university, it would be difficult to determine by just looking at the frequencies in **Table 2.2** because the classes are not the same size. Therefore, it is useful to standardize the frequencies based on the size of the distribution.

Table 2.2 *Students' Class Rank at Local and Out-of-State University*

Class Rank	University	
	Local	Out-of-State
Sophomore	2	2
Junior	7	10
Senior	11	14
Total	20	26

A **relative frequency distribution** shows the relative, as opposed to the absolute, number of times an outcome of a variable occurs in the data set, thereby standardizing each frequency based on the total number of cases, which is represented by N. Proportions or percentages are used to show the relative outcomes. One could think of proportions and percentages as mathematical first cousins, since they are closely related.

A **proportion** is calculated by dividing the frequency of an outcome by the total number of cases:

$$p = \frac{f}{N}$$

f = the frequency of an outcome

N = the total number of cases

Using the data for students' class rank from the local university in Table 2.2, we can calculate the proportion of sophomores by dividing the number of sophomores by the total number of students. We find that 2 out of 20 students are sophomores, meaning that the proportion is:

$$p = \frac{2}{20} = .10$$

Since proportions are always less than 1.00, many researchers prefer to express the relative measure as a **percentage,** which is the proportion multiplied by 100. If the proportion is not known, we divide the frequency of an outcome by the total number of cases and then multiply by 100:

$$\% = \frac{f}{N} \times 100$$

The percentage of students of who are sophomores is calculated below:

$$\% = \frac{2}{20} \times 100 = 10\%$$

Table 2.3 shows the absolute and relative frequency of students' class rank at the local university.

Table 2.3 *Distribution of Students' Class Rank at Local University*

Class Rank	f	p	%
Sophomore	2	.10	10
Junior	7	.35	35
Senior	11	.55	55
	$N = 20$	1.00	100

Now that we have reviewed how to calculate proportions and percentages, let's return to the original question: Do seniors comprise a higher composition of the class at the local university versus the out-of-state university? By transforming the data in **Table 2.2** from an absolute frequency distribution into a relative frequency distribution, we will be able to answer this question. The relative frequency distributions in **Table 2.4** allow us to compare the composition of the classes at the local and out-of-state universities quite easily. We can see that the composition of seniors is similar, with the local university having just a slightly higher composition of seniors than does the out-of-state university (55% compared to almost 54%).

**Table 2.4 *Percentage of Students' Class Rank at Local
and Out-of-State Universities***

Class Rank	University	
	Local (N = 20)	Out-of-State (N = 26)
Sophomore	10.00	7.69
Junior	35.00	38.46
Senior	55.00	53.85
Total	100.00	100.00

Relative frequency distributions are also useful when one needs to generalize to the entire population from a sample. Since samples are a subset of the population, the absolute frequencies of a sample are not indicative of the number of cases in the population. However, if the sample is representative of the population, then the percentages in a sample can be generalized to the population.

Ratio and Rates

Two other measures that are standardized by size are ratios and rates. A **Ratio** directly compares the number of cases falling in one category with the number of cases falling into another category. Because the frequency of the first category is often greater than the frequency of the second category, their quotient is multiplied by a constant (for example, $k = 100$) so that the ratio is not less than 1:

$$\text{ratio} = \frac{f_1}{f_2} \times k$$

Demographers often calculate the sex ratio to compare the sex composition of different populations. To compute the sex ratio:

$$\frac{f_{males}}{f_{females}} \times 100$$

If there are 21 million females compared to 15.8 million males, then the sex ratio is 75.24, meaning that for every 100 females in the country, there are approximately 75 men.

$$\frac{15.8}{21.0} \times 100 = 75.24$$

A **Rate** compares the actual number of events or occurrences to the potential frequency of events or occurrences (sometimes called the population at risk):

$$\text{Rate} = \frac{f_{(actual)}}{f_{(potential)}} \times k$$

In 2006, the number of births, deaths, and the population in the United States (U.S. Census Bureau, 2009) were:

Number of births = 4, 266,000
Number of deaths = 2,426,000
Population = 299,157,000

To calculate the crude birth rate for the United States in 2006:

$$\frac{\#\text{ of births}}{\text{total population}} \times 1000 = \frac{4,266,000}{299,157,000} \times 1000 = 14.26$$

The crude birth rate of 14.26 indicates that there were approximately 14 births for every 1000 people in the country.

To calculate the crude death rate in the United States in 2006:

$$\frac{\#\text{ of deaths}}{\text{total population}} \times 1000 = \frac{2,426,000}{299,157,000} \times 1000 = 8.11$$

The crude death rate of 8.11 indicates that there were approximately 8 deaths for every 1000 people in the country.

Grouped Frequencies Distributions

Interval/ratio variables, such as test scores and credit hours in the mini-dataset of university students, could have numerous outcomes. For example, some students could have failed the test with very low scores, while others students could have gotten very high scores and others earned scores between these extremes. The results would create a very long and difficult-to-read frequency distribution, as depicted in **Table 2.5.**

Table 2.5 *Frequencies of Students' Test Scores*

Test Scores	f	Test Scores	f	Test Scores	f
56	1	70	1	84	2
57	0	71	3	85	0
58	0	72	0	86	2
59	0	73	1	87	0
60	0	74	0	88	0
61	0	75	1	89	0
62	0	76	0	90	0
63	1	77	0	91	0
64	0	78	0	92	2
65	0	79	0	93	0
66	0	80	2	94	1
67	1	81	0	95	0
68	0	82	0	96	0
69	0	83	1	97	1
					N = 20

To deal with such varied and multiple outcomes, researchers typically present such data in a grouped frequency distribution. In a grouped frequency distribution, instead of listing each score, the categories represent a range of values. Each category of a group frequency distribution is referred to as a **class interval.** The first class interval in **Table 2.5** is 56–60, and the class interval with the highest scores is 96–100. There are five scores in each class interval (for example in the first class interval, 56, 57, 58, 59, and 60); thus, the class interval size is 5.

Table 2.6 *Grouped Frequency Distribution of Students' Test Scores*

Test Scores	f	%
56 – 60	1	5
61 – 65	1	5
66 – 70	2	10
71 – 75	5	25
76 – 80	2	10
81 – 85	3	15
86 – 90	2	10
91 – 95	3	15
96 – 100	1	5
Total	20	100

Class Limits

Stated limits are those shown in the table, while the true limits are used to classify observations into the proper class intervals. It is important to distinguish between true and stated limits because the values of interval/ratio-level variables found in grouped frequency distributions do not have to be whole numbers. For example, suppose one of the students who earned a score of 86 on the test had not scored 86 but instead 85.5. Each class interval has an upper and lower true limit. The upper true limit is the upper stated limit plus ½ the unit of measurement, while the lower true limit is the lower stated limit minus ½ the unit of measurement.

It is important to determine the true limits for multiple reasons. First, we need to know the true limits in order to calculate some statistics when data are in a grouped frequency distribution. We will encounter this in later chapters. Also, we use the true limits to classify cases in the proper class interval. For example, which class interval would we place a score of 90.25 in **Table 2.7?** Looking at the stated limits, would we place it in the class interval 86–90 or 91–95? The true limits show us that 90.25 would be classified in the class interval 86–90 with true limits of 85.5–90.5, since 90.25 falls between those two values.

Table 2.7 Class Limits and Midpoints of Test Scores

Test Scores	True Limits	m
56–60	55.5–60.5	58
61–65	60.5–65.5	63
66–70	65.5–70.5	68
71–75	70.5–75.5	73
76–80	75.5–80.5	78
81–85	80.5–85.5	83
86–90	85.5–90.5	88
91–95	90.5–95.5	93
96–100	95.5–100.5	98

It may seem that the true limits are not mutually exclusive, since the upper true limit of one class interval is the same value as the lower true limit of the next class interval (*e.g.,* 60.5, 65.5, etc.). However, the lower class interval actually contains values that are less than the upper true limit. Therefore, a test score of 60.5 is classified in the class interval 61–65 with the true limits of 60.5–65.5, not in 56–60 with the true limits of 55.5–60.5.

The Midpoint

As we have seen in group frequency distributions, we no longer know the exact number of cases for each outcome, but instead, the number of cases within a range of values that define each class interval. At times, instead of using the range of values that comprise the class interval, researchers need a single value to represent the value of the class interval. However, the scores of the cases could have high values, low values, or be distributed throughout the class interval. The **midpoint,** which is the middle score in a class interval, provides us with the best single number to represent the class interval. The midpoint of the interval 71–75 is:

$$m = \frac{\text{lower limit} + \text{upper limit}}{2} = \frac{71 + 75}{2} = 73$$

Cumulative Distribution

Researchers may also want to present the data in a cumulative way that shows the relative position of cases compared to the overall distribution. **Cumulative frequencies** (*cf*) refer to the number of cases that have a given score *or less*. Either simple or grouped frequency distributions can include cumulative frequencies. If we wanted to know how many students scored 70 or lower, we would add the frequencies in that class interval with the frequencies in class intervals that have scores that are less than 70.

Table 2.8 *Cumulative Distribution of Students' Test Scores (Based on Table 2.6)*

Test Scores	f	cf	%	$c\%$
56–60	1	1	5	5
61–65	1	2	5	10
66–70	2	4	10	20
71–75	5	9	25	45
76–80	2	11	10	55
81–85	3	14	15	70
86–90	2	16	10	80
91–95	3	19	15	95
96–100	1	20	5	100
Total	20		100	

Cumulative distributions may have cumulative frequencies, cumulative percentages, or both. The formula that we used to calculate percentages is modified to calculate cumulative percentages:

$$c\% = \frac{cf}{N} \times 100$$

cf = the cumulative frequency in any category

N = the total number of cases

If we wanted to determine the percentage of students who scored a 70 or less, we first need to know the number of students who scored a 70 or less. Looking back to **Table 2.8,** we can see that we determined that four students scored a 70 or less. Next, we need to know what the N, or the total number, of cases is. By looking at the sum of the frequencies, we can determine that N = 20. Now we can plug these values into the cumulative percentage formula:

$$c\% = \frac{4}{20} \times 100$$

$$= .20 \times 100$$

$$= 20$$

As you might have noticed, if you have a percentage distribution, you can also obtain the cumulative percentages by adding the percentage in that class interval with the percentage in class intervals with lower scores. Therefore, to determine the cumulative frequency for 70, we would add the 5 + 5 = 10, and then 10 + 5 = 20. We could then say that 20% of the students scored 70 or less on the test.

Percentile Rank

Percentile rank is a number that indicates the percentage of cases in a distribution that fall at or below a given score. The percentile rank is different from a percentage. The percentage tells us how many times something occurred based on 100. If a person took a quiz and got 15 answers right out of 20, 75% of the questions were answered correctly. In contrast, the percentile compares how well one person did compared to others who took the test.

If we wanted to find the percentile rank for 75, we would have to find out how many cases fall below that score out of the total number of cases. Since 9 out of 20 fall below 75, its percentile rank is 45%:

$$56, 63, 67, 70, 71, 71, 71, 73, 75, 80, 80, 83, 84, 84, 86, 86, 92, 92, 94, 97$$

Thus, if we compare how one person did on the test in relation to others who took the test, we could come to different conclusions from the percentage and the percentile rank. For example, if 70–79% is a C, the person who got 15 out of 20 answers correct may have thought that he or she had performed about average on the test. However, the percentile rank of 45% tells us that the majority of the class did better on the test than that person did.

Cross-Tabulations

Up to this point, we have only analyzed one variable at a time. However, researchers are often interested in the relationship between variables. One of the most basic ways to look at the relationship between variables is a **cross-tabulation,** or **cross-tab,** which is a cross-classification of one variable by at least one other variable. The simplest form of a cross-tabulation is a two-by-two table, in which the two variables that are examined only have two categories, as in **Table 2.9.**

Table 2.9 *Cross-Tabulation of Graduating Students' Post-Graduation Plans by Honor Student Status*

Post-Graduation Plans	Honor Student		
	Yes	No	
Employment	9	55	64
Graduate School	13	14	27
	22	69	91

The intersections of the rows and columns are referred to as the cells of the cross-tab. Since each cell has its own frequency, it is useful to distinguish the different frequencies. The first subscript refers to the variable in the rows, for example, in **Table 2.9,** Post-Graduation Plans. The second subscript refers to the variable in the columns, in our example, Honor Student status. Therefore, $f_{12} = 55$ refers to the frequency in the cell of the first row and second column. What is the frequency for f_{21}?

The numbers that appear outside the cells are the marginals. The row marginal is the total of the frequencies in each row of the table, and the column marginal is the total of the frequencies in each column of the table. The frequency distribution of each of the variables is represented in the marginal distribution.

Just as we calculated percentages in our frequency distributions, we can also calculate percentagess with cross-tabs. However, there are three different types of percentages that can be calculated depending on whether we use the total N, the row totals, or the column totals. The total percentages are calculated by dividing each frequency by the total sample size:

$$\text{total\%} = \frac{f}{N_{\text{total}}} \times 100$$

The percentage for the first cell is calculated by dividing the f_{11} by the total number of cases; that is $(9 \div 91) \times 100 = 9.9\%$. This percentage tells us that 9.9% of the graduating seniors are honor students who plan to seek employment.

Table 2.10 *Cross-Tabulation of Graduating Students' Post-Graduation Plans by Honor Student Status with Total Percentages*

Post-Graduation Plans	Honor Student		
	Yes	No	
Employment	9 9.9%	55 60.4%	64 70.3%
Graduate School	13 14.3%	14 15.4%	27 29.7%
	22 24.2%	69 75.8%	91 100.0%

Researchers typically find it more useful to calculate row or column percentages. To calculate the row percentages for the first row in **Table 2.10**, we would divide the number of graduating honor students who plan to seek employment by all graduating students who plan to seek employment, and the number of nonhonors graduating students who plan to seek employment by all graduating students who plan to seek employment. Then, we would move on to the second row by dividing the number of graduating honor students who plan to go to graduate school by all students who plan to go to graduate school, and the number of nonhonors graduating students who plan to go to graduate school by all students who plan to go to graduate school. If there were third, fourth, and additional rows, the process would be applied to those rows. The formula for row percentages is:

$$\text{row}\% = \frac{f}{N_{\text{row}}} \times 100$$

The row percentage for the first cell is calculated by divided the f_{11} by all the graduating students who plan to seek employment; that is $(9 \div 64) \times 100 = 14.9\%$. The interpretation of this percentage is that 14.1% of the graduating students who plan to seek employment are honor students. What percentage of the graduating students who plan to seek employment are not honor students?

Table 2.11 *Cross-Tabulation of Graduating Students' Post-Graduation Plans by Honor Student Status with Row Percents*

Post-Graduation Plans	Honor Student		
	Yes	No	
Employment	9 14.1%	55 85.9%	64 100.0%
Graduate School	13 48.1%	14 51.9%	27 100.0%
	22	69	91
	24.2%	75.8%	100.0%

Since **Table 2.11** has row percentages, except for rounding error, the row marginal percentages should total 100%. However, the column marginal percentages are not the sum of the percentages in the columns in a table with row percentages. The column marginal for "Yes," a student is an honor student, refers to the percentage of all students who are honor students. Therefore, the row marginal is calculated like any other row in the table, taking the frequency (22) divided by the total of the row (91) to obtain the row marginal percentage (24.2%). We would apply the same process to the column marginal for the "No" column for honor student status.

The last alternative involves using the column marginal totals to calculate total percentages. To calculate the percentages for the first column, we would divide the number of graduating honor students who plan to seek employment by all graduating honors students, and the number of graduating honors students who plan to go to graduate school by all graduating honors students. For the second column, we would divided the number of nonhonors graduating students who pland to seek employment by all nonhonors graduating students, and the number of nonhonors graduating students who plan to go to graduate school by all nonhonors graduating students. If there were third, fourth, and additional columns, the process would be applied to those columns. The formula for column percentagess is:

$$\text{column}\% = \frac{f}{N_{\text{column}}} \times 100$$

The column percentage for the first cell is calculated by dividing the f_{11} by all the students graduating with honors; that is, $(9 \div 22) \times 100 = 40.9\%$. The interpretation is that 40.9% of graduating honor students plan to seek employment. What percentage of graduating honor students plan to go graduate school?

Table 2.12 *Cross-Tabulation of Graduating Students' Post-Graduation Plans by Honor Student Status with Column Percentages*

Post-Graduation Plans	Honor Student		
	Yes	No	
Employment	9 40.9%	55 79.7%	64 70.3%
Graduate School	13 59.1%	14 20.3%	27 29.7%
	22 100.0%	69 100.0%	91 100.0%

Since **Table 2.12** has column percents, except for rounding error, the column marginal percentages should total 100%. However, the row marginal percentages are not the sum of percentages in the column in a table with column percentages. The row marginal for Employment for Post-Graduation Plans refers to the percentage of all students who plan to seek employment. Therefore, the column marginal is calculated like any other column in the table, taking the frequency (64) divided by the total (91) to obtain the column marginal percent (70.3%). We would apply the same process to the marginal for those students whose post-graduation plans are Graduate School.

With three different types of percentages, which do we use? As mentioned earlier, total percentages are seldom used. Total percentages are generally only used when neither of the variables appears to affect the other variable. In other words, an independent-dependent variable relationship cannot be identified. For example, if we were looking at respondents' views on abortion and the death penalty, an opinion on one of the issues does not necessarily cause a person to have an opinion on the other issue. Since cross-tabs involve cross-classification of one variable by at least one other variable, we often want to know percentages based on a cross-classification, not the entire group. We decide if we want to calculate row or column percentages based on which variables' categories we want to compare. If we can identify an independent variable, we will mostly likely compare those categories. In our example, honor student status may affect students' post-graduation plans. Therefore, honor student status is the independent variable. To show the possible effect of being an honor student on post-graduation plans, we would want to calculate column percentages.

Graphics

Graphics are an excellent way to communicate statistical information with clarity, precision, and efficiency. Large data sets can be presented coherently by displaying many numbers in a small space. Whereas columns of numbers may appear daunting, graphics can be used to encourage readers to compare different pieces of data. However, researchers should not assume that graphics can stand by themselves. Researchers should always provide a written description interpreting the statistical information displayed in a graphic.

Social scientists use a variety of different graphics in their work. Among the commonly used presentations are pie charts, bar graphs, histograms, frequency polygons, and line charts. The level of measure needs to be considered when choosing which graphic presentation to use. The **pie chart** is the simplest graphical display that is useful for showing the frequencies or percentages of a nominal-level or ordinal-level variable. The categories are shown as part of a circle whose segments total 100%. **Figure 2.1** shows the distribution of the types of previous jobs held by national sports broadcasters. Nearly half (49%) of broadcasters' previous jobs were at national networks, while 17% worked at regional networks and 28% came from local stations. Because of the relatively few broadcasters who previously worked elsewhere, they were combined into a single category of "Other" (6%).

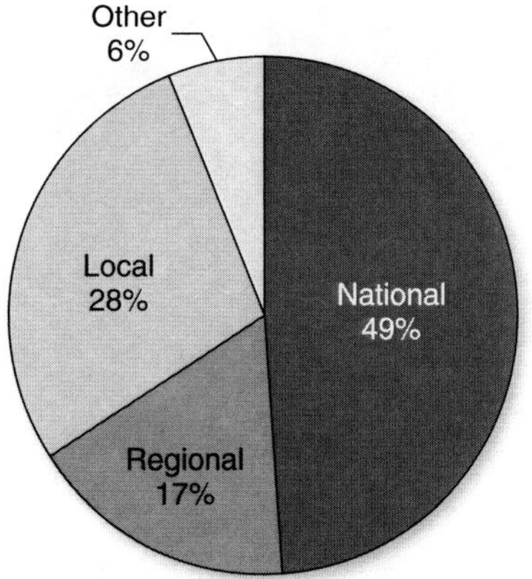

Figure 2.2 Type of Previous Broadcasting Job (N = 238)

The **bar graph** is used more often than pie charts since it can display a variable with more categories and multiple variables, which allow the comparison of how members of various groups differ on another variable. Along the horizontal (or *X*) axis of the bar graph are the categories of the variable, while the vertical (or *Y*) axis displays the frequency or percentage for each category of variables. This description assumes a horizontal construction of the graph. However, bar graphs can also be vertically constructed. Deciding which orientation to use is often determined by which design fits better on a page. The highest frequency of a variable is readily apparent to the viewer by identifying the highest bar in the graph.

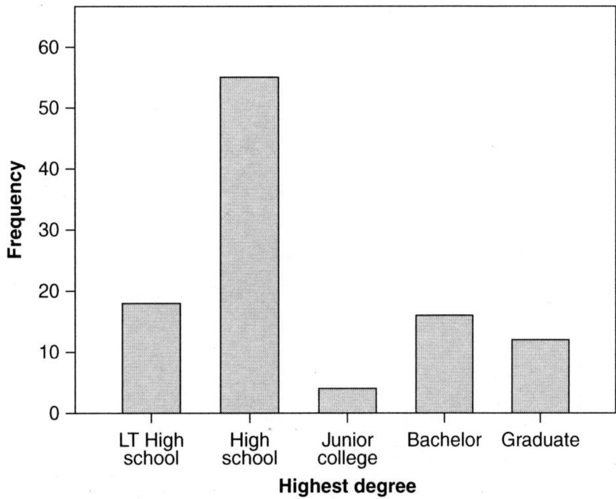

Figure 2.3 Bar Graph of Highest Degree (N = 105)

The **histogram** looks very similar in appearance to the bar graph. However, there is an important difference between the two graphs. In a bar graph, the bars depicting the different categories of the variable have spaces between each bar, while in the histogram, the bars touch each other. This difference in display is to emphasize the level of measurement of the variable. Nominal-level and ordinal variables are displayed in bar graphs to emphasize the distinct and separate nature of the categories of the variable. Researchers would graph an interval-level or ratio-level variable, on the other hand, as a histogram to emphasize the continuous nature of the variable's categories. The width of a bar corresponds to each class interval, with the bar's height representing the frequency or the percentage of the class interval.

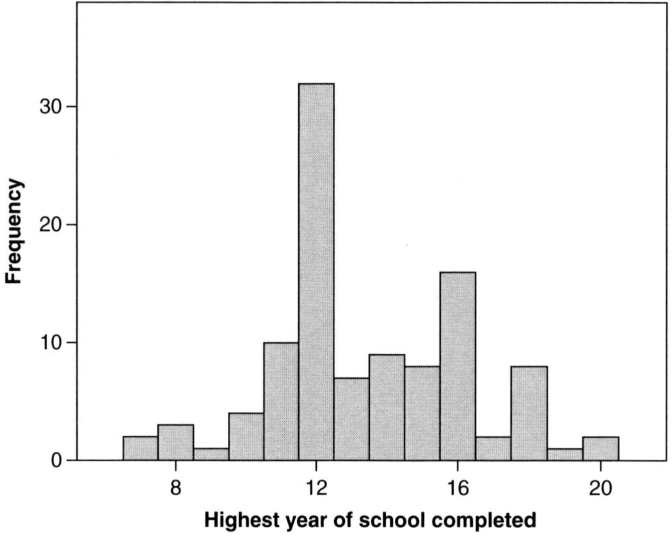

Figure 2.4 Histogram of Highest Year of School Completed (N = 105)

The **frequency polygon** is another graphic that is often used to emphasize the continuous nature of the outcomes of a variable. In a frequency polygon, frequencies are displayed by points over the score values or the midpoint of each class interval, with each point connected with a straight line. Whereas in a bar graph or histogram, the highest frequency of a variable is readily apparent by the highest bar in the graph, in a frequency polygon, the highest point in the graph represents the highest frequency. **Figure 2.5a** displays the points over each score, and **Figure 2.5b** displays the midpoint of each class interval.

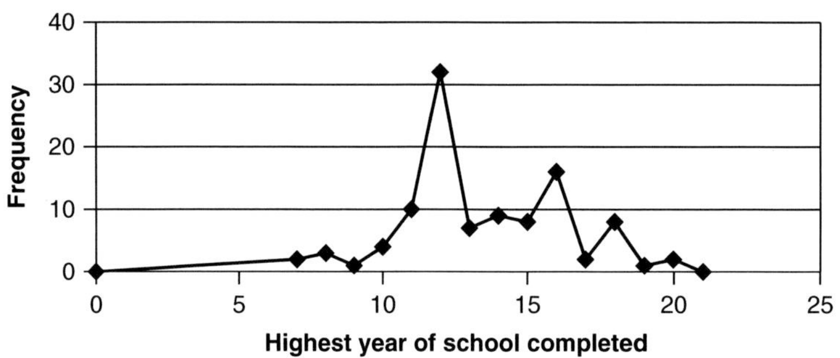

Figure 2.5a Frequency Polygon of the Highest Year of School Completed (N = 105)

Just as we can graph the frequencies or percentages of a variable, we can also graph the cumulative frequency or percentages in a **cumulative frequency (or percentage) polygon.** However, a cumulative frequency polygon differs from a regular frequency polygon in some important ways. Given the additive nature of cumulative frequencies, the succeeding points are never lower and typically higher than the previous point, and the highest point of a cumulative frequency is always the last point in the graph. When graphing grouped data, midpoints are used to graph regular frequency polygons, while the upper limit of the class interval must be used in a cumulative distribution, since cumulative frequencies indicate how many cases had that score or lower.

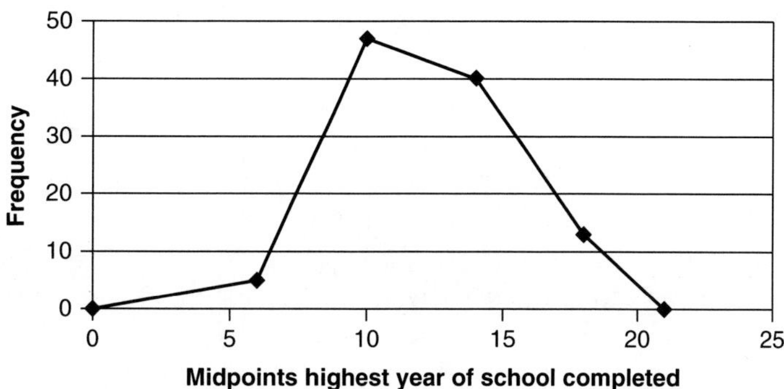

Figure 2.5b Frequency Polygon of the Highest Year of School Completed (N = 105)

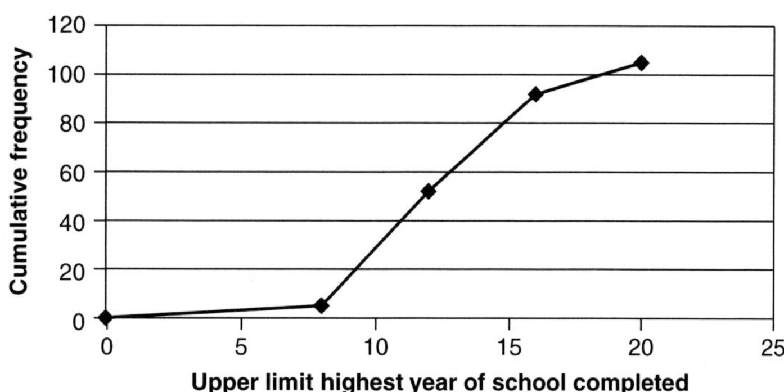

Figure 2.6 Cumulative Frequency Polygon of the Highest Year of School Completed (N = 105)

Researchers often use scatterplots to view the shape of the distribution of a variable or the relationship between variables. The outcomes of some variables are sometimes symmetrical, with each half of the curve being the mirror image of the other half of the curve. Skewed distributions are the result of outliers in the data; an **outlier** is a data point that differs significantly from the rest of the data. In positively skewed distributions, the extreme values are relatively high values compared to most of the scores in the distribution. In negatively skewed distributions, the outliers or extreme values are relatively low values compared to the rest of the scores in the distribution.

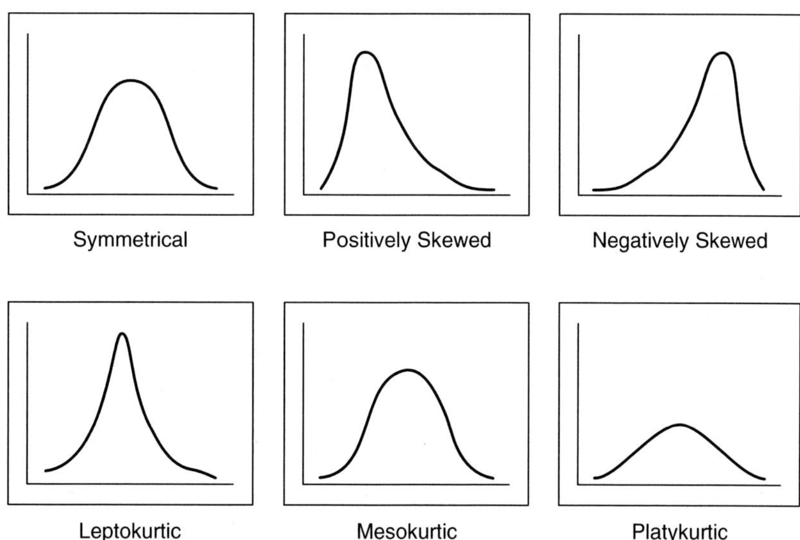

Figure 2.7 Distributions Depicting Symmetry, Skewness, and Kurtosis

Symmetrical distributions can vary in terms of their height (or kurtosis). Some can be quite tall (known as leptokurtic), while other are rather flat (platykurtic). Distributions that are not extremely high or flat are referred to as mesokurtic. The normal curve is a mesokurtic distribution.

Scatterplots may also be used to show the relationship between variables. They can help identify whether any outliers exist in the data. It is important to know if there are outliers because they can reduce the magnitude of the statistics that the data produce. Also, as we will see later on, some statistics assume a linear relationship between variables, such as the one depicted in **Figure 2.6.** However, not all relationships are linear, as seen in **Figure 2.7.** It is important to know whether the relationship between variables is linear in order to choose the correct statistics for the analysis.

Name: ______________________________________   Date: _________________

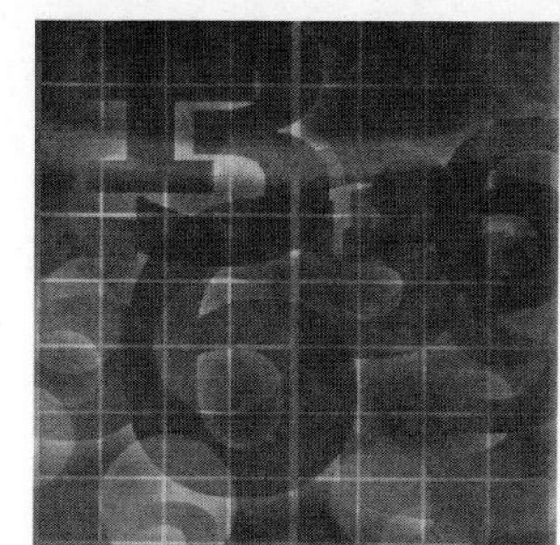

Exercises and Notes

- **Outcome** is a ______________________ or ______________________ of a variable.

- **Simple frequency distribution** is a table that contains a ______________________
  ____.

- **Relative frequency distribution** is a table that shows the ______________________ rather than the
  ______________________ outcomes of the variable.

1. Using the data from the spreadsheet below,

    (a) determine the frequencies for each category of the variable "Major" and place them in **Table 2.1.1**.
         Then, calculate the statistics need to complete the table.

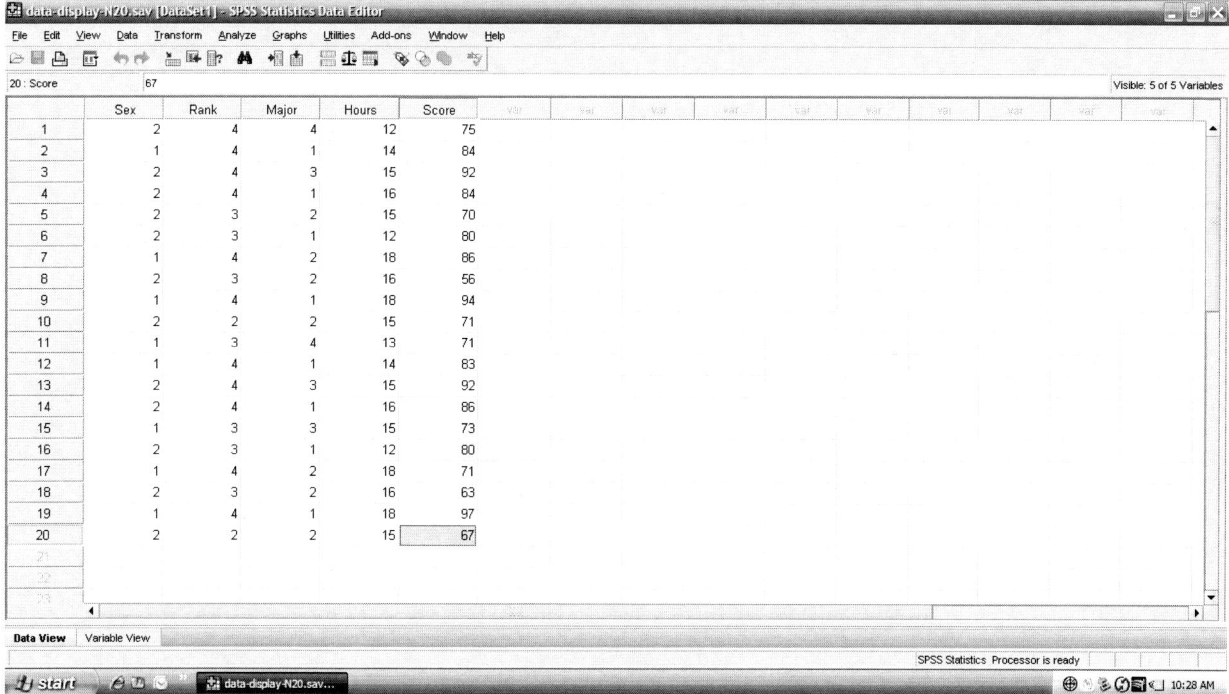

Table 2.1.1 *Distribution of Students' Major*

Major	f	p	%
Sociology			
Social Work			
Nursing			
Criminology			

(b) How many students were majoring in sociology?

(c) What percentage of students were social work majors?

(d) Which major had the smallest proportion of students?

Ratios and Rates

- **Ratios** directly compare ___________________ with the ___________________.

  2. If there are 21 million females compared to 15.8 million males, the sex ratio is

  $$\underline{\hspace{4cm}} \times k$$

  Interpretation: ____

  ____

  Calculate the sex ratio of the students in your classroom.

  In this class, the sex ratio is

  $$= \underline{\hspace{4cm}} \times k$$

  Interpretation: ____

  ____

- **Rates** compare the ___________________ of events or occurrences to the
  ___________________ of events or occurrences (sometimes called the population at risk).

  Rates =

  3. **Table 2.1.2** shows the reasons for workers' job loss or displacement and the total number of workers
  employed in the labor force.

  Table 2.1.2 Displaced and Employed Workers

# of workers whose plants closed	=	540,000
# of workers whose plants lacked work	=	288,000
# of workers whose positions were abolished	=	372,000
Total # of displaced workers	=	1,200,000
# of employed workers	=	145, 300,000

  (a) What is the ratio of workers whose positions were abolished to those whose plants closed?

  $$\underline{\hspace{4cm}} \times \underline{\hspace{1.5cm}}$$

  Interpretation: ____

  ____

  (b) What is the rate of job loss?

  $$\underline{\hspace{4cm}} \times \underline{\hspace{1.5cm}}$$

  $$\underline{\hspace{4cm}} \times \underline{\hspace{1.5cm}} =$$

  Interpretation: ____

  ____

Group Frequency

- ___________________ variables could have ___________________ outcomes.
- Researchers typically present variables with ___________________ outcomes in a
  ___________________ frequency distribution.
- **Class interval** refers to a category of a group frequency distribution that contains a
  ___________________, not ___________________.

4. How do you know that **Table 2.1.3** is a group frequency distribution?

Table 2.1.3. *Grouped Frequency Distribution of Hours Spent Watching TV Per Week*

Hours Watching TV	True Limits	f	Midpoints	cf	$c\%$
0–9		13			
10–19		15			
20–29		9			
30–39		7			
40–49		6			
Total		50			

- **Stated limits** are the ______________________________.
- **True limits** are the ______________________________ in a group frequency distribution.
  - **Upper true limit** = the upper stated limit ______________________________ the unit of measurement.
  - **Lower true limit** = the lower stated limit ______________________________ the unit of measurement.
- **Midpoint** is the ______________________________.
- The midpoint is the ______________________________ to ______________________________.
- **Cumulative frequencies** (*cf*) refer to the number of cases that ______________________________.
- **Cumulative percentage** (c%) is the percentage of cases that ______________________________.
- Cumulative frequencies or percentages can be calculated for ______________________________.

What if the scores in **Table 2.1.3** were organized from high to low, as in **Table 2.1.4**?

Table 2.1.4 *Grouped Frequency Distribution of Hours Spent Watching TV Per Week*

Hours Watching TV	True Limits	f	Midpoints	cf	$c\%$
40–49		6			
30–39		7			
20–29		9			
10–19		15			
0–9		13			
Total		50			

5. In which class interval would you classify someone who watched

  (a) 19.7 hours of TV?

  (b) 12.6 hours of TV?

  (c) 39.4 hours of TV?

  (d) 29.5 hours of TV?

6. What is the best single number to represent those who watched

  (a) 10–19 hours of TV?

  (b) 30–39 hours of TV?

7. What does the cumulative frequency that you calculated for the class interval…

    (a) 0–9 mean?

    (b) 20–29 mean?

8. What does the cumulative percentage that you calculated for the class interval.

    (a) 10–19 mean?

    (b) 40–49 mean?

Percentile Rank

• Percentile rank is a number that indicates the ________________________ in a distribution that ________________________ .

9. Given the following data:

$$94, 91, 85, 80, 77, 74, 71, 65, 56, 48$$

    (a) What is the percentile rank for 80?

    (b) What is the percentile rank for 65?

Cross-tabulations

• Cross-tabulations are tables that show ____

____

10. The subscript referring to the first variable is usually "i" and the subscript for the second variable is usually "j."

f_{ij} refers to the frequency in the ith row and the jth column.

f_{12} refers to the frequency in the _________ row and the _________ column.

f_{21} refers to the frequency in the _________ row and the _________ column.

Given the data on the following 20 students, create a cross-tab table.

	Sex	Major		Sex	Major
1	Female	Criminology	11	Male	Criminology
2	Male	Sociology	12	Male	Sociology
3	Female	Nursing	13	Female	Nursing
4	Female	Sociology	14	Female	Sociology
5	Female	Social Work	15	Female	Nursing
6	Female	Sociology	16	Female	Sociology
7	Male	Social Work	17	Male	Social Work
8	Female	Social Work	18	Female	Social Work
9	Male	Sociology	19	Male	Sociology
10	Female	Social Work	20	Female	Social Work

Title ____

	SEX		
Major	Male	Female	TOTAL
Criminology			
Nursing			
Social Work			
Sociology			

TOTAL

11. What is

(a) the marginal frequency for Social Work? _________________________

(b) the frequency for $cell_{21}$? _________________________

(c) the frequency for $cell_{22}$? _________________________

Total Percentages *are calculated by dividing each* ________*by the* _______________ _________________
and _____________________________________.

12. Using the data from the table in question 11, create a table with total percentages.

Title ____

	SEX		
Major	Male	Female	TOTAL
Criminology			
Nursing			
Social Work			
Sociology			

TOTAL

What does

(a) the marginal 65% mean?

(b) the marginal 15% mean?

(c) the percentage in $cell_{12}$ mean?

(d) the percentage in $cell_{41}$ mean?

Row Percents *are calculated by dividing each* _____________________ *by the* __________________
_____________________ *and* ___________________.

13. Using the data from the table in question 11, create a table with row percentages.

Title ____

Major	SEX		TOTAL
	Male	Female	
Criminology			
Nursing			
Social Work			
Sociology			

TOTAL

What does

   (a) the percentage in $cell_{12}$ mean?

   (b) the percentage in $cell_{31}$ mean?

14. Using the data from the table in question 11, create a table with column percentages.

Title ____

Major	SEX		TOTAL
	Male	Female	
Criminology			
Nursing			
Social Work			
Sociology			

TOTAL

What does

   (a) the percentage in $cell_{12}$ mean?

   (b) the percentage in $cell_{31}$ mean?

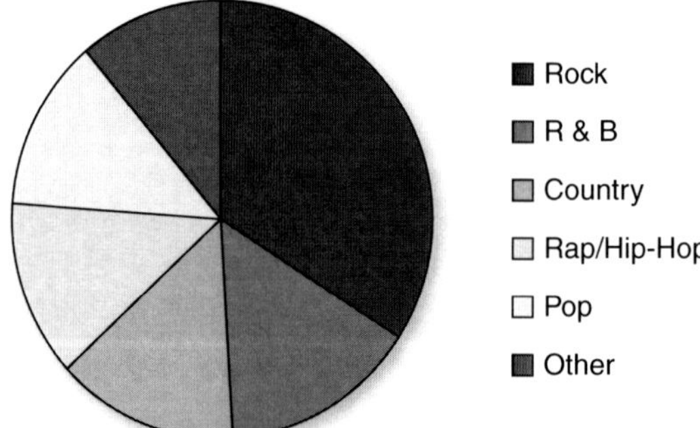

Identify the above type of graphic. _____________________________

Write a conclusion about the data presented in the above graphic.

____

____

Top Music Genres by Precentage of Sales

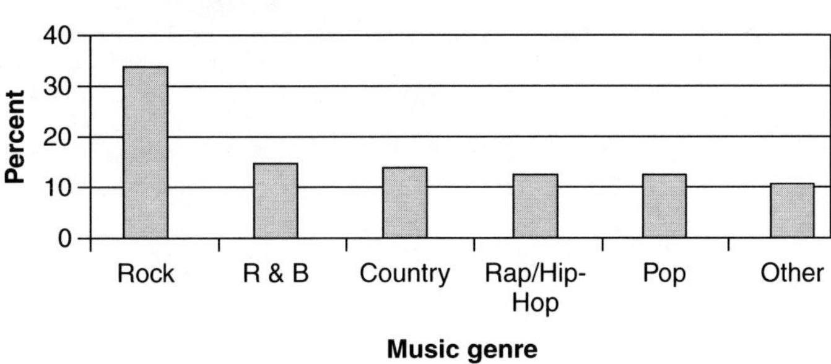

Identify the above type of graphic. ________________________

Write a conclusion about the data presented in the above graphic.

____

____

Average Hourly Compensation of Full-Time Employees at Acme Records

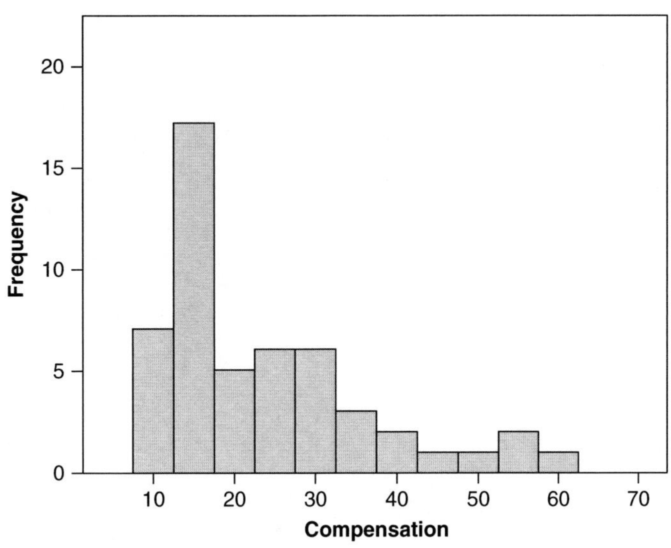

Identify the above type of graphic. ________________________

Write a conclusion about the data presented in the above graphic.

____

____

Average Hourly Compensation of Full-Time Employees at Acme Records

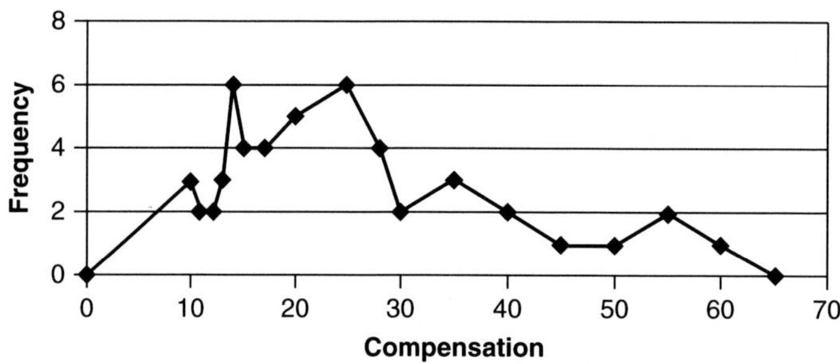

Identify the above type of graphic. __________________________________

Write a conclusion about the data presented in the above graphic.

____

____

Average Hourly Compensation of Full-Time Employees at Acme Records

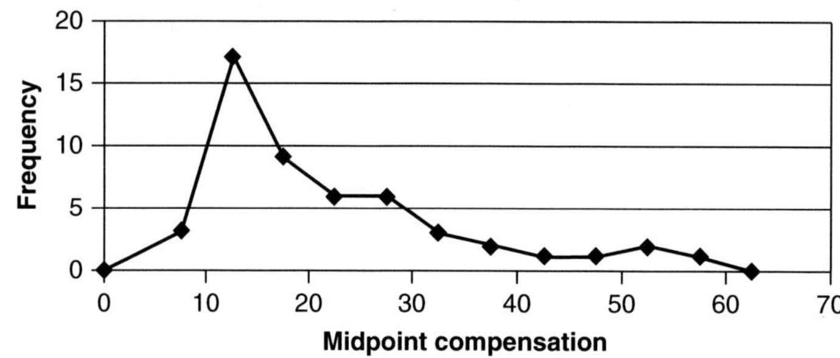

Identify the above type of graphic. __________________________________

Write a conclusion about the data presented in the above graphic.

____

____

Average Hourly Compensation of Full-Time Employees at Acme Records

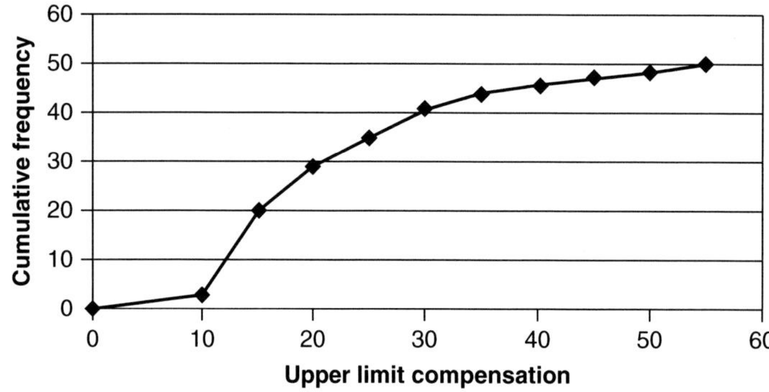

Identify the above type of graphic. __________________________________

Write a conclusion about the data presented in the above graphic.

____

____

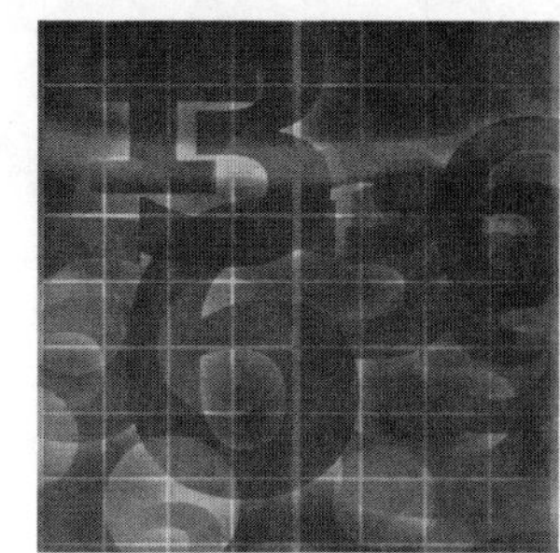

Learning Check

1. A researcher observes children at recess. The following symbols were used to record their behaviors: M = playing marbles; H = playing hopscotch; K = playing kickball; T = talking; O = other.

   The raw observations for the students are:

H	K	K	T	M	H	K	K	M	H	M	M
T	K	M	O	M	K	T	H	K	M	K	O

   Complete the following table.

   Students' Recess Activities

Activity	f	p	%
Marbles			
Hopscotch			
Kickball			
Talking			
Other			
Total			

   (a) How many children were playing hopscotch?

   (b) What percent of children were talking?

   (c) In which activity did the greatest proportion of children participate?

2. From the data in question 1, what is the ratio of children playing kickball to hopscotch? Interpret the ratio in a sentence.

3. There are 11 right-handers and 4 left-handers on the baseball team. What is the ratio of right-handers to left-handers? Interpret the ratio in a sentence.

4. If 44,000 people die in automobile accidents in a population of 3 million, what is the death rate due to auto accidents (per 1000 people)?

5. The following table provides application, admission, enrollment, and retention records for State U.

   Student Applications, Admission, Enrollment, and Retention Numbers for State U.

Number of applicants for fall 2010	7400
Number of students admitted in fall 2010	5200
Number of students enrolled in spring 2010	18,000
Number of students who were retained after spring 2010	11,500

   (a) Calculate the admission rate.

   (b) Calculate the retention rate.

6. Given the following IQ scores:

$$81, 88, 95, 103, 106, 112, 127$$

(a) What is the percentile rank for someone who has an IQ of 95?

(b) What is the percentile rank for someone who has an IQ of 106?

7. Given the following 40-yard dash times:

$$4.49, 4.47, 4.46, 4.45, 4.43, 4.42, 4.41, 4.28$$

(a) What is the percentile rank for someone whose time was 4.41?

(b) What is the percentile rank for someone whose time was 4.47?

8. Create a group frequency distribution from the following raw data. Create five class intervals with a class size of 10, with 99 as the upper stated limit of the highest class interval. Calculate true limits, midpoints, and cumulative percentages.

80	98	72	65	77
85	85	70	65	70
77	75	82	80	80
60	53	55	93	88
85	90	87	60	80
90	68	82	77	75
80	73	62	92	65
65	60	75	70	68
73	85	77	87	87
70	95	82	90	80
75	85	50	87	75
83	80	55	57	95
50	82	50	65	55

Test Scores of Students

Test Scores	True Limits	f	Midpoints	cf
Total				

9. Complete the following table. Calculate true limits, midpoints, and cumulative percentages.

IQ Scores of 50 People

IQ Scores	True Limits	f	Midpoints	cf	$c\%$
56–70		1			
71–85		7			
86–100		17			
101–115		17			
116–130		7			
131–145		1			
Total		50			

10. The following is a cross-tabulation of test results by studied for the test for a random sample of students.

Test Results	Studied for the Test		
	Yes	No	
Passed	34	3	37
Failed	5	8	13
	39	11	50

a. Are there independent and dependent variables in this table? If so, what are they? If not, why not?

b. Based on your answer for part a, calculate the most appropriate percents for the cross-tabulation.

Test Results	Studied for the Test	
	Yes	No
Passed		
Failed		

c. What percentage of the students passed the test?

d. What percentage of the students who did not study failed the test?

e. What percentage of the students who studied failed the test?

f. What percentage of students who studied passed the test?

g. What can be concluded about the relationship between the variables?

11. The following is a cross-tabulation of people attitudes regarding government funding for protecting the environment and improving the educational system.

Protecting the Environment by Improving Education

Protecting Environment	Improving Education			
	Too Little	About Right	Too Much	
Too Little	26	12	4	42
About Right	11	13	1	25
Too Much	8	2	3	13
	45	27	8	80

a. Are there independent and dependent variables in this table? If so, what are they? If not, why not?

b. Based on your answer for part a, calculate the most appropriate percents for the cross-tabulation.

Protecting the Environment by Improving Education

Protecting Environment	Improving Education		
	Too Little	About Right	Too Much
Too Little			
About Right			
Too Much			

c. What percentage of people thought government funding for protecting the environment was about right?

d. What percentage of people thought government funding for improving the educational system was too little?

e. Interpret the percentage that you called for $cell_{12}$.

f. Interpret the percentage that you called for $cell_{31}$.

12. (a) Construct a pie chart (using percentages) using the data from question 1.

(b) Construct a bar graph (using frequencies) using the data from question 1.

13. In a survey of 80 Americans, respondent were asked their opinion regarding government efforts to protect the environment. Construct (a) a pie chart (as percentages) and (b) a bar graph (as frequencies) for the following results: 42 said "too little," 25 said "about right," and 13 said "too much."

14. Construct a (a) histogram, (b) a frequency polygon, and (c) a cumulative frequency polygon using the data in question 8.

Chapter Three

Measures of Central Tendency

Researchers often want a single number to represent all of the cases of a variable in a sample or an entire population. Measures of central tendency often serve this purpose because measures of central tendency refer to the most common or typical value that represent a distribution. These values are generally located in the middle, or center, of the distribution. While there are many measures of central tendency, we will focus on the three most commonly used measures: the mode, median, and arithmetic mean. Each of these measures is calculated differently, given the level of measurement of the variable. The shape of the distribution (symmetrical versus skewed) can affect the appropriateness of using these various measures of central tendency.

The Mode

The **mode** is the value that occurs most frequently in a distribution. For example, since more adult Americans currently are married than having any other marital status, the modal marital status is currently married. Similarly, if more students in your class are majoring in sociology than in any other discipline, then the mode for students' major is sociology. However, in another section of the course, there could be just as many social work majors as sociology majors. In that case, we would have a bimodal distribution with two modes: sociology and social work.

The mode is appropriate for all levels of measurement. As the previous example shows, we can determine the mode for nominally measured variables such as marital status and academic major. We could also determine the mode for an ordinal, interval, or ratio measured variable. For example, if more student workers make $7.75 an hour than any other wage, then the modal wage of students would be $7.75. Since the mode can be calculated for any level of measurement, it may seem to be a useful statistic. However, it is limited because it only tells us that more cases in a data set fall in that category of a variable than in any other category of the variable. In some instances, the mode can represent less than half the cases.

The Median

The **median** is the middle point in an ordered raw-data array or frequency distribution. Since the median is the middle point, 50% of the cases fall below the median and 50% of the cases fall above the median. Because the data must be ordered, the variable must be measured at the ordinal, interval, or ratio level.

Suppose we want to find the median of seven people's responses to the question: To what social class do you belong? Their responses were:

Middle Class	Working Class	Middle Class	Upper Class
Working Class	Working Class	Middle Class	

First, the responses need to be ordered from highest to lowest (or lowest to highest). Since we have an odd number of cases ($N = 7$), the median is the value of the middle case. In this instance, case 4 is the middle case and its value is middle class. Therefore, the median social class of the respondents is middle class.

Upper Class	Middle Class	Middle Class	**Middle Class**
Working Class	Working Class	Working Class	

Let's consider an example in which the number of cases was even. What if another respondent who was working class had been included in the above survey? The median would fall between middle class and working class. If we had interval-level data, we could average the two scores. However, it is not appropriate to average ordinal scores. The median simply falls between the two responses of working class and middle class.

With interval-level or ratio-level data and an even number of cases, we can average the two middle scores. Suppose we have a sample comprised of six students of the following ages: 20, 20, 21, 22, 29, and 41. The middle point is between 21 and 22, or 21.5. Thus, the median age of the students is 21.5 years old. Similarly, if the distribution of ages were 20, 20, 21, 23, 28, and 41, then the median age would be 22 years of age.

With so few cases, it is easy to look at the data and determine which value is the middle point of the distribution. However, with larger data sets, this method would not be realistic. The position of the median can be located by:

$$\text{Position of the median} = \frac{N+1}{2}$$

If we apply this formula to the first example above, with an odd number of cases, we find that $(7 + 1) \div 2 = 4$. Therefore, the median is equal to the value of the fourth case in the distribution, or middle class. In the example with an even number of cases, we had an N of 8, and $(8 + 1) \div 2 = 4.5$. Therefore, the median falls between the fourth and fifth cases and is between working and middle classes. Using the ratio-level data, the median falls between the third and fourth cases and is 21.5 years of age. In the last example, that would be between 21 and 23 years of age. Thus, the median age of that distribution is 22 years.

An advantage of the median is that it is unaffected by extreme cases. If we return to the data set with ages of 20, 20, 21, 23, 28, and 41, all the students are in their twenties except one, who is 41 years old. However, the fact that this one student is older does not affect the median. Moreover, that student could be 101 years old and this still would not affect the median, since the value of the middle point does not change. Therefore, when there are extreme cases, the median can give us a better understanding of central tendency.

The Mean

The **mean** is considered the most powerful measure of central tendency because it takes every case into account. However, it can only be calculated using interval-level or ratio-level variables. The mean is viewed as the balancing point of the distribution, compensating, if need be, for cases at the extremes. While there are other means (such as the harmonic mean), we will focus on the arithmetic mean.

To calculate the mean (using $\overline{X}$, referred to as X bar) we sum all the scores and divide by the total number of cases:

$$\overline{X} = \frac{\Sigma X_i}{N}$$

If we return to our example of six cases of varying ages, we add 20, 20, 21, 23, 28, and 41 and then divide by the number of cases. Therefore, the mean age of the students is 25½ years old:

$$\overline{X} = \frac{\Sigma X_i}{N} = \frac{153}{6} = 25.5$$

Locating Measures of Central Tendency

If observations of a variable are symmetrically distributed, then all three measures of central tendency have the same value. As shown in **Figure 3.1(a),** the mode is the highest point since the mode is the value that occurs most often in the distribution. The median also falls at the highest point, with half of the cases falling at the median or above and the other half of the cases falling at the median or below. Since the cases are normally distributed, there are no extreme cases for which the mean must compensate. A symmetrical distribution is balanced with the mean located at the center of the distribution.

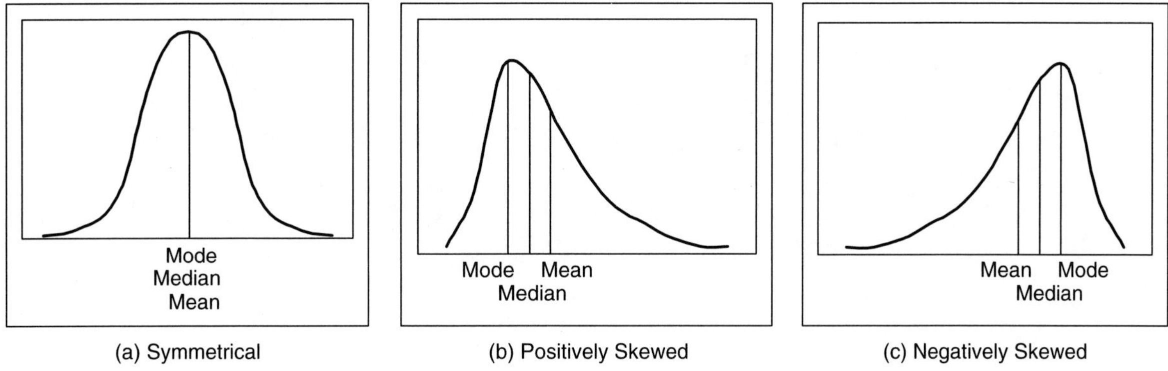

Figure 3.1 Measures of Central Tendency in Symmetrical and Skewed Distributions

In a skewed distribution, the mode again is the highest point. However, the value of the mean in a skewed distribution is never equal to the value of the mode. A skewed distribution has at least one outlier for which the mean must compensate. As the positively skewed distribution in **Figure 3.1(b)** shows, the distribution has a few high cases, which extend the tail, or end, in the positive direction. The mean moves towards the positive end or tail to compensate for those few high values. The opposite occurs in a negatively skewed distribution. There are relatively few low cases, producing a tail at that end of the distribution. The mean has a relatively low value to balance the low-scoring outliers in the distribution. Since the mean takes all values in the distribution into account and is affected by extreme cases, it will be located closer to the tails of skewed distributions.

The median of the distribution may or may not equal the mode. **Figure 3.1(b)** and **(c)** show where the median falls relative to the other measures of central tendency when the value of the median and mode are not the same. If the values of the median and mode are not equal, then the median falls between the mode and the mean. However, in grouped data, which we cover later in this chapter, this may not occur, since we no longer know the precise value of each case. In group data, we will not know the exact number of times a value occurs but rather the number of times that cases fall within a range of values.

Determining Measures of Central Tendency from Simple Frequency Distributions

As we saw in Chapter 2, besides having raw data, data may be organized in simple or grouped frequency distributions. We can still obtain measures of central tendency from data in frequency distributions. However, the way that we obtain the measures may differ if we have raw data, a simple frequency, or a grouped frequency distribution.

Suppose we have data from a survey of high-school cell-phone users in which respondents were asked: At what age did you get your first cell phone? The following are a raw-data array of their responses:

7	8	9	9	9	10	10	10	10	10	10
10	10	10	11	11	11	11	11	11	11	12
12	12	12	12	12	13	13	13	13	13	13
13	13	13	14	14	14	14	14	14	14	15
15	15	15	15	16	16	16	16	17	17	18

There are two ages, 10 and 13, at which most respondents received their first cell phone. Nine students indicated that they received their first cell phone when they were 10 years old, while another nine students said that they were 13 years old when they got their first cell phone. Thus, the distribution is bimodal, as the mode equals 10 and 13.

These raw data could also be organized in a frequency distribution, as shown in **Table 3.1.** In a frequency distribution, the mode is determined by looking for the highest frequency, which indicates the value of the variable that occurs most often. By looking at the frequency, we see that the highest frequency is 9. Because that highest frequency occurs twice, we have a bimodal distribution. The values associated with the highest frequency are 10 and 13 years old. Thus, the mode is 10 and 13, and we can conclude that more than any other age, students got their first cell phone at ages 10 and 13.

**Table 3.1 *Worksheet for Identifying the Mode
from a Simple Frequency Distribution***

Age at First Cell Phone	f
7	1
8	1
9	3
10	9
11	7
12	6
13	9
14	7
15	5
16	4
17	2
18	1

(mode → 10, f = 9 ← highest frequency; mode → 13, f = 9 ← highest frequency)

We also can calculate the median from a simple frequency distribution. The median is the middle case in an ordered data array. In a frequency distribution, the values of the variable are ordered from lowest to highest or vice versa. We will use the formula that we used to find the position of the median of raw data:

$$\text{Position of median} = \frac{N+1}{2}$$

If we use the frequency distribution of "Age at First Cell Phone," the position of the median would be the 28th case, since the total number of cases (N) is 55.

$$\frac{55+1}{2} = 28^{th} \text{ case}$$

To determine what age is associated with the 28th case, we use the cumulative frequencies. The cumulative frequency tells us how many students got their first cell phone at that age or younger. For example, 14 students got their first cell phone at age 10 or younger. However, we are not looking for the 14th case but rather the 28th case. As we continue down the cumulative frequency column, we see that 27 students got their first cell phone when they were 12 or younger. Therefore, cases 27 through 36 got their first cell phones when they were 13. Since the 28th case first got a cell phone at 13 years old, the median is 13.

Table 3.2 *Worksheet for Identifying the Median*
from a Simple Frequency Distribution

Age at First Cell Phone	*f*	*cf*
7	1	1
8	1	2
9	3	5
10	9	14
11	7	21
12	6	27
13	9	36
14	7	43
15	5	48
16	4	52
17	2	54
18	1	55
	55	

value of median ⟶ 13   36 ⟵ position of median

To calculate the mean from a simple frequency distribution, we need to modify the formula that we used to calculate the mean using raw data. When we had raw data, we added up all the scores and divided by the total number of cases. When using the data from the frequency distribution of "Age at First Cell Phone," we must take into account that three students got their first cell phone when they were age 9 and that nine students got their first cell phone when they were age 10, and so on. Thus, in order to take into account the number of times each category of the variable (in this instance, each age) occurs, we multiply it by the frequency of the occurrence:

$$\overline{X} = \frac{\Sigma f x}{N}$$

Table 3.3 *Worksheet for Calculating the Mean*
from a Simple Frequency Distribution

Age at First Cell Phone	*f*	*fx*
7	1	7
8	1	8
9	3	27
10	9	90
11	7	77
12	6	72
13	9	117
14	7	98
15	5	75
16	4	64
17	2	34
18	1	18
	55	687

$$\overline{X} = \frac{687}{55} = 12.49$$

The mean age at which students got their first cell phone was nearly 12½ years old.

Determining Measures of Central Tendency from Grouped Frequency Distributions

As we saw in Chapter 2, it is often impractical to display some interval-level or ratio-level variables in a simple frequency distribution because they have too many outcomes. Variables with numerous outcomes are combined into a grouped frequency distribution. Although the age at which students got their first cell phone did not have an unmanageably large number of outcomes, we will use it to illustrate how the measures of central tendency are obtained from a grouped frequency distribution. Thus, we will be able to compare the measures of central tendency obtained from raw, simple frequency, and group data.

Table 3.4 *Group Frequency Distribution of the Age at which Students Got their First Cell Phone*

Age at First Cell Phone	f
7−9	5
10−12	22
13−15	21
16−18	7
	55

To calculate the measures of central tendency from grouped data, we need to modify the procedures we followed when using simple frequency data. To determine the mode of grouped data, we look for *the* value associated with the highest frequency. In our example, the highest frequency is 22. However, when we go to the age at which students got their first cell phone, the age is not a single value but instead a range from 10 to 12 years old. To arrive at a value, we use the midpoint of that class interval because the midpoint is the best single number to represent the class interval. The midpoint is the upper limit plus the lower limit divided by 2. In this instance, we calculate $10 + 12 = 22 \div 2 = 11$. Therefore, the mode at which students got their first cell phone is 11 years old. When we grouped the data, we lost the precise age at which students got their first cell phone. Therefore, our mode of 11 approximates the actual bimodal distribution (Mo = 10 and 13) of the raw data and also that found in the simple frequency distribution.

The first step in finding the median from a grouped frequency distribution is the same as it is with raw data and simple frequency data; we use the following formula to find the location of the median:

$$\text{Position of the median} = \frac{N+1}{2} = \frac{55+1}{2} = 28^{\text{th}} \text{ case}$$

Once we know that the position of the median is the 28th case, we use the cumulative frequencies to determine the class interval in which the median is located. We can see that 27 students reported getting their first cell phone when they were 12 years old or younger. Therefore, the 28th case is located in the class interval of 13–15 years old.

Table 3.5 *Worksheet for Identifying the Median from a Group Frequency Distribution*

Age at First Cell Phone	f	cf
7−9	5	5
10−12	22	27
13−15	21	48
16−18	7	55
	55	

After we determine the critical interval in which the median is located, we use the following formula to calculate the median from a grouped frequency distribution:

$$\text{Median} = L + \left(\frac{\frac{N}{2} - cf_l}{f} \right) i$$

L = Lower true limit of cases in the critical interval

N = Number of cases

cf_l = cumulative frequency that is less than lower true limit of the critical interval

f = frequency of the critical interval

i = size of the class interval

$$\text{Median} = 13.5 + \left(\frac{\frac{55}{2} - 27}{21} \right) 3 = 13.57$$

Thus, our approximate median calculated from the grouped data is very close to the actual median of 13 obtained from the raw data or a simple frequency distribution.

The calculation of the mean from grouped data also involves modifying the formula used to calculate the mean. When we calculated the mean using simple frequency data, we modified the formula for the mean by multiplying each outcome by its frequency. However, with grouped data, we do not know how many times each outcome occurs but we do know the number of times a range of outcomes (the class interval) occurs. Since the midpoint is the best number to represent the class interval, we replace the value of each outcome (x) with the midpoint (m). Therefore, the formula used to calculate the mean from a grouped frequency distribution is:

$$\overline{X} = \frac{\Sigma fm}{N}$$

To calculate the mean from group data, we must do a similar procedure to that which we did when we computed the mean from simply frequency data. As the formula above indicates, we must obtain the Σfm. Thus, we multiply each frequency (f) by each corresponding midpoint (m) and then sum all of the products to obtain Σfm.

Table 3.6 *Worksheet for Calculating the Mean from a Group Frequency Distribution*

Age at First Cell Phone	f	m	fm
7−9	5	8	40
10−12	22	11	242
13−15	21	14	294
16−18	7	17	119
	55		695

$$\overline{X} = \frac{\Sigma fm}{N} = \frac{695}{55} = 12.64$$

When we calculate the mean from the grouped data on the age at which students got their first cell phone, we find that the mean is 12.64. Thus, on average, students got their first cell phone when they were a little older than 12½ years old. This means that what we approximated from the group data ($\overline{X} = 12.64$) is very close to the actual mean calculated from the raw data or simple frequency ($\overline{X} = 12.49$).

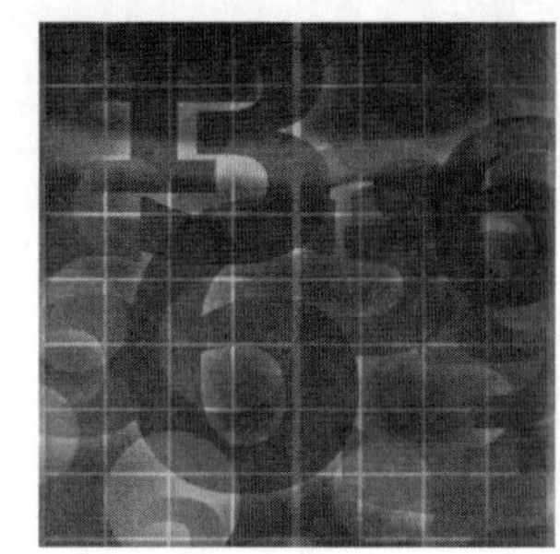

Exercise and Notes

- **Mode**
  - The value of the category or class interval that occurs ___________________. In a frequency distribution, the mode has the ___________________ frequency.
  - Limited—only tells us that more cases fall in that category than in any other.

- **Median**
  - Middle point in an ordered raw data array or frequency distribution; thus, ___________________ of the cases fall below the median and ___________________ of the cases fall above the median.
  - Appropriate for only ordinal and interval/ratio variables, since the data must be ordered.

- **Mean**
  - The arithmetic mean is commonly thought of as the ___________________, determined by summing all the scores and divide by the total number of cases.
  - A powerful measure, but it can only be calculated using interval-level or ratio-level variables.

Determining Measures of Central Tendency Using Raw Data

1. Look around the classroom. What is the modal sex of students attending class today?

   ____

   ____

2. Can we determine the median or mean sex of the students in the classroom?

   ____

   ____

   ____

3. Can we find the mode and median of the following students' responses to the question: Are you satisfied with the food service on campus?

STUDENT	RESPONSE
John	Very Satisfied
Lee	Somewhat Dissatisfied
Sue	Very Dissatisfied
Chris	Very Dissatisfied
Taylor	Somewhat Satisfied

4. What if another student were surveyed? Include in the data set for question 3 Morgan, who indicted that she was "Somewhat Satisfied." What is the median?

STUDENT	RESPONSE
John	Very Satisfied
Lee	Somewhat Dissatisfied
Sue	Very Dissatisfied
Chris	Very Dissatisfied
Taylor	Somewhat Satisfied

5. The IQs of 11 students are:

$$94, 97, 99, 102, 105, 105, 105, 107, 111, 111, 129$$

Calculate the measures of central tendency for these scores.

Determining Measures of Central Tendency Using Simple Frequency Data

6. The GPAs of 60 students applying for graduate programs at State University are listed in the table below. Calculate the measures of central tendency.

GPAs of 60 Graduate School Applicants to State University

GPA	f		
3.8	3		
3.7	4		
3.6	5		
3.5	6		
3.4	6		
3.2	7		
3.1	9		
3.0	8		
2.9	5		
2.8	4		
2.6	2		
2.5	1		

Determining Measures of Central Tendency Using Grouped Data

7. Calculate the measures of central tendency of the GPAs of 60 students applying for graduate programs at State University from the following grouped frequency distribution.

GPAs of 60 Graduate School Applicants to State University

GPA	f		
3.6−3.8	12		
3.3−3.5	12		
3.0−3.2	24		
2.7−2.9	9		
2.4−2.6	3		
Total	60		

8. Are the measures of central tendency for the students' GPAs computed from the raw data and simple frequency the same as the measures calculated from the grouped data?

Measure of Central Tendency	Raw Data/ Simple Frequency	Grouped Data
Mode		
Median		
Mean		

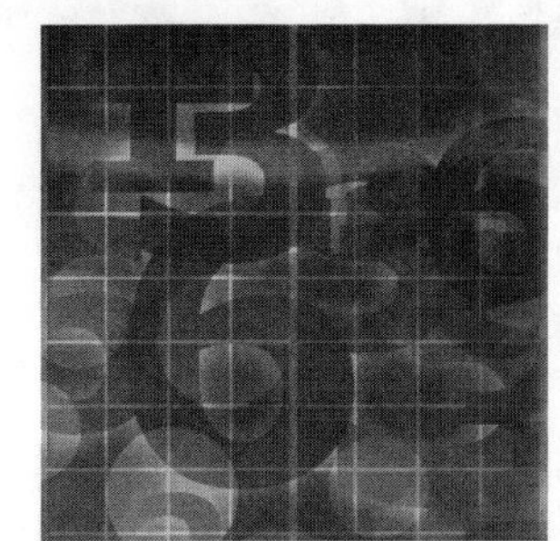

Learning Check

1. Find the measures of central tendency for each of the following variables (sex, ethnicity, age, and health).

Name	Sex	Ethnicity	Age	Health
Bob	Male	Euro-American	72	Good
Phyllis	Female	Euro-American	87	Fair
Karen	Female	African American	77	Poor
Mike	Male	Latino	66	Poor
Cathy	Female	Euro-American	81	Fair
Don	Male	African American	69	Good
Maria	Female	Latino	66	Fair

2. (a) Determine the measures of central tendency from the following ages of students.

   17   17   18   18   18   20   21   21   21

   21   21   23   23   24   25   26   26   27

   (b) Given the values of the measures of central tendency that you calculated, how would you describe the shape of the distribution?

3. A group of college students were asked: How old were you when you went out on your first date? Calculate the measures of central tendency from the following data.

Age at First Date	f
12	1
13	3
14	8
15	9
16	6
17	4
18	2

4. The number of televisions in 100 American households is displayed below. Calculate the measures of central tendency.

Number of TVs per Household	f
0	1
1	23
2	11
3	60
4	5

5. Suppose we ask a group of customers at a local bar: How many days in the past month have you had an alcoholic drink? Calculate the measures of central tendency.

Days of Alcohol Use	f
1−5	18
6−10	11
11−15	6
16−20	11
21−25	4
26−30	5

6. The grade distribution for a math class is listed below. Calculate the measures of central tendency.

Grades	f
90−99	7
80−89	18
70−79	14
60−69	12
50−59	11

Chapter Four

Variability

Now that you understand the mean, median, and mode, you are on the way to understanding and making sense of how data are distributed across a sample. This provides us with important information about our data. The next concepts to understand are the measures of **variability** for the data. Variability refers to the amount of variation that we have across a sample. This chapter will discuss the **range**, **variance,** and **standard deviation**.

The **range** refers to the distance between the highest number and the lowest number in a data set. We find this difference by subtracting the lowest value from the highest value. Consider the following data:

Student #	Age
1	18
2	18
3	21
4	22
5	23
6	28

If we want to know the range, we subtract the lowest age from the highest age:

$$28 - 18 = 10$$

Thus, the range for this data set is 10.

Calculating the range is the first step in understanding how much variability there is within the data set. Ten years between the youngest and oldest student is moderate. We can tell by examining this number that there will be a generational divide between the students. An 18-year-old would be expected to be a freshman, likely away from home for the first time, and most likely not yet oriented to academic life. A 28-year-old is more likely to be a senior or graduate student and living independently from his or her parents. This student would also be likely to have a job and perhaps a family. If we understand this variability, it helps us to be better prepared for addressing the needs and potential problems that the students might have in class.

Calculating Variance and the Standard Deviation

The mean, median, mode, and range can give us useful information about our sample. What they fail to do, however, is tell us how wide the variation in our sample happens to be. For that, we turn to the variance and the standard deviation.

Imagine that you work for a marketing research firm and you are told to develop a marketing plan for two movies. Each of these movies has an average viewer age of 30 years old. It seems reasonable that you would want to target places where the average 30-year-old would see your advertisements. You might recommend that television ads run between 9:00 P.M. and midnight. You might also recommend running print ads in magazines such as *Cosmo, GQ* and *People.* You might be surprised to find that one of your marketing plans is highly successful while the other fails. It is possible for two widely different populations to have the same average age. Let's take a closer look at our movie goers.

	Theater One	Theater Two
Age	50	28
	48	32
	52	34
	12	26
	10	31
	8	29
Average Age	30	30

Both samples have the same average age of 30 years old. However, a closer inspection of the data shows that Theater One is comprised of children and older adults, and Theater Two is comprised of young adults. While the mean is mathematically accurate for both groups, it is only a good representation for one of the groups. We use the variance and the standard deviation to help us determine whether or not the mean is a good indicator of the sample.

The **variance** is the average of the squared distance from the mean, and the **standard deviation** is the average distance from the mean, or the square root of the variance. These two values are very closely related. The most commonly used of the two is the standard deviation. These measures allow us to determine how closely our population clusters around the mean. We are going to calculate both of these next. The calculations are numerous and may appear to be complex, but in reality they entail simple addition, subtraction, multiplication, and division. We are going to calculate the variance and the standard deviation for Theaters One and Two simultaneously, but the calculations are divided into easy steps.

The first thing that we need to discuss is the formula and a few facts about data. Social science students often look at complex formulas and get confused. This is because math uses its own shorthand language of symbols. Once you understand the language of the symbols, however, it is easier to make sense of the formulas. The formulas for variance and the standard deviation use the following symbols.

X — Any value of our variable. In the theater example X stands for age. Since we have two theaters we will use X_1 (Theater One) and X_2 (Theater Two).

$\overline{X}$ = The average of X for our sample

μ = The average of X for the population

sd = the standard deviation of X for the sample

σ = the standard deviation of X for the population

Sd^2 **or** σ^2 = the variance of X for the sample or population, respectively

$\Sigma = \text{sum}$

$N = \text{the number of cases in a data set}$

As you can see, there are several symbols to learn. Also, we will make references to the sample and the population. We use sample statistics in this course because we use sample data. How do we tell when to use a sample versus the population? In the example of the theaters, if we had the age of every single movie goer in the United States, then we would use population statistics. Since we are only interested in two theaters, we have a sample. The majority of studies use samples that are selected from the population.

The following are the formulas for variance:

$$\sigma^2 = \frac{\Sigma(x - \bar{\mu})^2}{N} \qquad\qquad sd^2 = \frac{\Sigma(x - \bar{x})^2}{N-1}$$

You will notice that there are only two differences in these formulas. The one on the left uses the population mean in the numerator and N in the denominator. The one on the right uses the sample mean in the numerator and $N - 1$ in the denominator. The -1 is the difference that has mathematical consequences. If we use the entire population, then we can say without a doubt that our data represents the group. When we take a random sample, we can say with confidence that our sample represents the group, but not with certainty. We acknowledge that our sample always has bias. For example, if we do a telephone survey, then we know that our sample is made up of people that have telephones. While this may apply to most people, for those who are in the percentage of people without a phone, our data may not reflect their reality. To correct for this bias, we subtract 1 from our sample size. Since we are using sample data, we will use the formula on the right. Remember, we are doing two simultaneous calculations for both theaters. We already know the average is 30 years of age for both groups, so we don't need to calculate this.

Data

Theater One	Theater Two
50	28
48	32
52	34
12	26
10	31
8	29

Step One: Find the difference between the age of each moviegoer and the mean age for the group:

$$(x - \bar{x})^2$$

Theater One	Theater Two
$50 - 30 = 20$	$28 - 30 = -2$
$48 - 30 = 18$	$32 - 30 = 2$
$52 - 30 = 22$	$34 - 30 = -4$
$12 - 30 = -18$	$26 - 30 = 4$
$10 - 30 = -20$	$31 - 30 = 1$
$8 - 30 = -22$	$29 - 30 = -1$

Step Two: Find the square of the difference. We square the differences before we sum them because if we don't, they will always add up to zero:

$$(x - \bar{x})^2$$

Theater One	Theater Two
$20^2 = 400$	$-2^2 = 4$
$18^2 = 324$	$2^2 = 4$
$22^2 = 484$	$4^2 = 16$
$-18^2 = 324$	$-4^2 = 16$
$-20^2 = 400$	$1^2 = 1$
$-22^2 = 484$	$-1^2 = 1$

Step Three: Sum the squares:

$$\Sigma(x - x)^2$$

Theater One	Theater Two
400	4
324	4
484	16
324	16
400	1
+484	+1
2416	42

Step Four: Divide the sum of the squares by $N - 1$. Since each theater has 6 attendees, we divide by 5.

Theater One	Theater Two
$2416/5 = 483.2$	$42/5 = 8.4$

The variance for Theater One is 483.2 and the variance for Theater Two is 8.4. This tells us the average of the sum of squares for the theaters. We can see that there is much more variability for Theater One than there is for Theater Two.

The variance is not as useful as the standard deviation is, however. The standard deviation tells us the average difference from the mean for the sample. We add one step to the variance formula to calculate the standard deviation. That formula (for the sample) is:

$$sd = \sqrt{\frac{\Sigma(x = \bar{x})^2}{N = 1}}$$

Once we have the variance, finding the standard deviation is as simple as finding the square root.

Step Five: Find the square root of the dividend:

$$\sqrt{483.2} = 22 \qquad\qquad\qquad\qquad \sqrt{8.4} = 2.9$$

Interpreting the Standard Deviation

Now that we have the standard deviation, how do we interpret it? We take the average and apply the standard deviation as $a +/-$ to it.

Theater One

Average age = 30

30 + 22 = 52

30 − 22 = 8

This means that the majority of people at Theater One are between the ages of 8 and 55.

Theater Two

Average age = 30

30 + 2.9 = 32.9

30 − 2.9 = 27.1

This means that the majority of people at Theater Two are between the ages of 27 and 33.

Now we can go back and look at our marketing plan. The successful marketing strategy was for Theater Two, where the mean of 30 years of age actually does reflect the sample. The unsuccessful strategy was for Theater One, where the sample was probably parents and grandparents taking children to see a movie. With the standard deviation, we can revise our marketing plan for Theater One and run ads in *Parenting Magazine*, on the Nickelodeon channel, and during daytime and early evening programming and Saturday morning cartoons.

Generally, the larger the standard deviation, the less the mean represents the sample. Keep in mind, though, that the definition of *large* is dependent upon the data. If we are looking at income, an average of $50,000 with a standard deviation of $10,000 is fine. While $10,000 is a large number, incomes within $40,000 and $60,000 are not that dissimilar. In order to accurately interpret the standard deviation, you need to know the data.

Using SPSS to Calculate the Variability Measures

The examples that we have used have extremely small samples, so it was easy to get the information that we needed to calculate the range, variance, and standard deviations by looking at the data and using a calculator. Normally, social scientists use large data sets for statistical analysis. We can get the same information easily from SPSS. Again, we will use the 2008 General Social Survey (GSS). First, we open the data file. Once it is open, we select Analyze → Descriptive Statistics → Descriptives on the tool bar at the top of the screen. (Figures for SPSS shown below.)

We are going to use the variable HRS2, which asks the respondents how many hours they normally work each week. We move that variable into the Analysis column.

Then we highlight the variable and click the arrow in the center of the box to move it into the Variable column, and then click OK.

You will notice that the variable moves to the right column when you click the arrow. Once you have moved it, click on the Options button in the upper righthand corner of the center box. This will bring up another box, where you can tell SPSS the tests that you want it to run for you.

We have selected the standard deviation, variance, range, minimum, and maximum. Mean is selected by default. Once we have selected our tests, we click continue and then click OK when we are taken back to the variable selection box.

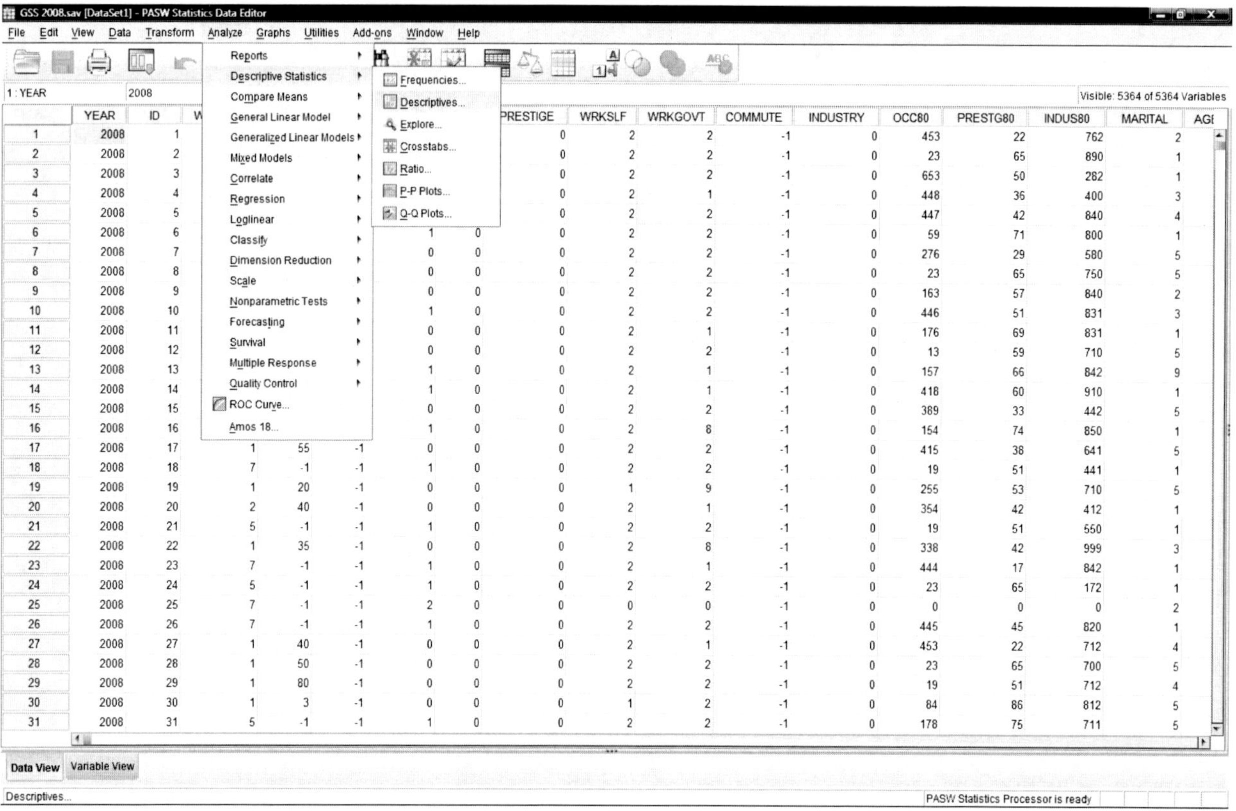

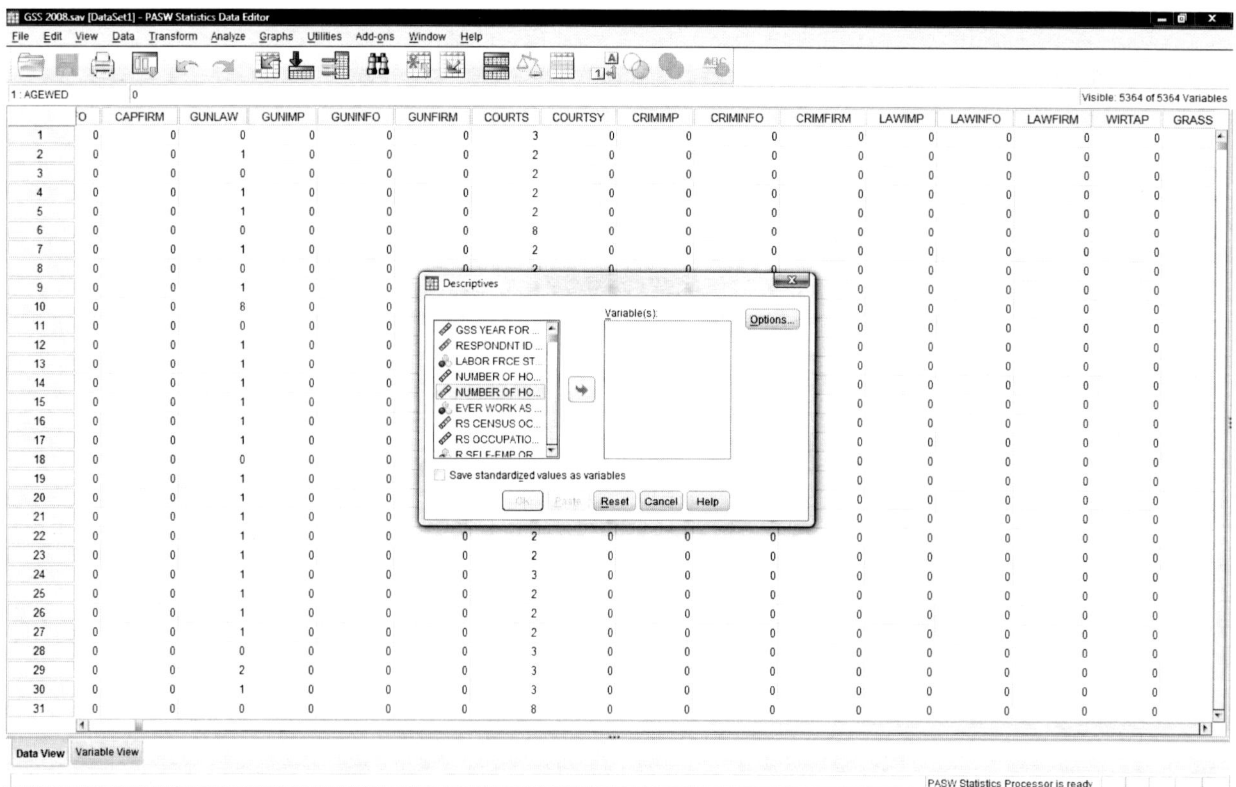

When we click OK, we see the SPSS wheels turning, and in a second, the output window opens. If it fails to open, look at the bottom of the screen. It should be flashing orange to get your attention. Click if necessary to open the window. You will see the following:

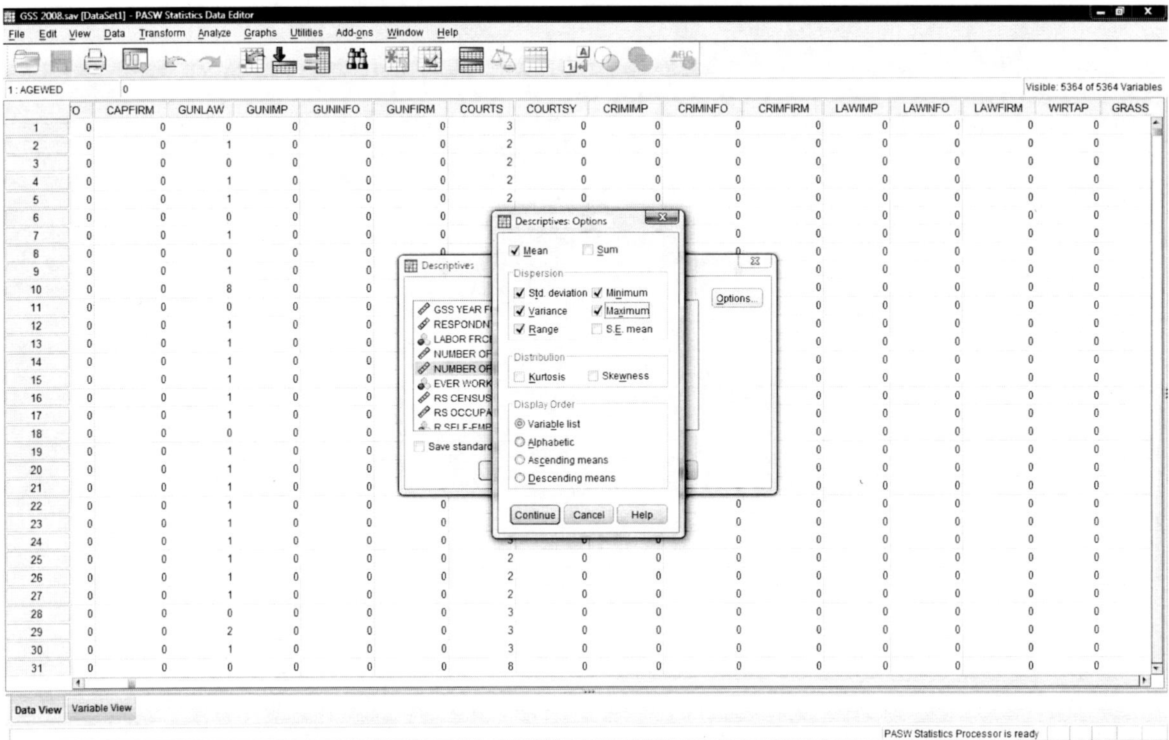

The center is where you will find your results. We'll get to that in a minute. The lefthand column is a table of contents for your output. Right now there is only one analysis, but if you use the program for a while, your output will grow. Each data run appends to the bottom of the file. You can find what you need easily by

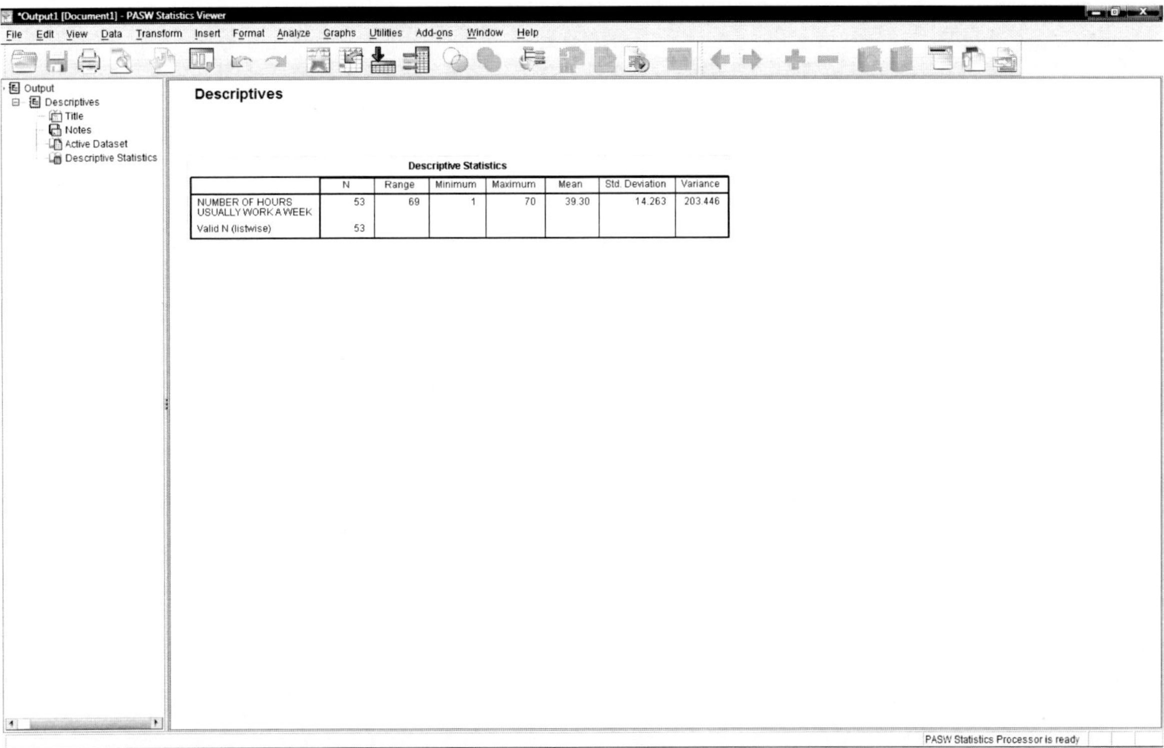

scrolling through this list and clicking on what you need. This will bring the selected analysis to the front of the screen. The top toolbar provides you with options for editing the output window to make it presentation ready. We will not cover those options in this class.

What we are really interested in is the analysis in the center of the window, shown in **Table 4.1.** This is the information that we asked SPSS to compute for us.

Table 4.1 *Descriptives*

	Descriptive Statistics						
	N	Range	Minimum	Maximum	Mean	Std. Deviation	Variance
Number of Hours Usually Work a Week	53	69	1	70	39.30	14.263	203.446
Valid N (listwise)	53						

As you can see, SPSS provides us with all of the information that we have learned to compute mentally in this chapter.

Name: ____ Date: _______________

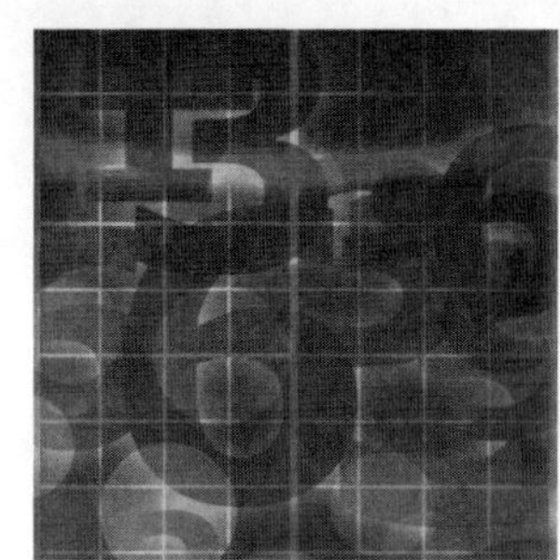

Exercises and Notes

Distribution of Grades
58
64
68
73
74
83
83
84
86
89
91
93
93
94
97
97
99
99
101

1. Find the range, variance, and standard deviation for the grades listed above. Does the average reflect the sample?

Name: ____   Date: _______________

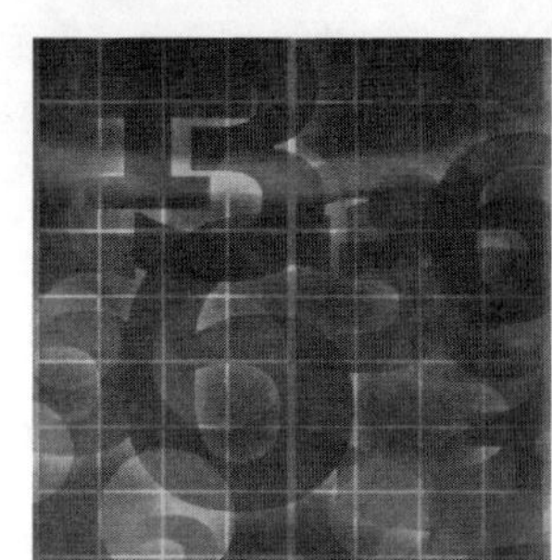

Learning Check

Number of Hours Worked Last Week
20
26
28
30
32
34
40
40
40
43
45
45
45
45
50
50
55
60
65

1. Find the range, variance, and standard deviation for the grades listed above. Does the average reflect the sample?

Chapter Five

Probability, the Normal Distribution, and Estimation

Up to this point we have dealt with **descriptive statistics,** measures used to summarize the main characteristics of a data set. Except when we discussed the variance, we have not distinguished between data from a population and from a sample. The **population** is the entire group of individuals, objects, or events that the researcher is interested in, while the **sample** is a subset of the population. As we progress toward using inferential statistics, such differences are important, since **inferential statistics** allow us to generalize to population by measuring properties of a sample and then deciding if those properties probably exist or do not exist in the population. In this chapter, we will explore topics that lay the foundation for steps in hypothesis testing that will be introduced in the following chapter.

Probability

Probability refers to the likelihood that an outcome will occur. There are two types of probabilities: observational and theoretical. Theoretical probabilities are based on theoretical mathematics, while observational probabilities rely on a very large number of observations. As an example of observational probability, we know that researchers have determined by examining records of millions of births that the probability of having twins has risen over the last 30 years because of the increased number of women taking fertility drugs. While observational probabilities rely on empirical data, theoretical probabilities represent the chance of events or outcomes based on certain assumptions. A simple example of a theoretical probability involves the flip of a coin. Let's say you choose heads for a coin flip. We know that you have a 50-50 chance of being right, assuming a fairly weighted coin is used and that it lands face up or facedown.

Probabilities range in value from 0 to 1.0, with zero suggesting that something is impossible and 1.0 indicating an absolute certainty. Probabilities are often reported as percentages to avoid decimals. For example, meteorologists will report that there is a 70% of rain rather than the calculated probability value of .70.

We calculate a probability as:

$$\text{Probability} = \frac{\text{number of times } \textit{a particular outcome} \text{ can occur}}{\text{number of times } \textit{all possible outcomes} \text{ can occur}}$$

By applying the probability formula to flipping a coin, we find that probability to be:

$$\text{Probability of a head} = \frac{\text{number of heads on a coin}}{\text{number of all outcomes of coin flip}} = \frac{1}{2} = .50$$

The Rules of Probability

The calculations of many probabilities are not as simple as the example of a coin flip. Different rules of probability may need to be invoked in order to determine the likelihood of some events or outcomes. Let's use as an example buying grab bags at a charity event to see how different rules of probability work. Suppose you buy a grab bag. Of the 25 grab bags, one of the bags contains \$100. Since there are only 25 grab bags and you pick one first, your friend thinks the chance of winning the \$100 is pretty good. You say that you think the chances of your getting something other than the \$100 are greater than 90%. We can see if you are right by using the **converse rule**—the probability of an event not occurring is 1 minus the probability of the event occurring. Therefore, we first need the probability of picking the bag with the \$100, which is:

$$\text{Probability of \$100} = \frac{\text{number of bags with \$100}}{\text{total number of grab bags}} = \frac{1}{25} = .04$$

By the converse rule, the probability of not choosing the \$100 is: $1 - .04 = .96$, or 96%. Therefore, you were correct that the chance of your picking something other than the \$100 was greater than 90%.

Sometimes people are interested in what the probability of multiple outcomes occurring is. In addition to the \$100, suppose that among the other prizes there are also three bags with movie passes, one bag with football tickets, and two bags with a certificate for dinner for two at the best restaurant in town in the grab bags. What is the probability that you would pick one of those four items? Since each grab bag includes only one prize, you cannot get two or more prizes with one pick. Therefore, we are dealing with distinctly different, or mutually exclusive, outcomes. With mutually exclusive outcomes, we use the **addition rule:** the probability of multiple distinct outcomes is the sum of their individual probabilities. Therefore, to find the probability of getting one of those four desired items, we need to find the separate probabilities for each of the four different prizes and then add them together:

$$\text{Probability of \$100} = \frac{\text{number of bags with \$100}}{\text{total number of grab bags}} = \frac{1}{25} = .04$$

$$\text{Probability of movie passes} = \frac{\text{number of bags with movie passes}}{\text{total number of grab bags}} = \frac{3}{25} = .12$$

$$\text{Probability of football tickets} = \frac{\text{number of bags with football tickets}}{\text{total number of grab bags}} = \frac{1}{25} = .04$$

$$\text{Probability of dinner at restaurant} = \frac{\text{number of bags with dinner at a restaurant}}{\text{total number of grab bags}} = \frac{2}{25} = .08$$

The probability that you would pick one of the four desired prizes (either the \$100, or the movie passes, or the football tickets, or the dinner at a restaurant) is .28 ($.04 + .12 + .04 + .08 = .28$).

Other times we may be interested in probabilities that have multiple outcomes that are not mutually exclusive but instead are a combination of two or more outcomes. Suppose that at the charity event, they are also selling 50 mystery boxes. One of the boxes contains \$500. What is the probability that you could win both cash prizes by choosing only one bag and one box? We know that the probability of choosing the \$100 in the grab bag is .04. We can calculate the probability of choosing the \$500 mystery box as:

$$\text{Probability of \$500} = \frac{\text{number of boxes with \$500}}{\text{total number of boxes}} = \frac{1}{50} = .02$$

Since we are interested in determining the probability of choosing the cash amounts in the bag and box, both must occur. Therefore, we apply the **multiplication rule of probability:** the probability of a combination of independent outcomes is the product of their separate probabilities. Therefore, the probability that you will pick both the bag with the \$100 and the box with \$500 is .0008 ($.04 \times .02 = .0008$). While you had a

1 in 25 chance of picking the bag with the $100 and a 1 in 50 chance of choosing the box with $500, the likelihood that you would pick both cash amounts is 1 in 1250.

The probabilities that we have encountered until now have assumed that the total number of possible outcomes (the number in the denominator) has not changed. However, this is not always the case. Suppose that you decided to pick two grab bags and there were four items you were interested in obtaining: the $100 in one bag, movie passes in three bags, football tickets in one bag, and a certificate for dinner in two bags. The probability that you would pick one of these items from the first bag is .28, since 7 out of the 25 bags has a desired prize (7 ÷ 25 = .28). However, you open the first bag and find that it contains a certificate for 10 gallons of gasoline—a useful item but not what you had hoped to have picked. However, you still have your second bag. The probability that it has one of the four items that you desire is not the same as it was with the first bag. It is now .29 (7 ÷ 24 = .292). The probability changed because after you opened the first bag, you did not put it back before picking the second bag. This distinction is called with or without replacement. **With replacement**, when an item is withdrawn, it is then replaced before another item is withdrawn. **Without replacement**, when an item is withdrawn, it is not replaced before another item is withdrawn.

Let's look at another example to clarify the distinction between with or without replacement. Suppose you need a quarter for a parking meter. You have 16 coins in your pocket: three quarters, four dimes, two nickels, and seven pennies. Therefore, the probability of pulling a quarter out of your pocket in a single attempt is .19 (3 ÷ 16 = .1875). You reach in your pocket and pull out a nickel. You try to hold it in your hand while you reach for another coin, but it falls out of your hand and back into your pocket before you can pull out another coin. Therefore, your second attempt will be with replacement and the probability of drawing a quarter is still .19 (3 ÷ 16 = .1875). However, on this second attempt, you get a penny. You decide to hold the penny in your other hand so it won't fall back into your pocket. Now, your third attempt without replacement has a .20 probability of successfully drawing a quarter (3 ÷ 15 = .20). As you can see, when multiple drawings are taking place, it is important to determine whether they are done with or without replacement.

Probability Distribution

In Chapter 2, you learned how empirical data collected in the real world can be organized in a frequency distribution. In this chapter, you will learn how to construct a **probability distribution:** a table that contains the possible categories for a variable and the likelihood, based on probability theory, that each category will occur. Suppose you have two dice and you want to determine the likelihood of rolling one pip (a dot on the die) on each die. You could roll one pip on one die, two dice, or neither of the dice. Therefore, this probability distribution will have three outcomes: 0, 1, and 2.

To end up with one pip on each die, you have to get one pip on Die 1 and one pip on Die 2. First, let's find the probability of getting one pip on Die 1 and then Die 2.

$$p \text{ of } \textbf{one pip} \text{ on Die 1} = \frac{\text{number of sides with one pip}}{\text{total number of sides}} = \frac{1}{6} = .167$$

$$p \text{ of } \textbf{one pip} \text{ on Die 2} = \frac{\text{number of sides with one pip}}{\text{total number of sides}} = \frac{1}{6} = .167$$

In order to roll two 1s, you must get one pip on Die 1 *and* on Die 2. Therefore, we use the multiplication rule to obtain the probability of getting two 1s: (.167 × .167 = .028).

Now that we know the probability of getting one pip on each die, we can use the converse rule to find the probability of getting no dice with one pip. In other words, for neither dice to have one pip, each dice would have to have two, three, four, five, or six pips. The probabilities for each die are:

$$p \text{ of } \textbf{two or more pips} \text{ on Die 1} = 1 - .167 = .833$$

$$p \text{ of } \textbf{two or more pips} \text{ on Die 2} = 1 - .167 = .833$$

In order to get zero dice with one pip, we must not roll a 1 on Die 1 *and* Die 2. Therefore, we use the multiplication rule to obtain the probability of getting zero dice with one pip, (.833 × .833 = .694).

The process of finding the probability of getting a single die with one pip is slightly different than the methods previously used because there are two ways to obtain one pip on one and only one of the two dice. The one could occur on Die 1 *or* Die 2. Therefore, if Die 1 has one pip, then the side showing for Die 2 must have two, three, four, five, or six pips. As we discovered earlier:

$$p \text{ of } \textbf{one pip} \text{ on Die } 1 = .167$$

$$p \text{ of } \textbf{two or more pips} \text{ on Die } 2 = .833$$

Since both of these events must occur to get a single die with one pip, we use the multiplication rule to get the product of these separate probabilities ($.167 \times .833 = .139$). Therefore, the probability of getting one pip on Die 1 is .139. However, the single pip could have occurred on Die 2. Therefore, we need to calculate this second way of obtaining a single pip on the two dice.

$$p \text{ of } \textbf{two or more pips} \text{ on Die } 1 = .833$$

$$p \text{ of } \textbf{one pip} \text{ of Die } 2 = .167$$

To obtain the probability of obtaining single pip, this time appearing on Die 2, we must multiply these separate probabilities together ($.833 \times .167 = .139$). The last step is to add the separate probabilities on the two different ways that we could obtain a single pip on one die. We add the two probabilities because the two ways are distinctly different, mutually exclusive ways of achieving *a die* with one pip. Thus, the probability of a die with one pip is .278 ($.139 + .139 = .278$).

Just as we organized observed frequencies in a frequency distribution, we can organize these theoretical probabilities in a probability distribution, as shown in **Table 5.1.** Because the sum of the probabilities of all possible outcomes equals 1.0, the probability distribution indicates that all possibilities in the distribution are included. The sum of our three included outcomes do indeed sum to 1.0.

Table 5.1 *Probability Distribution of Rolling 1s with a Pair of Dice*

# of 1s on a Die	Probability (*p*)
0	.694
1	.278
2	.028
Total	1.000

The Normal Curve

The best-known continuous probability distribution is the normal curve. People often refer to the normal curve as the bell curve because of its shape. The normal curve is a theoretical probability distribution that is symmetrical and unimodal—having one peak. The mean and median have the same value as the mode. Another property of the normal curve is that it gradually decreases from its highest point, extending in both directions infinitely, with 100% of the area located under the curve. Thus, the curve never actually touches the baseline. Although the normal curve is a theoretical distribution, researchers have found that some variables, such as standardized test scores, height, and some attitudinal measures, appear to correspond to this ideal model. We can use the normal curve to describe observed distributions, interpret the standard deviation, and determine probabilities.

Researchers can use the normal curve as a probability distribution for generalizing observed sample results to the population. Because we need to distinguish between sample and population measures in inferential statistics, we use different notation to distinguish between the measures. For example, the use of the notation of $\bar{X}$ to represent the mean refers to the mean as a sample statistic, while μ indicates the mean as a population parameter. **Table 5.2** contains the notation for some sample statistics and the population parameters used in this chapter. Some additional notations will be introduced in subsequent chapters. You may have already noticed a pattern in the notations. Letters from the Roman alphabet that are used in the English language are used to represent sample statistics, while Greek letters are used to denote population parameters.

Table 5.2 *Sample and Population Notations*

Measure	Sample Statistic	Population Parameter
Mean	$\overline{X}$	μ
Standard Deviation	S	σ
Variance	S^2	σ^2
Proportion	P	Π

The normal curve as a theoretical probability distribution also uses Greek letters to represent measures in order to distinguish the mean and standardized deviation in this idealized model from observed measures of the mean and standard deviation from actual data. **Figure 5.1** displays the normal curve with the mean and standard deviations. Because the distribution is symmetrical, the mean (μ) is located at the center of the distribution, with half the cases falling above the mean and half falling below the mean. Three standard deviations (σ) above and below the mean are also shown. Positive standard deviations appear above the mean, while negative standard deviation fall below the mean.

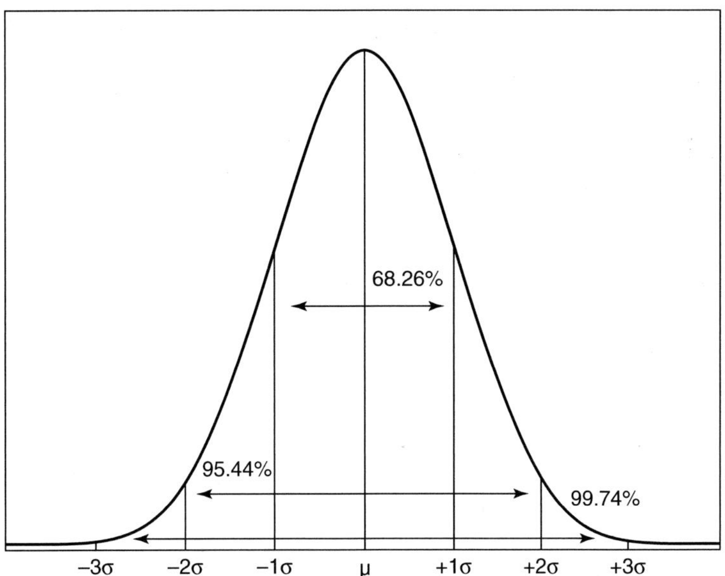

Figure 5.1 Percent of Area under the Normal Curve

Figure 5.1 also displays the area under the normal curve. The graph shows that 68.26% of the distribution falls within one standard deviation (-1σ to $+1\sigma$) of the mean. Because the curve is symmetrical, we also can determine that 34.13% of the distribution (68.26% ÷ 2 = 34.13%) lies between the mean and one standard deviation below *or* above the mean. Using the same logic, we can also conclude that 47.72% of the cases (95.44% ÷ 2 = 47.72%) lie between the mean and two standard deviations below *or* above the mean and that 49.87% (99.74% ÷ 2 = 49.87%) of the distribution is between the mean and three standard deviations below *or* above the mean.

The Standard Z Score Table

Often, we want to know the percentage of cases that lie somewhere between, for example, a standard deviation and the mean. Suppose we want to know the percentage of cases that occur within 1.19σ of the mean. We know that 68.26% of the cases lie between -1σ and $+1\sigma$ and 95.44% fall between -2σ and $+2\sigma$. Because of the shape of the curve, we know that more cases fall closer to 1σ from the mean than to 2σ from the mean.

To determine the exact percentage that falls between -1.19σ and $+1.19\sigma$ (see **Figure 5.2**), we use the standard normal table. While the entire table can be found in Appendix A, a portion is presented here for illustrative purposes (see **Table 5.3**). The entire standard Z Table contains hundreds of Z scores and the corresponding areas associated with the Z score. To find the percentage of cases that lies within $+1.19$ standard deviations of the mean, we look in Column A for the Z score 1.19. Next, we look at the corresponding value in column B and find 38.30. This means that 38.30% of cases fall between the mean and 1.19 standard deviations above the mean. However, we were interested the area *within* 1.19σ standard deviations of the mean. A score could be within 1.19 and be below the mean. Therefore, we must also find the percentage of cases that lies within -1.19 standard deviations of the mean. The standard Z Table does not need to contain any negative Z scores because the normal curve is symmetrical. The percentages of cases that lie between $+1.19\sigma$ and -1.19σ are the same. Therefore, to determine the area within 1.19 standard deviations of the mean, we add $38.30 + 38.30 = 76.60$ and conclude that 76.60% of cases fall within 1.19 standard deviations of the mean.

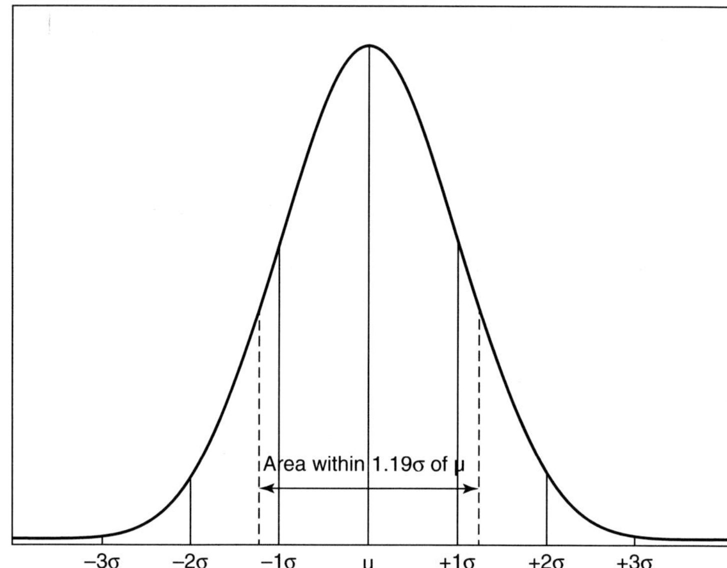

Figure 5.2 Position of a Score within 1.19σ of $\overline{X}$.

Converting a Raw Score into a Z Score

The normal curve can be useful in interpreting the standard deviation of real-world data, as long as the variables are approximately normally distributed. For example, over 1 million students take the ACT college entrance exam each year. The scores on the exam have been found to approximate the normal curve with a mean of 21 and a standard deviation of about 5. Therefore, we can conclude that about 68% of the test takers scored between 16 (1 standard deviation below the mean) and 26 (1 standard deviation above the mean) on the ACT exam.

Suppose you scored a 28 on the ACT exam. You know your score falls between 1 and 2 standard deviations above the mean, but you do not know that exact point or the exact percentage of cases that fall below your score. We can convert your test score of 28 to a Z **score**—the number of standard deviations that a given raw scores falls above or below the mean. Z scores are also referred to as **standardized scores** because the variable is no longer measured in the raw units of the variable—in this case ACT scores—but are standardized based on the standard deviation. **Figure 5.3** illustrates the conversion of the raw ACT scores to their corresponding Z scores.

We now need to mathematically convert your ACT score into a Z score. The formula for a Z score is:

$$Z = \frac{X - \mu}{\sigma}$$

Table 5.3 *Abridged Standard Z Score Table*

Column A Z	Column B Area between Mean & Z	Column C Area beyond Z
.00	.00	50.00
.01	.40	49.60
.02	.80	49.20
.03	1.20	48.80
.04	1.60	48.40
.05	1.99	48.01
. . .	. . .	. . .
1.15	37.49	12.51
1.16	37.70	12.30
1.17	37.90	12.10
1.18	38.10	11.90
1.19	38.30	11.70
1.20	38.49	11.51
1.21	38.69	11.31
. . .	. . .	. . .
3.80	49.99	.01
3.90	49.995	.005
4.00	49.997	.003

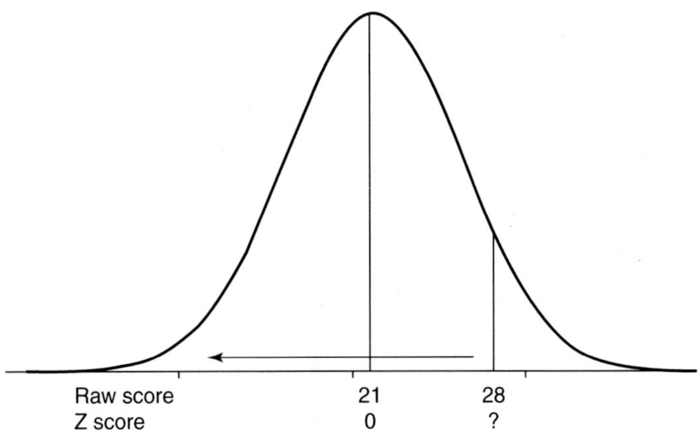

Figure 5.3 Position of ACT Raw Score of 28

Thus, we standardize your ACT score of 28 as:

$$Z = \frac{28 - 21}{5} = 1.4$$

We now know that your ACT score of 28 is 1.4 standard deviations above the mean. However, in order to determine the percentage of cases that fall below your score, we need to use the standard Z score table. To find the percentage of cases that are lower than your standardized ACT score (or Z score) of 1.4, we look in Column A for the Z score 1.4. Next, we look at the corresponding value in column B and find 41.92. This means that 41.92% of cases fall between the mean and above the mean at your raw score of 28, or standardized score of 1.4. However, to determine how many scores are lower than your score, we need to account for the cases that fall below the mean. Because the mean divides this symmetrical distribution in half, 50% of the cases lie below the mean (see **Figure 5.4**). Therefore, we add 50% to the 41.92% that lies between the mean and your score to determine that 91.92% of cases have lower ACT scores than your raw score of 28, or standardized score of 1.4.

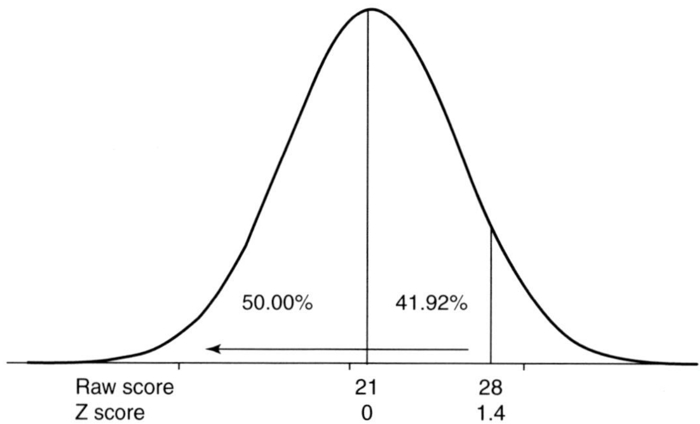

Figure 5.4 Percentage of Cases that Score Less Than the ACT Raw Score of 28

Some scores, particularly test scores, are often presented in terms of their **percentile rank:** the percentage of cases in a distribution that fall at or below a given score. In a normal distribution, the percentile rank refers to the percentage of cases that lie to the left of a particular score. When we just found the percentage of cases that scored less than 28 on the ACT, we were in fact finding the percentile rank. Thus, someone, like yourself, who scored a 28 on the ACT achieved the 92nd percentile.

On the other hand, you might be interest in finding the percentage of students who scored higher than you on the ACT exam. **Figure 5.5** illustrates the area of the curve that we wish to find. Since the portion of the curve is to the *right* of your score (28), we will *not* be finding the percentile rank. We can determine the percentage of scores higher than 28 in a variety of different ways. Since we know that the percentage of cases that lie below your raw score of 28 (Z score of $+1.4\sigma$) is 91.92%, by the converse rule, 8.08% have a higher score ($100 - 91.92 = 8.08$). However, if we did not know that 91.92% of cases fell below your score, then the easiest way to determine the percentage of cases that scored higher than your standardized score of 1.4σ would be to use the standard Z Table. We would look in Column A for the Z score 1.4, as we did before. However, this time, we look at the corresponding value in column C and find 8.08. Thus, using either method, we conclude that 8.08% of cases have higher ACT scores than your raw score of 28, or standardized score of 1.4.

Estimating Population Means and Population Proportions

As mentioned earlier, researchers often use sample data rather than population data because of time and monetary limitations. Researchers typically employ methods of **probability sampling** that specify that each case in the population has the same probability of being included in the sample. Probability sampling methods are used to ensure that the sample represents the population and allow researchers to use **inferential statistics**

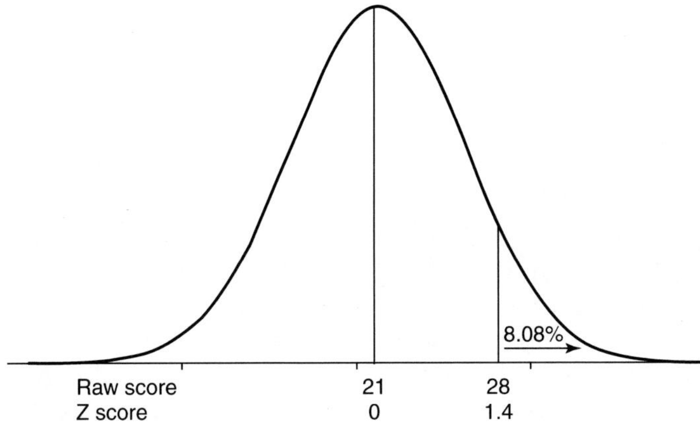

Figure 5.5 Percentage of Cases that Score Higher Than the ACT Raw Score of 28

to generalize to the population. The simple random sample, the stratified random sample, and the systematic random sample are examples of probability sampling.

Data analyses from probability samples often require **estimation,** which involves using a sample statistic drawn from a random sample to approximate the population parameter. Even with probability sampling techniques, it is likely that there would be a discrepancy between the population parameter and the sample statistics, known as **sampling error.** A mean drawn from sample data, $\overline{X}$, probably will not equal the population mean, μ, indicating the need for estimation. In this chapter, we will estimate both population means and proportions. First, we will examine how to do a confidence interval around a point estimate of a mean when the population variance is both known and unknown. Then, we will explore how to construct a confidence interval around the point estimate of a proportion.

A crucial concept in estimation is the **sampling distribution,** a theoretical probability distribution of all possible values for a particular statistic. The sampling distribution allows researchers to indicate how much confidence they have in their estimates and provide them with a basis to infer from their samples to the population. Let's focus on the **sampling distribution of the means,** a theoretical probability distribution of sample means drawn from all possible samples of the same given size. Suppose our population is the test scores of students in a diversity class. We draw a random sample of five test scores and calculate the mean. We replace the scores, draw another random sample of five test scores, and calculate the mean. Since the sampling distribution is theoretical and we would have to draw an infinite number of samples and compute an infinite number of means, for this example, we will draw 100 samples and calculate the means.

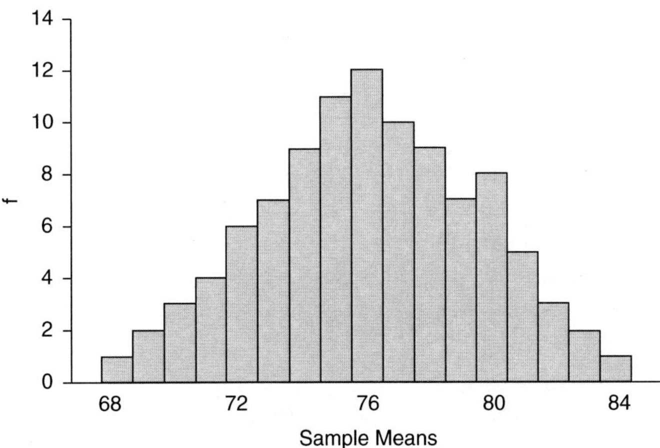

Figure 5.6 Sampling Distribution of a 100 Sample Means of Sample Size N = 5 Drawn from a Population of 65 Students' Test Scores

Figure 5.6 shows the results of the 100 sample means. Some very important properties of the sampling distribution of the mean are beginning to emerge. The **Central Limit Theorem** emphasizes three important principles that deal with the shape, central tendency, and variability of the sampling distribution. First, with sufficient sample size, the sampling distribution of the means will be normal, even if the shape of the distribution in the population is not normal. The 100 sample means contained in **Figure 5.6** are beginning to resemble a normal distribution. If more and more samples were drawn, the shape of the distribution would become the normal curve. Second, the mean of the sample distribution equals the population mean. The distribution of the sample means centers around the true population mean and is normally distributed with low sample means offsetting high sample means. Thus, with a normal distribution around the population mean, the mean of the sampling distribution would equal the true population mean. Third, the standard deviation of the sampling distribution of means, known as the **standard error of the mean** ($\sigma_{\bar{X}}$), is smaller than the standard deviation in the population distribution. The standard error of the mean equals:

$$\sigma_{\bar{X}} = \frac{\sigma}{\sqrt{N}}$$

The standard deviation of the population is measuring the standard distance between a score (X) and the population mean, while the standard error of the mean measures the standard distance between a sample mean ($\bar{X}$) and the population mean (μ). The standard error of the mean is extremely useful because it indicates how well a sample mean ($\bar{X}$) estimates the population mean (μ).

Confidence Intervals for Means When σ is Known

Sample statistics, such as the sample mean ($\bar{X}$), are **point estimates,** which use a single number to estimate the population parameter. However, since sample means are affected by sampling error, you may be wondering how confident you should be in a point estimate. One way to demonstrate the likelihood that the sample data represent the population is to establish a **confidence interval:** a range of values within which the population may fall. Instead of using the single sample mean as an estimate of the population mean, with a confidence interval, we will have a range of values around the sample mean that allows us to state the probability that the interval contains the population mean. The lowest value of a confidence interval is known as the **lower confidence limit,** while the highest value of a confidence interval is the **upper confidence limit.**

Let's look at an example of a confidence interval for a sample mean. Suppose a mayor wants to determine whether city employees maintain competency levels for their jobs without having to test the entire city's workforce. He selects 50 workers at random and gives them competency exams. He finds that the mean is 80 for the sample of 50 workers. The mayor realizes that he cannot be certain that the sample mean reflects the population mean, given the likelihood of sampling error. However, the mayor knows that the population standard deviation equals 12, giving the history of use in other cities. Therefore, based on one of the principles of the Central Limit Theorem, we can calculate the standard error as follows:

$$\sigma_{\bar{X}} = \frac{\sigma}{\sqrt{N}} = \frac{12}{\sqrt{50}} = 1.70$$

We can use the sample mean of 80 and establish a range of values around it that allows us to state the probability that the interval contains the population mean. One of the acceptable probabilities used by researchers is the 95% level of confidence, which means that there are 95 chances out of 100 that the true population mean falls within the interval. Since the standard deviation is known, we construct the confidence interval using the following formula:

$$CI = \bar{X} \pm Z\,\sigma_{\bar{X}}$$

$\bar{X}$ = the sample mean

Z = the Z score associated with the confidence level (in this case, 95% confidence)

$\sigma_{\bar{X}}$ = the standard error of the mean

We can determine the Z score associated with the 95% confidence level by using the standard Z Table. However, instead of using a Z score to find a probability, we use the probability of 95% to find a Z score. **Figure 5.7** illustrates that since the standard Z Table deals with half of the curve, we divide 95% in half and get 47.5%. If we look for 47.5 in Column B of the standard Z Table, in Column A we find the Z score of 1.96.

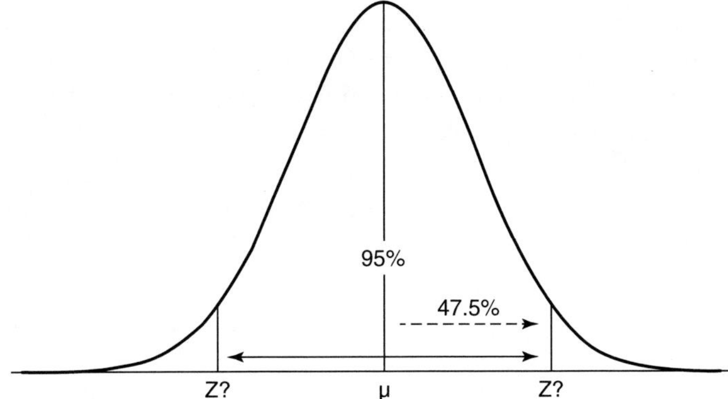

Figure 5.7 Normal Curve with a 95% Confidence Interval.

If we apply the 95% confidence level to our example of 50 city workers, we find:

$$CI_{95\%} = \overline{X} \pm Z\,\sigma_{\overline{X}}$$
$$CI_{95\%} = 80 \pm (1.96)\,1.70$$
$$= 80 \pm 3.33$$
$$= 76.67 \text{ to } 83.33$$

Note that we add and subtract ($\pm$) the product of a value associated with the level of confidence (Z) and the standard error ($\sigma_{\overline{X}}$) to the point estimate ($\overline{X}$). Thus, from the calculated confidence level, the mayor can be 95% confident that all city workers' test average (the population mean) lies between 76.67 and 83.33.

If the mayor wanted to be even more confident in stating the range of values in which the population mean lies, we could calculate a 99% confidence interval. The only value in the confidence interval formula that would change is the Z score.

$$CI = \overline{X} \pm Z\,\sigma_{\overline{X}}$$
$$CI_{99\%} = 80 \pm (2.58)\,1.70$$
$$= 80 \pm 4.39$$
$$= 75.61 \text{ to } 84.39$$

Therefore, the mayor can be 99% confident that the average of all city workers' test scores fall between 75.61 and 84.39. Because the mayor wanted to be more confident in estimating the true population mean, the confidence interval widened.

Confidence Intervals for Means When σ Is Unknown

In the previous section, when we calculated a confidence interval around a mean, we knew what the standard deviation in the population was. Therefore, we could calculate the standard error. However, in most cases, if we do not know the mean in the population, we do not know the standard deviation in the population since the mean is used in calculating the standard deviation. In these circumstances, we use the standard deviation of the sample (S) to estimate the standard error of the mean. We obtain the unbiased estimate of the standard error by replacing σ with S:

$$S_{\overline{X}} = \frac{S}{\sqrt{N}}$$

In addition to using the unbiased estimate of the standard error, our calculation of the confidence interval around a mean changes in yet another way when the standard deviation in the population is unknown. Since utilizing an estimated standard error adds uncertainty, instead of including the Z scores from the normal distribution in our confidence intervals, we use t scores that are derived from a family of distributions. The t distribution differs from the normal distribution in some important ways. Whereas the standard Z Table is based on one curve (the normal curve), the **t distribution** is a family of curves, each based on the number of **degrees of freedom** (*df*), a measure of the unique information available when conducting a statistical test. In addition, the standard error of a t distribution is larger than the standard error of the normal distribution. However, if the degrees of freedom are large enough (i.e., greater than 120, the probabilities associated with the t distribution are the same as those associated with the normal distribution. **Figure 5.8** illustrates the difference between two curves in the t distribution: one with 5 degrees of freedom and the other with a large enough number of degrees of freedom to be equivalent to the normal curve. As you can see, with fewer degrees of freedom (in this case, 5), the curve is flatter than the normal curve. Lastly, although the t distribution assumes that the variable is normally distributed in the population, research has shown that data from nonnormal distributions only slightly affect the computation of the test statistic. Therefore, unless the data severely violate the assumption of normality, we can use the t distribution even with relatively small samples.

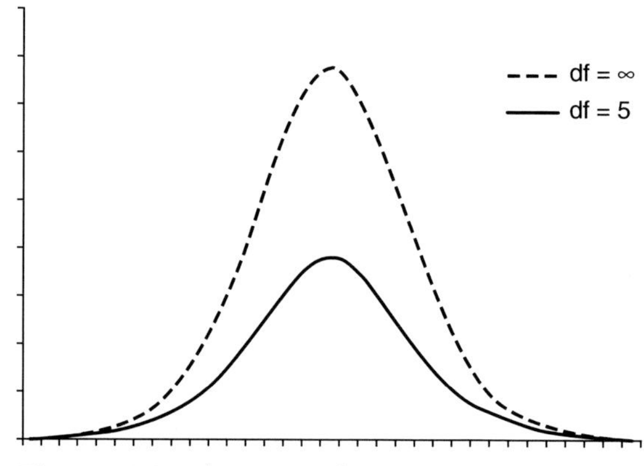

Figure 5.8 The t Distribution with ∞ and 5 Degrees of Freedom

Let's look at an example in which we do not know the population standard deviation. The average income for a random sample of 41 people in Citiopolis is $45,000, with a standard deviation of $20,000. What is the 95% confidence interval estimate for the mean income of the population of Citiopolis? Since we only have the sample standard deviation (S), and not the standard deviation in the population, we must estimate the standard error of mean ($S_{\bar{X}}$) and use the t distribution in constructing our confidence interval instead of Z. Therefore, the confidence interval when σ is unknown is:

$$CI = \bar{X} \pm t(S_{\bar{X}})$$

While we know that the sample mean income in Citipolis is $45,000, we don't know the t value or the standard error of the mean. However, we can use the sample standard deviation ($S = 20,000$) to estimate the standard error of mean ($S_{\bar{X}}$):

$$S_{\bar{X}} = \frac{S}{\sqrt{N}} = \frac{20,000}{\sqrt{41}} = 3123.48$$

Now let's determine the value of t for this example. Since we know the value of a sample mean ($45,000), the mean of one possible sample is no longer free to vary; we know it is $45,000. Therefore, the number of degrees of freedom is $N - 1$. We will learn in later chapters that the degrees of freedom will be calculated differently, because various statistical tests have different information available for the analysis. However, since we have information only from one sample, the degrees of freedom are 40 ($N - 1$ or $41 - 1 = 40$).

Although we have the degrees of freedom, we also need to know the alpha level (α) to find the value of t for our confidence interval. The alpha level is a function of the confidence level. In this case, we want to be 95% confident. Since alpha is stated as a probability and not a percentage, we change 95% to .95 and subtract it from 1.00. Thus, if we wish to be 95% confident, $\alpha = .05$ ($1.00 - .95 = .05$). If we want to be 99% confident, $\alpha = .01$ ($1.00 - .99 = .01$).

We can find the t value needed to calculate our confidence interval around the mean by using the two-tailed t-distribution table provided in Appendix A. We need to look down the column of degrees of freedom to find 40 degrees of freedom and across to find the .05 α. Thus, the t value for 40 degrees of freedom and an α of .05 is 2.021.

Now that we have our sample mean (45,000), the standard error (3123.48), and the t value (2.021), we can calculate our confidence interval:

$$CI = \bar{X} \pm t(S_{\bar{X}})$$

$$CI_{95\%} = 45,000 \pm (2.021)(3123.48)$$

$$CI_{95\%} = 38,687.45 \text{ to } 51,312.55$$

We can be 95% confident that the mean income of people in Citiopolis is between \$38,687.45 and \$51,312.55.

If we wanted to be 99% confident about the mean income of all people in Citiopolis, we need to modify our calculation of the confidence interval. Because we want to be more confident in our estimation, our alpha level changes from .05 to .01, which in turn alters the t value in our confidence interval. The two-tailed t-distribution table indicates that 40 degrees of freedom and an alpha level of .01 are associated with a t value of 2.704. Therefore, the 99% confidence level is:

$$CI_{99\%} = 45,000 \pm (2.704)(3123.48)$$

$$CI_{99\%} = 36,554.11 \text{ to } 53,445.89$$

We can be 99% confident that the mean income of people in Citiopolis is between \$36,554.11 and \$53,445.89. Comparing the 95% and 99% confidence levels, we see that the 99% confidence interval is wider. Because we wanted to be more confident in our estimation of the population mean, we needed to report a wider range of values.

Confidence Intervals for Proportions

The mean is just one point estimate that we can use when calculating a confidence interval. Confidence intervals are also often calculated around proportions. Many opinion polls report results in terms of the percentages of people that gave various responses. In Chapter 2, you learned that percentages are proportions that have been multiplied by 100 to avoid decimal places. Therefore, although the mathematical formula for the confidence interval requires a proportion, the question and the results of such confidence intervals are typically presented as percentages.

The formula for calculating a confidence interval around a proportion is similar to calculating a confidence interval around the mean. Each calculation starts with the point estimate (the mean or proportion), and then we add and subtract the product of a value associated with the level of confidence and the standard error. The formula for calculating a confidence interval around a proportion is:

$$CI = p \pm Z\,S_p$$

p = the observed sample proportion

Z = the Z score associate with a confidence level (*e.g.,* 95% or 99%)

S_p = the estimated standard error of the proportion

Since the population proportion is unknown, we estimate the standard error of proportion using the following formula to calculate the estimated standard error of the proportion:

$$S_p = \sqrt{\frac{P\,(1-P)}{N}}$$

S_p = the estimated standard error of the proportion

p = the observed sample proportion

N = the sample size

As mentioned earlier, opinion polls often calculate confidence interval around a proportion. If we look at poll data from a previous presidential election, we can see how pollsters use these statistics to make predictions about the outcome of coming elections. Days before the 2008 presidential election, the final Gallup Poll showed Barack Obama with a 53% to 42% advantage over John McCain among 1952 likely voters.[1] Let's construct a 95% confidence interval for Obama's estimated percentage, in other words, a 95% confidence interval around the proportion of .53. We learned earlier that the Z score associated with a 95% confidence level equals 1.96. Next, we need to calculate the estimated standard error of the proportion:

$$S_p = \sqrt{\frac{.53\,(1-.53)}{1952}} = \sqrt{\frac{.2491}{1952}} = .011$$

Now we can calculate the confidence interval:

$$CI_{95\%} = .53 \pm (1.96 \times .011)$$
$$CI_{95\%} = .53 \pm .02$$
$$CI_{95\%} = .51 \text{ to } .55$$

We would interpret these results after converting the proportions back to percentages. The final Gallup Poll before the 2008 presidential election indicated with 95% confidence that Barack Obama would receive between 51% and 55% of the popular vote. However, the Gallup Poll does not typically report results this way. It typically reports the predicted percentages (in this case, 53% for Obama and 42% for McCain) and a margin of error. The **margin of error** is the product of the Z score and the estimated standard error of the proportion, in this case, .02 or 2%. The Gallup Poll predicted that 53% of voters would cast their votes for Obama with a 2% margin of error.

Suppose the pollsters wanted to be even more confident; they wanted to have 99% confidence in their predictions. The proportion of the vote that we wish to construct the confidence interval around does not change, nor does the standard error of the proportion. However, the Z score would change, since it is associated with the level of confidence that we are increasing from 95% to 99%. Therefore, the Z score changes from 1.96 to 2.58. With this change in the Z score, the 99% confidence interval around Obama's proportion would be:

$$CI_{99\%} = .53 \pm (2.58 \times .011)$$
$$CI_{95\%} = .53 \pm .03$$
$$CI_{99\%} = .50 \text{ to } .56$$

These results show that the final Gallup Poll before the 2008 presidential election could have predicted with 99% confidence that Barack Obama would receive between 50% and 56% of the popular vote. Thus, the Gallup Poll would have predicted that 53% of voters would cast their votes for Obama with a 3% margin of error.

---

[1] Newport, Frank, Jeffrey M. Jones, and Lydia Saad. 2008. "Final Presidential Estimate: Obama 55%, McCain 44%: Independents Break for Obama, Boosting Obama's Broad Democratic base," Gallup, 2008. Retrieved on August 21, 2010, from www.gallup.com/poll/111703/final-presidential-estimate-obama-55-mccain-44.aspx.

Let's look at one more scenario regarding our confidence interval around votes for Obama. This time, let's consider a 95% confidence interval with the poll consisting of only 800 likely voters instead of 1952 likely voters. Because N has changed, the standard error of the proportion will change:

$$S_P = \sqrt{\frac{.53\,(1-.53)}{800}} = .018$$

The 95% confidence interval based on 800 voters would be:

$$CI_{95\%} = .53 \pm (1.96)\,(.018)$$
$$CI_{95\%} = .53 \pm .04$$
$$CI_{95\%} = .49 \text{ to } .57$$

Therefore, the poll of 800 likely voters before the 2008 presidential election could predict with 95% confidence that Barack Obama would receive between 49% and 57% of the popular vote. The poll could predict that 53% of voters would cast their votes for Obama with a 4% margin of error.

We calculated three confidence intervals around Obama's predicted votes that produced three different results. Why did the confidence intervals change? Changing the level of confidence alters the margin of error and thus the width of the confidence interval. To be more confident (99% versus 95%), the range of values in which we expect the population proportion (π) falls must be wider. Modifying the sample size also changes the margin of error and the width of the confidence interval. The smaller sample size increases the likelihood of sampling error, and the confidence interval widens given that uncertainty. Thus, the level of confidence and the sample size have an impact on the width of a confidence interval around a proportion.

In this chapter, you learned about some the basics of probability and about probability distributions, such as the normal curve with Z scores and the family of t distributions. We used both Z and t scores to construct confidence intervals around the point estimates of the mean and proportion. In the next chapter, we will use probability in hypothesis testing to determine whether we should or should not generalize findings from sample data to the population.

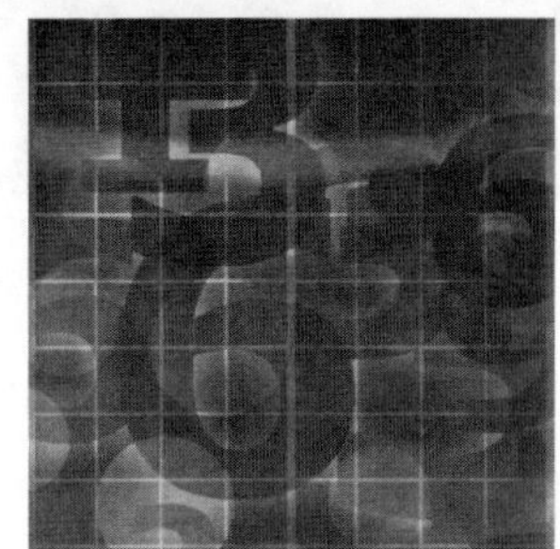

Exercises and Notes

Probability

Probability *refers to the* _______________ *that an* ___________________________.

$$\text{Probability} = \frac{\text{number of times} \underline{\hspace{3cm}} \textit{outcome } \text{can occur}}{\text{number of times} \underline{\hspace{3cm}} \textit{outcomes } \text{can occur}}$$

Probability of Drawing an Ace =

Probability of NOT Drawing an Ace =

_______________ ***Rule of Probability:*** *used to determine the probability of an* _______________ _______________.

Event _______________ = ___ − *the probability of the event* _______________.

$$p \text{ of No Ace} = \underline{\hspace{4cm}}$$

Probability of Drawing an Ace or King

$$\text{Probability of an Ace} = \underline{\hspace{4cm}}$$

$$\text{Probability of a King} = \underline{\hspace{4cm}}$$

_______________ ***Rule:*** *the probability of* ___________________ *outcomes is the* ______ *of their individual probabilities.*

$$\text{Probability of an Ace or King} = \underline{\hspace{4cm}}$$

Probability of Drawing Two Aces from Two *Separate* Decks =

___________________ ***rule:*** *the probability of a* _______________ *of independent outcomes is the* _______________ *of their separate probabilities.*

$$p \text{ of two Aces} = \underline{\hspace{4cm}}$$

Probability of Drawing Two Aces from the *Same* Deck

With replacement, *when an item is withdrawn, it is* ___________________ ___________________________.

$$p \text{ of first Ace} =$$

$$p \text{ of second Ace} =$$

$$p \text{ of two Aces} =$$

Without replacement, *when an item is withdrawn, it is* ___________________ ___________________________.

$$p \text{ of first Ace} =$$

$$p \text{ of second Ace} =$$

$$p \text{ of two Aces} =$$

Probability Distribution of Drawing Two Face Cards from Two Decks =

# of Face Cards	p

- p of **face cards from both decks** =
  - **Deck 1**
    - p of one face card =
  - **Deck 2**
    - p of one face card =

    p *of* **face cards from both decks** =
- p of **zero face cards from both decks**:
  - **Deck 1**
    - p of a nonface card =
  - **Deck 2**
    - p of a nonface card =
    - p of a nonface card =
- p of **one face card on one card** =
  - One Way: Face card from first deck but not from second deck =
  - Second Way: Face Card from second Deck but not first deck =

Combine the Ways:

Normal Curve

Properties

- ____________________ .
- ________ shaped, ________________________________ .
- ________________ of the area is under the curve.
- The tails extend __________________ .

Sample and Population Notations

Measure	Sample Statistic	Population Parameter

Area under the Normal Curve

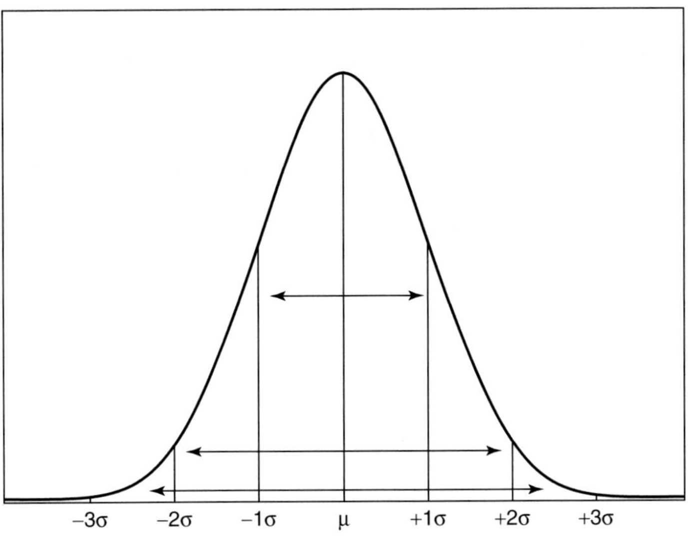

Using the *Z* Table to Find Area

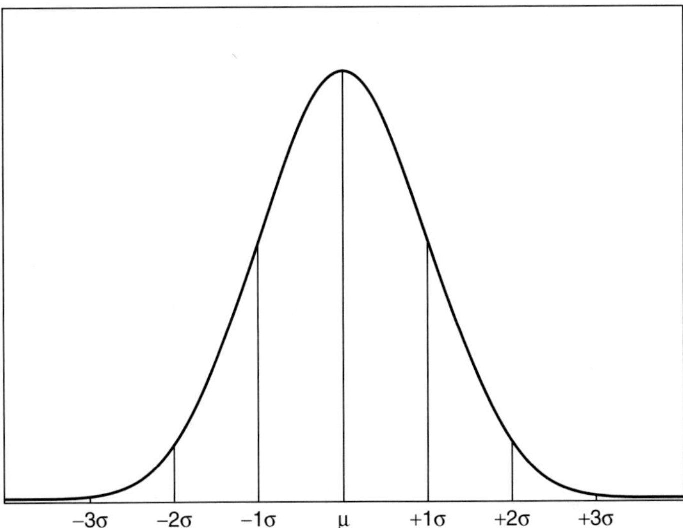

Abridged Standard Z Score Table

Column A Z	Column B Area between Mean & Z	Column C Area beyond Z
.00	.00	50.00
.01	.40	49.60
.02	.80	49.20
.03	1.20	48.80
.04	1.60	48.40
.05	1.99	48.01
...	...	...
1.15	37.49	12.51
1.16	37.70	12.30
1.17	37.90	12.10
1.18	38.10	11.90
1.19	38.30	11.70
1.20	38.49	11.51
1.21	38.69	11.31
...	...	...
3.80	49.99	.01
3.90	49.995	.005
4.00	49.997	.003

Converting Raw Scores to Z Scores

$$Z = \frac{x - \mu}{\sigma}$$

IQ scores are normally distributed with a mean of 100 and a standard deviation of 15. The qualifying IQ to join Mensa is 135. What is the likelihood of random selecting a prospective Mensa member from the population?

Percentile Ranks and the Normal Distribution

What would be the minimum IQ percentile rank of a Mensa member?

Percentile rank: the percentage of cases in a distribution that _____________________ a given score.

If people with IQs below 70 are considered to have an intellectual disability, what % of the population is NOT considered to have an intellectual disability?

Populations and Samples

_______________ sampling specifies the _____________ of each case in the ____________ of being included in the _______________.

- For example,

- For example,

Estimation: using a _______________________ drawn from a _______________________ to approximate the _______________________.

- What sample statistic estimates the population mean, μ?

- What sample statistic estimates the population proportion, π?

Sampling Error: the _______________________ between the _______________________ and the _______________________. e.g., the difference between _____ and _____ or _____ and _____.

Sampling Distribution: a _______________________ distribution of _______________ for a particular _______________.

Central Limit Theorem

1. The _______ of the sampling distribution of mean is _______________________.

2. The _______________ of the sampling distribution of mean, ___, _______________________
    _______________________.

3. The distribution is _______________________.

4. All of this is _______, even if the distribution of _______________________ in the
    _______________.

What is the standard deviation of the sampling distribution of the mean called?

Sample Point Estimates and Confidence Intervals

Confidence Interval (Population Standard Deviation **Known**)

Confidence Interval: the range of values around a _______________________ that allows one to state the _______________________ that the interval contains the _______________________.

$$CI = \bar{X} \pm Z\,\sigma_{\bar{X}} \qquad\qquad \sigma_{\bar{X}} = \frac{\sigma}{\sqrt{N}}$$

Students' test scores in an introductory course are normally distributed with a standard deviation of 13.65. A professor takes a sample of 100 students whose mean score was 71. Construct a 95% confidence interval of the mean test scores in the population.

- Calculate a 95% confidence interval around the sample mean.

$\bar{X} =$

$Z =$

$\sigma_{\bar{X}} =$

$CI =$

Interpretation:_______________________

____

____

- If we wanted to be 99% confident, how would the confidence interval change?

$\bar{X} =$

$Z =$

$\sigma_{\bar{X}} =$

$CI =$

Interpretation:____

____

____

Confidence Interval (Population Standard Deviation Unknown)

We can obtain an unbiased estimate of the standard error of the mean by replacing $\sigma_{\bar{X}}$ with S.

$$S_{\bar{X}} = \frac{S}{\sqrt{N}} \qquad\qquad CI = \bar{X} \pm tS_{\bar{X}}$$

With an estimated standard error of the mean, _______________________.

The **t distribution:** a family of curves, each based on _______________________________

____.

School officials are concerned about freshmen's adjustment to college life. One indicator that they decide to use in their study is the number of times out-of-town students go home for the weekend during their first semester. The sample of 27 had a mean of 5 and a standard deviation of 3.

- Construct a 95% confidence interval to estimate the number of times students go home for the weekend in the population.

$\bar{X} =$

$t =$

$S_{\bar{X}} =$

$CI =$

Interpretation:____

____

____

- If we wanted to be 99% confident, how would the confidence interval change?

Interpretation:____

____

____

Confidence Interval for Proportions

$$CI = p \pm Z\,S_P$$

$CI =$ the confidence interval

$p =$ the sample proportion

$Z =$ Z score corresponding to confidence level

$S_P =$ the standard error of the proportion

The estimated standard error of the proportion is:

$$S_P = \sqrt{\frac{P\,(1-P)}{N}}$$

A political science class wants to predict the winner of the election of the student senate president. They find from 1000 randomly selected students intending to vote that Smith is ahead with 44%, followed by Jackson with 36%, and Davis with 20%. Construct a 95% confidence interval for Smith's estimated percentage.

- Construct a 95% confidence interval for Smith's estimated percentage.

    $p =$

    $Z =$

    $S_P =$

    $CI =$

Interpretation:____

____

____

____

- Construct a 99% confidence interval for Smith's estimated percentage.

    $CI =$

Interpretation:____

____

____

____

- Suppose the poll had consisted of 400 likely voters. Construct a 95% confidence interval for Smith's estimated percentage.

    $CI =$

Interpretation:____

____

____

____

- In the three confidence intervals just calculated, why did the confidence intervals change?

____

____

____

____

Learning Check

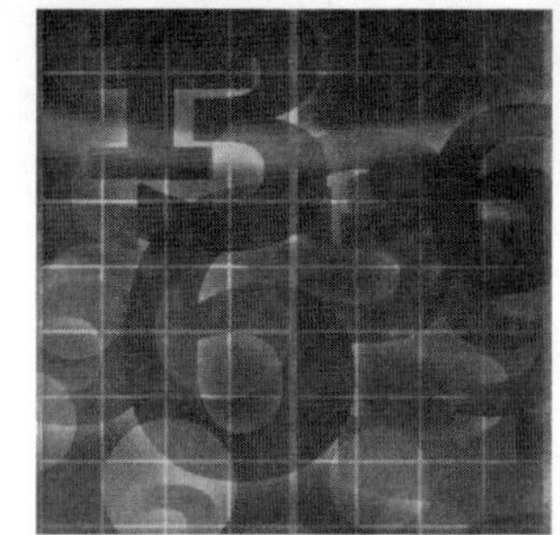

1. Suppose a university committee is randomly selecting a member of student senate members to serve on the committee. The 50 member student senate has the following composition: 14 seniors, 16 juniors, 13 sophomore, and 7 freshmen.

   (a) What is the probability that the student will be a junior?

   (b) What is the probability that the student will not be a senior?

   (c) What is the probability that the student will be a sophomore or freshman?

2. You have been doing laundry and you have to match a pile of socks.

Number of Socks	Color	Type
8	Black	Crew
2	Blue	Crew
4	White	¼ Crew
6	White	Footies
2	Blue	¼ Crew
4	Black	Footies

   (a) What is the probability that you randomly grab a crew sock?

   (b) What is the probability that you randomly grab a blue sock?

   (c) What is the probability that you randomly grab a sock that is not black?

   (d) What is the probability that you randomly grab a white footie?

   (e) Suppose you grab one sock and it is a black crew. What is the probability that you randomly pick a matching sock?

3. If about 10% ($p = .10$) of the American population is left-handed, what is the probability of electing three consecutive left-handed presidents?

4. Alexis is enrolled in a Spanish course of 20 students, 12 women and 8 men. When she arrives at class, there are three students in the room, two women and one man. What is the probability that the next student to come through the door will be a woman?

5. About 36% of Americans have heart disease and 8% of Americans have diabetes. What is the probability that a randomly chosen American has both heart disease and diabetes?

6. At Paul's summer job, some of workers order carry-out for lunch. Paul orders a ham sandwich and potato chips. The other workers order four more ham sandwiches, three turkey sandwiches, four roast beef sandwiches, and two tuna sandwiches, with five others ordering potato chips, four ordering tortilla

chips, and two ordering multigrain chips. With the sandwiches and chips in separate bags, Paul reaches in and pulls out a sandwich and a bag of chips.

(a) What is the probability that Paul picks a ham sandwich?

(b) What is the probability that Paul picks potato chips?

(c) What is the probability that he picks both correctly?

7. There are two bowls of Halloween candy. One bowl has 10 pieces of candy (4 candy bars, 5 suckers, and 1 gumball), and the other bowl has 8 pieces of candy (3 candy bars, 2 suckers, and 3 gumballs). Construct a probability distribution of drawing candy bars from the two bowls (one drawn from each container).

8. Assuming that the variable is normally distributed, what is the probability of a case falling

(a) within 2.73σ of $\bar{X}$?

(b) greater than 3.02σ of $\bar{X}$?

(c) less than $+1.11\sigma$ of $\bar{X}$?

(d) less than $-.72$ of $\bar{X}$?

9. Women's heights are normally distributed with a mean of 64 inches and a standard deviation of 3 inches.

(a) Who is further from the mean, Robin, whose standardized Z score height is $-.83$, or Jennifer, who is 66 inches tall?

(b) Many female gymnasts are 5 feet (60 inches) or shorter. What is the probability of randomly selecting a female of this height or shorter from the population?

(c) Many female supermodels are at least 5 feet 10 inches tall. What is the probability of randomly selecting a female of this height or taller from the population?

(d) Can the answers in part (b), (c), both (b) and (c), or neither (b) nor (c) be expressed as a percentile rank? Explain your answer.

10. The exam scores in Introductory Sociology classes approximate the normal distribution with $\bar{X} = 75$ and $\sigma = 11$.

(a) If a failing grade on an exam was 59 or lower, what percentage of the class failed?

(b) What percentage of students earned a C $(70 - 79)$?

(c) What is the probability of randomly selecting a student who earned an A or B (a grade of 80 or higher)?

(d) Determine the percentile rank of a student who scored 68 on the exam.

(e) Determine the percentile rank of a student who scored 93 on the exam.

11. A doctor takes a sample of 12 year-old boys' heights $(N = 20)$ drawn from a national population. The sample mean was 59 inches, while the population standard deviation was 2.5 inches.

(a) Construct a 95% confidence interval of the mean of 12 year-old boys' heights in the population.

(b) Construct a 99% confidence interval of the mean of 12 year-old boys' heights in the population.

12. A coffee-shop owner wants to determine the average number of cups of coffee drunk by her customers per day. A sample of 15 customers had a mean of 3.5 and a standard deviation of 1.73.

    (a) Construct a 95% confidence interval to estimate the average number of cups of coffee drunk per day by all of the customers.

    (b) Construct a 99% confidence interval to estimate the average number of cups of coffee drunk per day by all of the customers.

13. University advisors want to determine the number of hours students are studying per week. Their sample of 121 full-time students produced a mean of 26 hours, with a standard deviation of 6.71 hours.

    (a) Construct a 95% confidence interval to estimate the average number of hours studied per week by full-time students in the population.

    (b) Construct a 99% confidence interval to estimate the average number of hours studied per week by full-time students in the population.

14. A school administration is interested in estimating the number of students who like the semester system, as opposed to the quarter system. From a random sample of 200 students, they found that 58% favored semesters.

    (a) Can the administration be 95% sure that a majority of the entire student body favors semesters?

    (b) Would the administration's conclusion change if they wanted to be 99% confident?

15. A city council wants to determine public support for the use of cameras at red lights. They found from a sample of 500 citizens that 62% support the use of the cameras.

    (a) Construct a 95% confidence interval around the percentage of citizens that support the use of cameras at red lights.

    (b) What is the margin of error?

16. A university wants to determine if students preferred classes taught 3 days, 2 days, or 1 day a week. A survey of 600 students produced the following results: 34% favored 3 days a week, 39% liked 2 days a week, and 27% preferred courses taught 1 day a week.

    (a) Construct a 95% confidence interval to estimate the proportion that favors courses taught 3 days a week in the population.

    (b) How would the class interval have been different if the sample size had been 300?

Chapter Six

Testing the Significance of Means

In the previous chapter, we discussed how we were moving from descriptive statistics to inferential statistics. With **descriptive statistics,** we learned how to calculate such statistics as the mean and the standard deviation, which we can use to summarize the main characteristics of a data set. In this and subsequent chapters, we will be working with **inferential statistics,** which allow us to test our hypotheses by measuring properties of a sample and deciding if those properties probably exist or probably do not exist in the population. In the previous chapter, we covered topics, such as probability and estimation, that laid the foundation for hypothesis testing. In this chapter, we will examine the basics of hypothesis testing and apply them to testing the significance of means. T-tests will be used to test the significance of a single sample mean, the significance of the difference between two independent sample means, and the significance of the difference between two dependent or paired means. We will also examine the concept of one-tail versus two-tail hypothesis testing.

Hypothesis Testing

In Chapter 1, we learned that a **hypothesis** is a prediction that is testable through data analysis. We use sample data to test a hypothesis and see if we can generalize our findings in a sample in a data set to the entire population. Because our hypotheses are predictions about what exists in the population, hypotheses are always about the population, not about the sample. Therefore, hypotheses always contain symbols of population parameters but never symbols of sample statistics. When doing statistical analyses, it is also important to understand that hypotheses always come in pairs. There are both the null hypothesis and the alternative (or research) hypothesis. Stating these hypotheses is the first step in hypothesis testing. The steps in hypothesis testing are summarized below.

Step One: State null and alternative hypotheses. The **null hypothesis** predicts that there is no difference between two quantities or values; in other words, the null hypothesis predicts that the quantities or values are equal. Any difference that occurs in the sample data is seen as the result of sampling error. The **alternative (or research) hypothesis** predicts a difference between two quantities or values; in other words, the alternative or research hypothesis predicts that the quantities or values are not equal. The alternative hypothesis considers any difference that exists in the sample data to be the result of a true difference in the population, and not sampling error.

We test the null hypothesis because statistical tests assume that the null hypothesis is true. Therefore, the only way that we can find support for our alternative hypothesis is to refute the null hypothesis. We cannot "prove" that the alternative hypothesis is true but can only invalidate the null hypothesis. Furthermore, the hypotheses cannot be "proven" to be true because decision making in hypothesis testing is based on probability, and there is always the possibility that we will make the wrong decision, as we will discuss later in this chapter.

Step Two: Determine the significance or alpha level that will be used in the statistical test. The alpha level is an important element in the decision making process regarding the null hypothesis. The significance level that is chosen by researchers may vary among disciplines. Alpha (α) levels of .05, .01, and .001 are typically seen in sociological research and are used in this book. Alpha levels play an important role in hypothesis testing, since they are needed to determine the critical region, which is used in Step 4 to make a decision about the null hypothesis.

Step Three: Calculate the statistical test. The statistical test is calculated from our sample data to decide whether or not the null hypothesis should be rejected. The test statistic compares a sample statistic, such as a mean, to the value stated in the null hypothesis. The standard error, which we first learned about in the previous chapter, plays an important role in statistical tests because it accounts for the variability in sample statistics that occurs when smaller samples are drawn from a population.

Step Four: Make a decision regarding the null hypothesis. Depending on whether we are calculating the test statistic from a formula or whether we are using a statistical program, such as SPSS, the procedure will vary. If we are using a formula, we use the alpha level and the number of **degrees of freedom** (*df*)—the number of pieces of unique information used in calculating a test statistic—to find the critical value from a table, such as a *t* table (provided in Appendix AB). Once we have the critical value, we can determine the **critical region,** which is the area, or areas, of the sampling distribution that indicates that we should reject the null hypothesis. The critical region is based on the probability of a sample statistic falling in the critical region when the null hypothesis is true. Therefore, if the calculated test statistic falls in the critical region (*i.e.,* is greater than the critical value as an absolute value), the null hypothesis is rejected. If we are using a computer, SPSS provides us with the exact probability that the null hypothesis is true. Therefore, we do not need to use a table to determine a critical value and region. We can just compare the α with the probability (*p*) on the computer printout. If *p* is less than α, then we reject the null hypothesis.

If we reject the null hypothesis, the result is considered statistically significant. If we cannot reject the null hypothesis, many statisticians prefer to report that they fail to reject the null hypothesis rather than to say that they accept it. They make this distinction because acceptance suggests that the null hypothesis is true, whereas failure to reject implies that there was not sufficient data to convince us to prefer the alternative hypothesis. As we will see after our discussion of the steps of hypothesis testing, there are two types of errors that can occur when making a decision about the null hypothesis.

Step Five: Interpret the results. We conduct statistical tests to be able to make a conclusion about what exists in the population. Therefore, the final step in hypothesis testing is to make inferences about the population. We need to make statements about what we believe exists or does not exist in the population, based on our statistical analysis of sample data.

Error in Hypothesis Testing

The decision that we make about whether or not to reject the null hypothesis is based on probability. Although we make a decision based on what is most likely, there is still a chance that we can make the wrong decision. When deciding whether or not to reject the null hypothesis, four different outcomes are possible. As **Table 6.1** shows, only two of those outcomes involve truly correct decisions. If the null hypothesis is true and we decide not to reject it, we have made a correct decision. Similarly, if the null hypothesis is false and our decision is to reject the hypothesis, we have made a correct decision.

Table 6.1 *Types of Error in Testing Hypotheses*

	Decision Based on Sample Data	
	Reject H_0	**Fail to Reject H_0**
H_0 is true.	Type I	Correct
H_0 is False	Correct	Type II

Unfortunately, there are two ways that we could make incorrect decisions. If we reject a true null hypothesis, we have committed what is known as a **Type I error.** The likelihood of committing a Type I error is related to the level of significance. If we choose an alpha level of .05 and reject the null hypothesis, there is a .05 probability, or 5% chance, that we are wrong. We can reduce the likelihood of a Type I error by reducing the alpha level. If we choose an alpha level of .01, we reduce the probability of a Type I error to .01, or 1%. An alpha level of .001 gives us an even smaller chance of a Type I error, with a probability of .001, or 0.1%.

You may be wondering why we don't just choose the alpha level of .001, if that level gives us the smallest chance of a Type I error. Unfortunately, reducing the likelihood of a Type I error does not occur without risk. As previously mentioned, there are two ways that we can make an incorrect decision regarding a null hypothesis. If we reduce the chance of a Type I error, we increase the likelihood of a **Type II error**—failing to reject a false null hypothesis. Whereas the probability of a Type I error is equal to alpha, the probability of a Type II error is equal to beta (β). If α equal .05, β does *not* equal $1.00 - .05 = .95$; instead, β is equal to 1 minus the power of the statistical test. The power of a statistical test involves a more in-depth discussion of statistics than is needed here. However, we will discuss the power of statistical tests in Chapter 7, when we compare parametric and nonparametric tests. What you should always be mindful of is that these two types of errors exist in hypothesis testing and that reducing the likelihood of one increases the probability of the other.

Test of a Single Sample Mean

Let's start our application of hypothesis testing with a test of a single mean. In a test of a single mean, we compare a hypothesized value to a mean calculated from sample data. The objective of the test is to determine whether the sample data support the probability that the true population mean is equal to the hypothesized value. For example, a film student believes that the average movie is 110 minutes in length. He takes a random sample of 30 films with a mean length of 104 minutes and a standard deviation of 17 minutes. If he knew what the standard deviation in the population is, he could compute the test statistic, Z. However, just as we indicated in the previous chapter when dealing with confidence intervals, in practice, we seldom, if ever, know the population standard deviation and do not know the population mean. When we do not know the population standard deviation and therefore the standard error of the sampling distribution of the mean ($\sigma_{\bar{x}}$), we must estimate the standard error ($S_{\bar{x}}$), and use t as our test statistic. Therefore, in this chapter, we will focus on testing hypotheses using t, not Z.[1]

Step One: State null and alternative hypotheses. The null hypothesis is a statement of no difference; the alternative hypothesis indicates that there would be a difference. Since the question in our example is whether movies last 110 minutes or not, the null hypothesis indicates that there is no difference between the population mean and the hypothesized value. Therefore, the null and alternative hypotheses are:

H_0: The average length of movies in the population is 110 minutes ($\mu = 110$).

H_1: The average length of movies in the population is *not* 110 minutes ($\mu \neq 110$).

Step Two: Determine the significance or alpha level that will be used in the statistical test. Before the film student tests the null hypothesis, he chooses a level of significance. As earlier discussed, he can choose among the alpha levels of .05, .01, and .001, and he picks .05.

Step Three: Calculate the statistical test. When determining which test statistic to calculate, we need to assess the information that we do and do not have. The film student does not know the standard deviation

---

[1] If the standard deviation in the population (σ) is known, Z can be calculated as $Z = \dfrac{\bar{X} - \mu}{\dfrac{\sigma}{\sqrt{N}}}$ or $Z = \dfrac{\bar{X} - \mu}{\sigma_{\bar{x}}}$.

in the population. Therefore, he must estimate the standard error and use t as his test statistic. Since he is comparing a single mean to a hypothesized value, he uses the following formulas:

The t-Test of a Single Mean

$$t = \frac{\overline{X} - \mu}{S_{\overline{X}}}$$

Estimated Standard Error

$$S_{\overline{X}} = \frac{S}{\sqrt{N}}$$

$\overline{X}$ = the sample mean

μ = the population mean

S = the sample standard deviation

$S_{\overline{X}}$ = the estimated standard error of the mean

N = the total number of cases

Since the standard deviation (S) of lengths of movies for the sample of 30 movies is 17, the film student can estimate the standard error as:

$$S_{\overline{X}} = \frac{17}{\sqrt{30}} = 3.10$$

The film student now has the estimated standard error of 3.10 and the mean of the sample ($\overline{X}$), which is 104 minutes. However, he does not know the mean in the population; he is hypothesizing about what the mean in the population is. Since statistical tests assume that the null hypothesis is true, the film student can use the hypothesized value of 110 minutes that is stated in the null hypothesis (H_0: μ = 110). Therefore, the calculated t value is obtained as:

$$t = \frac{104 - 110}{3.10} = 1.935$$

Step Four: Make a decision regarding the null hypothesis. The film student needs to compare the calculated t of -1.935 with the critical value of t, which he finds in a t table. To find the critical value of t, he needs the number of degrees of freedom and the significance level (from step 2, α = .05). The way that the degrees of freedom are calculated varies by the statistical test. In this case, there are 30 different movies that provide unique information and the one sample mean of 104 minutes. Therefore, the formula for the degrees of freedom for the test of a single mean is $N - 1$, or in this case, $30 - 1 = 29$, where 1 is subtracted from the total number of cases because knowing the sample mean reduces the amount of unique information by 1. The film student uses the two-tailed t table (provided in Appendix A) and finds that the critical value of t, 2.045, is associated with an alpha level of .05 and 29 degrees of freedom. **Figure 6.1** shows the critical value of t, 2.045, which determines a critical region that designates whether the null hypothesis can be rejected. Since the calculated t of -1.935 is less than the absolute value of the critical t of 2.045, -1.935 falls in the area of the curve in which the null hypothesis is not rejected, as **Figure 6.1** shows. Thus, the film student would fail to reject the null hypothesis.

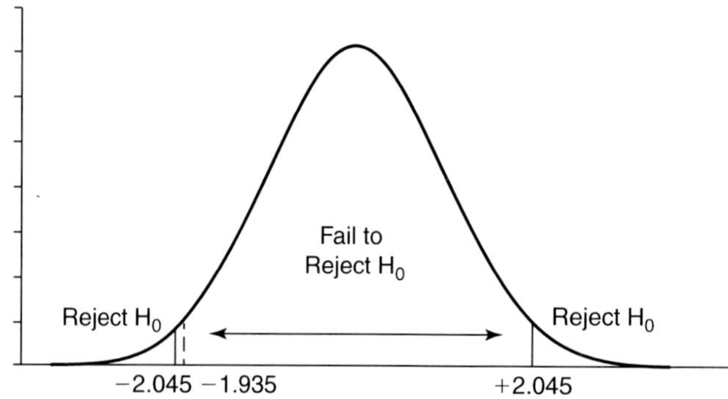

Figure 6.1 The t Curve with df = 29 and α = .05.

Step Five: Interpret the results. Since the film student failed to reject the null hypothesis, he did not have enough evidence in his sample of 30 movies to refute the null hypothesis. The discrepancy between the sample mean of 104 minutes and the population mean of 110 minutes is not statistically significant. Therefore, he concludes that the average length of movies in the population appears to be 110 minutes. However, the film student should be cognizant of the fact that he could be wrong. The average length of movies may not be 110 minutes. Despite correctly conducting the statistical test, he cannot be 100% certain of his results. In this case, he may have still have committed a Type II error—failing to reject a false null hypothesis.

Test of the Difference between Two Independent Sample Means

In the social sciences, researchers often are interested in comparing groups. For example, a sociologist may want to determine whether stores in urban areas charge different prices than stores in the suburbs. He then buys the same items at both the urban and suburban stores and analyzes the data to determine whether urban and suburban stores' prices differ for the same items. The data he collected are summarized in **Table 6.2,** followed by his analysis of the data.

Table 6.2 *Data for Urban and Suburban Stores*

	Urban	Suburban
Average Cost	$93.12	$90.67
S^2	6.25	4.41
N	21	21

Step One: State null and alternative hypotheses. Since the question here is whether two groups—urban and suburban stores—have different prices, the sociologist compares the mean cost of the same items from these different stores. The null hypothesis indicates that there is not a difference between the groups, while the alternative hypothesis indicates that there is a difference. Therefore, the hypotheses regarding the stores are:

H_0: The mean prices of the same items in urban and suburban stores' prices do not differ:

$$(\mu_1 - \mu_2 = 0 \text{ or } \mu_1 = \mu_2)$$

H_1: The mean prices of the same items in urban and suburban stores' prices do differ

$$(\mu_1 - \mu_2 \neq 0 \text{ or } \mu_1 \neq \mu_2)$$

Step Two: Determine the significance or alpha level that will be used in the statistical test. The sociologist decides to use the .01 level of significance for his statistical test ($\alpha = .01$).

Step Three: Calculate the statistical test. Since the sociologist does not know the standard deviations in the populations of urban or suburban stores,[2] he must estimate the standard error and use t as his test statistic.

---

[2] On the rare occasions in which the true population standard deviations of the two groups (σ_1 and σ_2) are known, the true standard error of the difference can be computed and used to calculate a Z statistic, using the following formulas:

$$\sigma_{\bar{X}_1 - \bar{X}_2} = \sqrt{\frac{\sigma_{X_1}^2}{N_1} + \frac{\sigma_{X_2}^2}{N_2}} \quad \text{and} \quad Z = \frac{(\bar{X}_1 - \bar{X}_2 - \mu_1 - \mu_2)}{\sigma_{\bar{X}_1 - \bar{X}_2}}$$

If he can assume that the two population variances are equal, he can use the following formula, which combines information from the two samples:

Estimated Standard Error of the Difference between Means (Assumed Equal Population Variances)

$$S_{\bar{X}_1 - \bar{X}_2} = \sqrt{\frac{[(N)_1 - 1)s_1^2 + [(N)_2 - 1)s_2^2}{N_1 + N_2 - 2}} \sqrt{\frac{1}{N_1} + \frac{1}{N_2}}$$

N_1 = the total number of cases from sample 1
N_2 = the total number of cases from sample 2
S_1^2 = the variance from sample 1
S_2^2 = the variance from sample 2

However, if the two sample variances are very different—for example, one is more than twice the size the other—we should not assume that the variances are equal in the population. With variance so dissimilar, it does not make sense to pool the sample variances together to estimate the standard error. Therefore, when sample variances are very different, we use the following formulate to estimate the standard error:

Estimated Standard Error of the Difference between Means (Assumed Unequal Population Variances)

$$S_{\bar{X}_1 - \bar{X}_2} = \sqrt{\frac{S_1^2}{N_1} + \frac{S_2^2}{N_2}}$$

The sociologist computed the variances from the sample of products from the urban stores and suburban stores, as **Table 6.2** shows. He found that the variance of the cost of items from the urban stores was 6.25, while the variance of items from the suburban stores equaled 4.41. Since one variance was not double other, the sociologist can assume that the variances for the two types of stores are equal in the population. Therefore, he uses the first formula to estimate the standard error of the difference:

$$S_{\bar{X}_1 - \bar{X}_2} = \sqrt{\frac{(21-1)6.25 + (21-1)4.41}{21 + 21 - 2}} \sqrt{\frac{1}{21} + \frac{1}{21}}$$

$$= \sqrt{\frac{125 + 88.2}{40}} \sqrt{.048 + .048}$$

$$= \sqrt{5.33} \sqrt{.096}$$

$$= 2.309 \times .310$$

$$= .716$$

The sociologist has the estimated standard error, so he can calculate the t statistic to test the null hypothesis. Since he is comparing two means from two groups (urban and suburban stores), he will use the following formula, which includes the two sample means and the two populations means:

$$t = \frac{(\bar{X}_1 - \bar{X}_2) - (\mu_1 - \mu_2)}{S_{\bar{X}_1 - \bar{X}_2}}$$

$\bar{X}_1$ = the sample mean of group 1
$\bar{X}_2$ = the sample mean of group 2
μ_1 = the population mean of group 1
μ_2 = the population mean of group 2
$S_{\bar{X}_1 - \bar{X}_2}$ = the estimate the standard error of the difference

While the sociologist has both sample means and the estimated standard error of the difference, he does not have the two population means. In fact, he is doing a statistical test to make an inference about the equality of those population means. However, his null hypothesis indicates that $\mu_1 - \mu_2 = 0$. Therefore, since we do statistical tests under the assumption that the null hypothesis is true, we can always assume that $\mu_1 - \mu_2 = 0$ when calculating the t statistic for the difference between two means. Therefore, the sociologist's calculated t statistic is:

$$\frac{93.12 - 90.67 - 0}{.716}$$

$$= \frac{2.45}{.716}$$

$$= 3.422$$

Step Four: Make a decision regarding the null hypothesis. To make a decision about the null hypothesis, the sociologist needs to compare his calculated t statistic with the critical value of t. However, he needs to determine the degrees of freedom in order to find the critical value of t. When testing the difference two means, the degrees of freedom are:

$$df = N_1 + N_2 - 2$$

In this case, the degrees of freedom equal 40 (21 + 21 − 2 = 40). Since the sociologist had decided on a significance level of .01, the critical value of t from the two-tailed t table (provided in Appendix A) is 2.704. He uses the critical value of 2.704 to identify the critical region as shown in **Figure 6.2.** He sees that the calculated t value of 3.422 is greater than the critical value of 2.704, falling in the right-hand tail of the curve. Since the calculated t value falls in the tail of the curve, the sociologist rejects the null hypothesis.

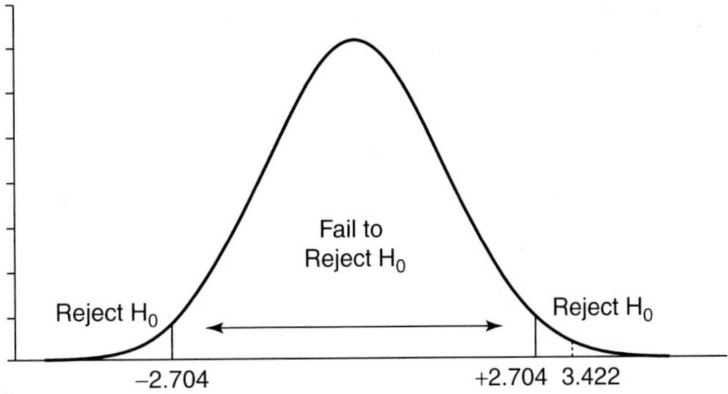

Figure 6.2 The t Curve with df = 40 and α = .01.

Step Five: Interpret the results. The sociologist rejects the null hypothesis; he has enough evidence in his samples of 42 stores to refute the null hypothesis. The difference between the two sample means of $93.12 and $90.67 is statistically significant. Therefore, he concludes that urban and suburban stores in the population probably do charge different prices for the same items. However, the sociologist should be aware that he could be wrong. The average price of the same items at all urban and suburban stores may actually be the same, but he cannot know this since he does not have the prices from the entire population of urban and suburban stores. In this case, he may have still committed a Type I error—rejecting a true null hypothesis.

Test of the Difference between Dependent or Paired Sample Means

The previous two sample t-test is referred to as an independent test because the analysis involves comparing means drawn from two independent groups. However, researchers also conduct dependent or paired sample t-tests because they are also interested in comparing the same sample at two points in time. The dependent sample *t*-test has the advantage that it has the exact same subjects for both time periods. A classic application of the dependent sample *t*-tests is the pretest and posttest situations, in which a sample is measured at one point in time prior to a "treatment" and then after the "treatment." The dependent sample *t*-test measures the effect of the "treatment" on the sample. Researchers may also pair or match subjects on a variable or variables that they would like to control in order to approximate the advantage of the dependent sample means *t*-test.

Suppose we are interested in testing the effect that hypnosis has on cigarette smoking. We survey 10 smokers to see how many cigarettes they smoke daily. The subjects are then given a posthypnotic suggestion that each time they light a cigarette, they will experience a horrible taste and feel nauseated. **Table 6.3** shows the number of cigarettes smoked daily before and after the posthypnotic suggestion. The table is followed by an analysis of the data.

Table 6.3 *Cigarettes Smoked by Subjects Before and After a Posthypnotic Suggestion*

Subject	Before	After
1	19	13
2	35	37
3	20	14
4	31	31
5	23	15
6	11	5
7	32	28
8	24	21
9	18	17
10	27	25

Step One: State null and alternative hypotheses.

H_0: The mean number of cigarettes smoked before and after hypnosis are the same

H_1: The mean number of cigarettes smoked before and after hypnosis are not the same

Step Two: Determine the significance or alpha level that will be used in the statistical test. Let's use the .01 level of significance for this statistical test ($\alpha = .01$).

Step Three: Calculate the statistical test. The formula used to test the difference between dependent or paired sample means is:

$$t = \frac{\overline{X}_1 - \overline{X}_2}{S_{\overline{D}}}$$

$\overline{X}_1$ = the sample mean at time 1

$\overline{X}_2$ = the sample mean at time 2

$S_{\overline{D}}$ = the standard error of dependent/paired means

As you can see, this is similar to the other *t*-tests that we have covered in this chapter, with sample means in the numerator and a standard error in the denominator. However, in this case, the standard error ($S_{\overline{D}}$) and standard deviations (S_D) are calculated using D, which refers to the difference between the value at time 1 and time 2.

$$S_{\bar{D}} = \frac{S_D}{\sqrt{N}}$$

$$S_D = \sqrt{\frac{\Sigma D^2 - \frac{(\Sigma D)^2}{N}}{N-1}}$$

Therefore, we must calculate the means before and after hypnosis and the difference between each subject's before and after values.

Subject	Before	After	D	D²
1	19	18	1	1
2	35	37	−2	4
3	20	18	2	4
4	31	31	0	0
5	23	24	−1	1
6	11	6	5	25
7	32	28	4	16
8	24	21	3	9
9	18	17	1	1
10	27	25	2	4
	240	225	15	65

We can find the "before" mean and "after" mean by dividing the sum of their values by the number of cases ($240 \div 10 = 24.0$ and $225 \div 10 = 22.5$, respectively). Next, we calculate D by subtracting each "after" value from each "before" value for each case. Thus, for subject 1, D equals 1 because $19 - 18 = 1$, with the process repeated for each case. The sum of the difference (ΣD) for before and after hypnosis is 15. The last column of work shows that each value of D is squared and those squared values are summed, with $\Sigma D^2 = 65$. The calculation of the standard deviation is:

$$S_D = \sqrt{\frac{65 - \frac{(15)^2}{10}}{10-1}} = 2.173$$

Now that we have the standard deviation, we can calculate the standard error as:

$$S_{\bar{D}} = \frac{2.173}{\sqrt{10}} = .687$$

With the standard error calculated, we now have all the values needed to calculate the t statistic:

$$t = \frac{24.0 - 22.5}{.687} = 2.183$$

Step Four: Make a decision regarding the null hypothesis. To make a decision about the null hypothesis, we need to compare the calculated t with the critical value of t. We need to determine the degrees of freedom in order to find the critical value of t. With N equal to the number of differences, or the number of *pairs* of observations, the degrees of freedom equal:

$$df = N - 1$$

In this case, the degrees of freedom equal 9 ($10 - 1 = 9$). Since we decided on a significance level of .01, the critical value of t from the two-tailed t table (provided in Appendix A) is 3.250. We can see in **Figure 6.3** that the calculated t of 2.183 is less than the critical value of 3.250, falling in the area in which we fail to reject the null hypothesis. Thus, we fail to reject the null hypothesis.

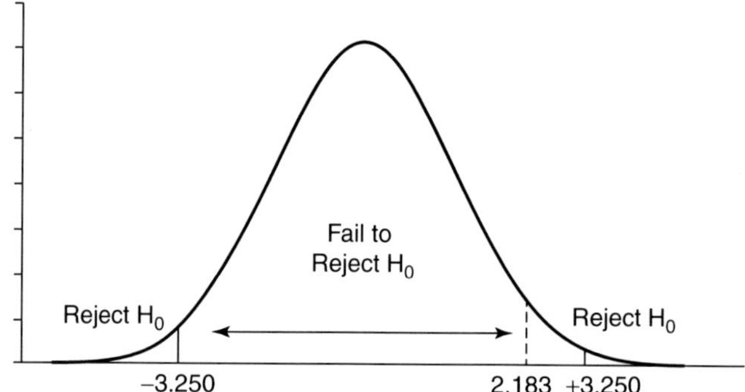

Figure 6.3 The t Curve with df $= 9$ and $\alpha = .01$.

Step Five: Interpret the results. Since we failed to reject the null hypothesis, we did not have enough evidence in our sample to refute the null hypothesis. Therefore, we conclude that the difference between the mean numbers of cigarettes smoked before and after hypnosis is the same. The observed difference in the sample was not statistically significant. In other words, based on the sample data, we believe that the number of cigarettes smoked before and after hypnosis is the same in the population. However, we should be aware that a difference might exist in the population; the average number of cigarettes smoked after hypnosis may actually be different in the population. In this case, we may have still committed a Type II error—failing to reject a false null hypothesis.

One-Tailed Versus Two-Tailed Tests

Up to this point, we have stated our alternative hypotheses in terms of a difference between the population mean and the hypothesized value in a test of a single mean or a difference between two means. These are referred to as two-tailed tests because they allow researchers to reject the null hypothesis in either the positive or negative tail. However, researchers may hypothesize not simply a difference but a direction to the difference, that is, one value is greater or less than another. Hypotheses that specify a direction require one-tailed tests that allow for the rejection of the H_0 in only one of the tails. **Figure 6.4** shows two one-tailed curves, one with a negative and the other with a positive test.

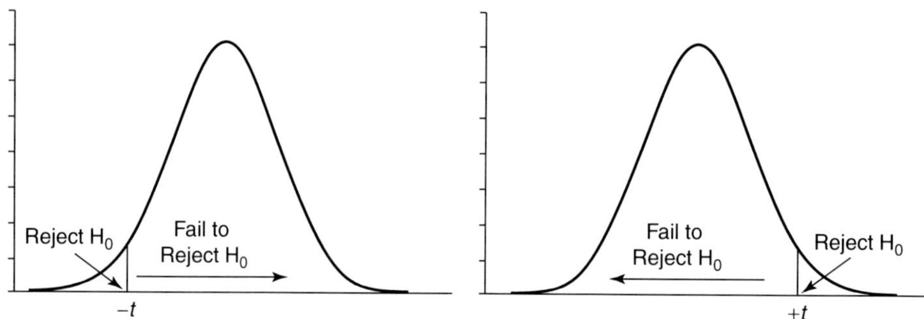

Figure 6.4 The One-Tailed t Curves

Let's revisit the previously covered two-tailed tests and formulate one-tailed tests. As discussed in an earlier example, the film student stated the alternative hypothesis simply as a difference from the hypothesized value of 110 minutes:

H_0: The average length of movies in the population is 110 minutes ($\mu = 110$).

H_1: The average length of movies in the population is *not* 110 minutes ($\mu \neq 110$).

These hypotheses reflect a two-tailed test, because if the population mean is significantly longer than *or* shorter than 110 minutes, the film student could reject the null hypothesis. The *t* curve would have critical regions in both of the two tails, indicating that if the calculated *t* falls in either tail, the null hypothesis should be rejected. Since critical regions exist in both tails, it is a two-tailed test.

Alternatively, the film student could have stated his hypotheses differently. If he thought that movies were *shorter* than 110, the hypotheses would be:

H_0: The average length of movies in the population *is not shorter than* 110 minutes ($\mu \geqslant 110$).

H_1: The average length of movies in the population is *shorter than* 110 minutes ($\mu < 110$).

These hypotheses indicate a one-tailed test, since the film student expects the true population mean to be less than the hypothesized value. Because a specific direction is specified in the hypothesis, there is only one rejection region, in this case, in the negative tail. Also, note that the critical value of *t* differs in a one-tailed test compared to a two-tailed test. When the film student used the two-tailed test, the critical value of *t* of 2.045 was associated with an alpha level of .05 and 29 degrees of freedom. However, the one-tailed *t* table indicates that the critical value of *t* associated with an alpha level of .05 and 29 degrees of freedom is 1.699, as shown in **Figure 6.5**. Since the entire probability for rejecting the null hypothesis ($\alpha = .05$) is in one tail as opposed to being split into two tails, the critical value of *t* in the one-tailed test is smaller than in the two-tailed test—in this case, 1.699 as opposed to 2.045. Due to the smaller critical value of *t* in the one-tailed test, the film student now would reject the null hypothesis and conclude that the average length of movies in the population appears to be shorter 110 minutes. Thus, we see that the one-tailed test is associated with a critical value of *t* that is lower, making it easier to reject the null hypothesis. However, this is not done without risk. Any calculated *t* of the opposite sign (in this case, positive), regardless of how large, would not allow for the rejection of the null hypothesis.

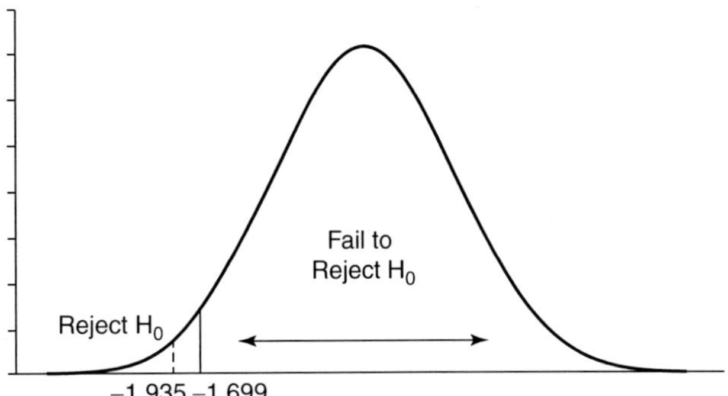

Figure 6.5 The One-Tailed *t* Curve with df = 29 and $\alpha = .05$.

Let's look at the other examples previously presented as two-tailed tests and see how they might differ under a one-tailed test. In the test of two independent means, the sociologist who wanted to study the possible inequality in the pricing of items in urban and suburban stores could have stated the hypotheses to indicate a direction. He suspects that urban stores charge more than suburban stores for the exact same items. Therefore, his hypotheses are:

H_0: The mean price of the items at urban stores *is not higher* than the mean price of those same items at suburban stores ($\mu_1 \leqslant \mu_2$) or ($\mu_1 - \mu_2 \leqslant 0$).

H_1: The mean price of the items at urban stores *is higher* than the mean price of those same items at suburban stores ($\mu_1 > \mu_2$ or ($\mu_1 - \mu_2 > 0$).

With the change in the stating of the hypotheses, the sociologist is no longer conducting a two-tailed test but rather a one-tailed test. Given the one-tailed test of the null hypothesis, the critical value of *t* will change. As we saw before, in a two-tailed test with 40 degrees of freedom and an alpha level of .01, the critical value of *t* was 2.704. However, the one-tailed *t* table indicates that the critical value of *t* with 40 degrees of freedom and an alpha level of .01 is 2.423. In this case, since the sociologist hypothesized that mean prices at urban stores are greater than mean prices at suburban stores, the entire rejection region is in the positive tail, as shown in **Figure 6.6.** Since the calculated *t* (3.436) is greater than the critical value of *t* (2.423), the sociologist would reject the null hypothesis, just as he had done in the two-tailed test. However, this time, his conclusion would be that the mean price of the items at urban stores *is higher* than the mean price of those same items at suburban stores.

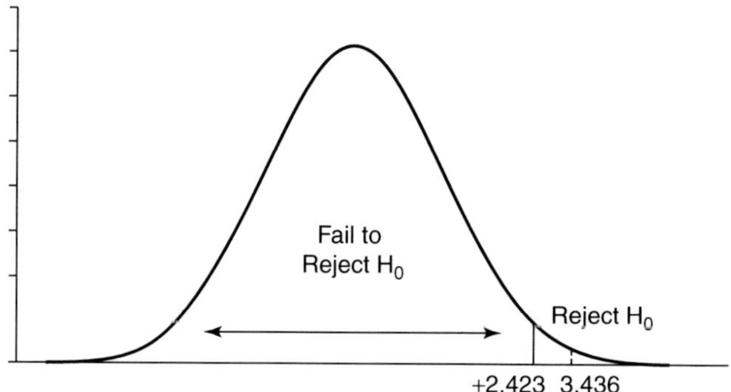

Figure 6.6 The One-Tailed *t* Curve with df $= 40$ and $\alpha = .01$.

Similarly, the hypotheses regarding hypnosis and cigarette smoking can be restated to indicate a direction. If we expected hypnosis to reduce the number of cigarettes smoked, we would hypothesize:

H_0: The mean number of cigarettes smoked before hypnosis *is not greater* than the mean number smoked after hypnosis.

H_1: The mean number of cigarettes smoked before hypnosis *is greater* than the mean number smoked after hypnosis.

Since the hypotheses indicate a direction, we are doing a one-tailed test. The one-tailed *t* table indicates that the critical value of *t* with 9 degrees of freedom and an α of .01 is 2.821. Although the critical value of *t* is smaller in the one-tailed test compared to the two-tailed test (2.821 compared to 3.250, respectively), the calculated *t* value is still smaller than the critical value of *t*, as shown in **Figure 6.7.** In this case, our decision remains the same; we fail to reject the null hypothesis but we now conclude that the mean number of cigarettes smoked before hypnosis *is not greater* than the mean number smoked after hypnosis.

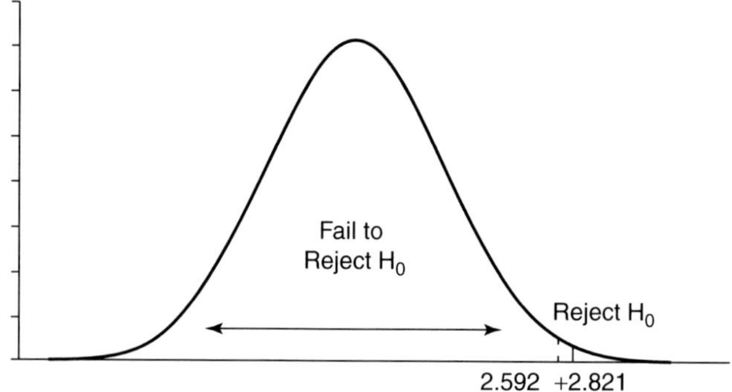

Figure 6.7 The *t* Curve with df $= 9$ and $\alpha = .01$.

Many types of statistical analysis in software programs, such as SPSS, automatically produce two-tailed tests. However, because theory must inform our hypothesis making, the decision to do a one-tailed or two-tailed test is a theoretical decision and should not be determined by a computer program. The decision to state our hypothesis as having a difference (two-tailed test) versus specifying the direction of the difference (one-tailed test) should be based on theoretical considerations, such as the expected relationship between variables. If SPSS produces a two-tailed test, we can take the *p* value from the output and divide it by 2 to obtain the *p* value for a one-tailed test. For example, if the *p* value is .072 for a two-tailed test, the one-tailed *p* value would be .036. Therefore, theory should continue to direct us in the formulation of our hypotheses, which in turn determines whether we will do a one-tailed or two-tailed test.

T-Tests in SPSS

Let's look at how one sample and two independent sample tests would be done using SPSS. Suppose Shannon and Taylor get into a discussion regarding the average hours of TV that American adults watch. Shannon thinks that the average American watches 5 hours of TV. She does a statistical test to see if the average number of TV watched is 5. She begins by stating her hypotheses:

H_0: The average American watches 5 hours of TV.

H_1: The average American does not watch 5 hours of TV.

In SPSS, Shannon first checks the variable for missing values, Hours Per Day Watching TV (TVHOURS). First, she needs to be sure that she is in Variable View, as shown in **Figure 6.8.** If you find that you are not in Variable View, click on the tab labeled Variable View that the solid arrow is point at. Find TV HOURS in the first column. In the row for TV HOURS, go to the column labeled Values, click on the cell, and box that the dashed arrow is pointed at will appear. Click on it.

GSS2008-subset-2.sav [DataSet1] - SPSS Statistics Data Editor

File   Edit   View   Data   Transform   Analyze   Graphs   Utilities   Add-ons   Window   Help

	Name	Type	Width	Decimals	Label	Values	Missing	Columns	Align	Measure
242	SUICIDE3	Numeric	1	0	SUICIDE IF DIS...	{0, IAP}...	0, 9	10	Right	Nominal
243	SUICIDE4	Numeric	1	0	SUICIDE IF TIR...	{0, IAP}...	0, 9	10	Right	Nominal
244	POLHITOK	Numeric	1	0	EVER APPRO...	{0, IAP}...	0, 9	10	Right	Nominal
245	POLABUSE	Numeric	1	0	CITIZEN SAID ...	{0, IAP}...	0, 9	10	Right	Nominal
246	POLMURDR	Numeric	1	0	CITIZEN QUES...	{0, IAP}...	0, 9	10	Right	Nominal
247	POLESCAP	Numeric	1	0	CITIZEN ATTE...	{0, IAP}...	0, 9	10	Right	Nominal
248	POLATTAK	Numeric	1	0	CITIZEN ATTA...	{0, IAP}...	0, 9	10	Right	Nominal
249	FEAR	Numeric	1	0	AFRAID TO W...	{0, IAP}...	0, 9	6	Right	Nominal
250	OWNGUN	Numeric	1	0	HAVE GUN IN ...	{0, IAP}...	0, 9	8	Right	Nominal
251	PISTOL	Numeric	1	0	PISTOL OR RE...	{0, IAP}...	0, 9	8	Right	Nominal
252	SHOTGUN	Numeric	1	0	SHOTGUN IN ...	{0, IAP}...	0, 9	9	Right	Nominal
253	RIFLE	Numeric	1	0	RIFLE IN HOME	{0, IAP}...	0, 9	7	Right	Nominal
254	ROWNGUN	Numeric	1	0	DOES GUN BE...	{0, IAP}...	0, 9	9	Right	Nominal
255	HUNT	Numeric	1	0	DOES R OR S...	{0, IAP}...	0, 9	6	Right	Nominal
256	NEWS	Numeric	1	0	HOW OFTEN ...	{0, IAP}...	0, 9	6	Right	Nominal
257	TVHOURS	Numeric	2	0	HOURS PER D...	{-1, IAP}...	-1, 99	9	Right	Scale
258	PHONE	Numeric	1	0	DOES R HAVE...	{0, IAP}...	0, 9	7	Right	Nominal
259	COOP	Numeric	1	0	RS ATTITUDE ...	{0, IAP}...	0, 9	6	Right	Nominal
260	COMPREND	Numeric	1	0	RS UNDERST...	{1, GOOD}...	0, 9	10	Right	Nominal
261	FORM	Numeric	1	0	FORM OF SPL...	{0, NO SPLI...	0	6	Right	Nominal
262	FECHLD	Numeric	1	0	MOTHER WOR...	{0, IAP}...	0, 9	8	Right	Nominal
263	FEPRESCH	Numeric	1	0	PRESCHOOL ...	{0, IAP}...	0, 9	10	Right	Nominal
264	FEFAM	Numeric	1	0	BETTER FOR ...	{0, IAP}...	0, 9	7	Right	Nominal
265	RACDIF1	Numeric	1	0	DIFFERENCES...	{0, IAP}...	0, 9	9	Right	Nominal
266	RACDIF2	Numeric	1	0	DIFFERENCES...	{0, IAP}...	0, 9	9	Right	Nominal
267	RACDIF3	Numeric	1	0	DIFFERENCES...	{0, IAP}...	0, 9	9	Right	Nominal
268	RACDIF4	Numeric	1	0	DIFFERENCES...	{0, IAP}...	0, 9	9	Right	Nominal

Data View   **Variable View**

SPSS Statistics Processor is ready

Figure 6.8 SPSS Variable View Display

Another box opens that shows the value labels for the variable (see **Figure 6.9**). Computers read and analyze numbers. If a category of the variable is not a numeric value, researchers assign a numeric value to represent that category. The variable, Hours Per Day Watching TV (TV HOURS), has meaningful numeric categories, for example, someone may watch "2" hours of TV per day. Therefore, these meaningful numeric categories do not need value labels. However, as shown in **Figure 6.9,** the Value Labels box shows three value labels for Hours Per Day Watching TV (TV HOURS). IAP stands for "Inappropriate." This value is assigned when it does not make sense to ask a respondent a particular question, such as asking someone who has never married whether he or she is divorced. DK is the abbreviation for the response "Don't Know," while NA stands for "No Answer."

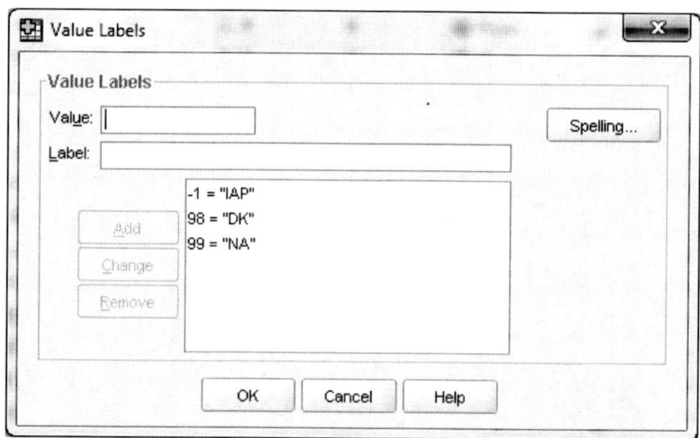

Figure 6.9 SPSS Value Labels for Hours Per Day Watching TV (TVHOURS)

These responses of "Inappropriate," "Don't Know," and "No Answer," do not represent how many hours of TV respondents watch. Furthermore, Shannon does not want, for example, those respondents who responded "Don't Know" to be analyzed as if they watched 98 hours of TV, but 98 is the numeric value assigned to the response "Don't Know." Therefore, the solution is to exclude these results from the analysis. Shannon clicks on the Missing cell for the variable, Hours Per Day Watching TV (TV HOURS); note the cell that the dotted arrow is pointing at in **Figure 6.8.** She clicks on the box with the three dots and another box, the Missing Values box, opens. **Figure 6.9** shows the missing values for the variable, Hours Per Day Watching TV (TV HOURS). The General Social Survey (GSS) data already has "Inappropriate" coded as −1 and "No Answer," coded as 99, defined as missing. However, "Don't Know," which is coded as 98, is not currently defined as missing. Therefore, Shannon enters 98 in the box that the arrow is pointing at in **Figure 6.10** and clicks OK.

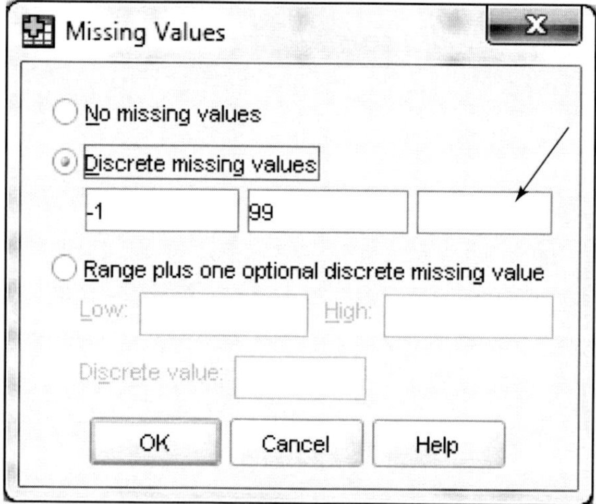

Figure 6.10 SPSS Missing Values for Hours Per Day Watching TV (TVHOURS)

Shannon now is ready to analyze the data. Shannon goes to Analyze=>Compare Means=>One Sample T Test and the box in **Figure 6.11** opens. She looks for the variable Hours Per Day Watching TV (TV HOURS) in the left-hand box, highlights the variable, and clicks on the arrow between the boxes so that the variables moves from the left-hand box into the right-hand box. Next, Shannon needs to change the Test Value from the 0 that SPSS automatically displays to the one she wants to use. Therefore, she enters 5 in the Test Value box and clicks OK.

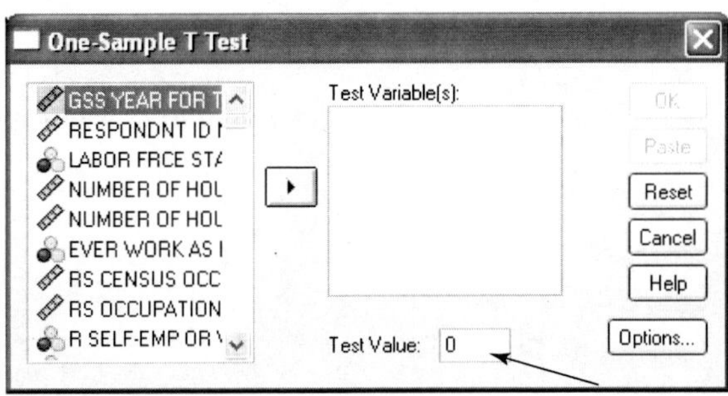

Figure 6.11 One-Sample T Test Box

SPSS produces the printout shown in **Table 6.4.** The first set of statistics displayed are descriptive statistics for Hours Per Day Watching TV (TV HOURS). The second box provides the *t* test. SPSS has calculated the *t*-test value of -27.615 and the degrees of freedom (*df*) of 1323. SPSS also provides the *p* value in the column labeled "Sig. (2-tailed)." The *p* is the exact probability at which we fail to reject the null hypothesis. Therefore, we do not need to find the critical value of *t* in a *t* table to compare with the calculate *t* score. Instead, we compare the α level with the *p* value in the column labeled "Sig. (2-tailed)." If the *p* value is less than the α value, then the null hypothesis is rejected. **Table 6.5** outlines some key differences between α and *p*.

Table 6.4 *SPSS Printout of the One-Sample Testing if Respondents Watch 5 Hours of TV Per Day*

One-Sample Statistics

	N	Mean	Std. Deviation	Std. Error Mean
Hours Per Day Watching TV	1324	2.98	2.659	.073

One-Sample Test

	Test Value = 5					
					95% Confidence Interval of the Difference	
	t	df	Sig. (2-tailed)	Mean Difference	Lower	Upper
Hours Per Day Watching TV	-27.615	1323	.000	-2.018	-2.16	-1.87

Table 6.5 *The Difference between α and p*

α	p
• α is the threshold probability at which H_0 is rejected.	• The *p* is the exact probability at which researchers fail to reject the H_0.
• The researcher determines α prior to testing hypotheses.	• The researcher does not determine *p*; it is determined by the data.
• We have used α and the *df* to obtain a critical value of *t* that we compared with our calculated *t*. If the calculated *t* was greater than the critical value of *t*, then H_0 is rejected.	• SPSS provides us with the *p* value. If the *p* is $< \alpha$, then H_0 is rejected. Therefore, we do not need to obtain a critical value of *t* from a *t* table to compare with the calculated *t* score.

To determine if respondents watch five hours of TV per day, Shannon examines the printout in **Table 6.4** and sees that the p value under "Sig. (2-tailed)" is .000. This probability is so small that at least the first three decimal place of the number are zeros. Therefore, Shannon realizes that p is less than α and decides that she should reject the null hypothesis. She concludes that the average number of hours of TV watched by adult Americans probably is not 5 hours.

Shannon and Taylor's discussion of TV watching habits continues. They begin to speculate on how the number of hours of TV watched might vary by age. Shannon and Taylor decide to test to see if those under 60 years old differ in the number of hours of TV watched compared to those who are age 60 and older. They begin by formulating the hypotheses:

H_0: Those 60 and older do not differ from those under 60 years old in terms of the number of hours of TV watched daily ($\mu_1 = \mu_2$).

H_1: Those 60 and older *differ* from those under 60 years old in terms of the number of hours of TV watched daily ($\mu_1 \neq \mu_2$).

Shannon and Taylor are analyzing the two groups in terms of their TV watching. To do the t-test, SPSS requires that they know how the Age variable is coded. Just as they did earlier, they can go to Variable View to Values to check to see how the Age variable is coded and check to be sure that missing values have been defined. To do the t-test in SPSS, Shannon and Taylor go to Analyze=>Compare Means=>Independent Samples T Test, and the box in **Figure 6.12** opens. Since they want to see if two age groups differ in the number of TV they watch, the Test Variable is Hours Per Day Watching TV (TV HOURS), and they search for it in the left-hand box and click on the arrow to move it into the Test Variable(s) box. The grouping variable is the Age variable; they want to compare two different age groups. After finding it in the left-hand box, they click the arrow to move the Age variable into the Grouping Variable box. The words "Define Groups" (see **Figure 6.12**) become darkened once a variable is moved into the Grouping Variable box. Taylor clicks on Define Group, and the box in **Figure 6.13** appears. If the groups that they were comparing were categories of a nominal variable, such as sex, they would need to know how males and females are coded, and they would enter those numeric values in the Group 1 and Group 2 boxes. However, Age is a ratio-level variable. Therefore, Taylor clicks on the Cut Point radial button (see the solid arrow in **Figure 6.13**) and is now able to type in the box to the right of the Cut Point radial button (see the dashed arrow in **Figure 6.13**). Since Shannon and Taylor want to compare those under 60 years old with those who are 60 and older, Taylor enters 60 in the box and then clicks Continue. The Define Groups box closes and Taylor clicks OK in the Independent-Samples T Test box.

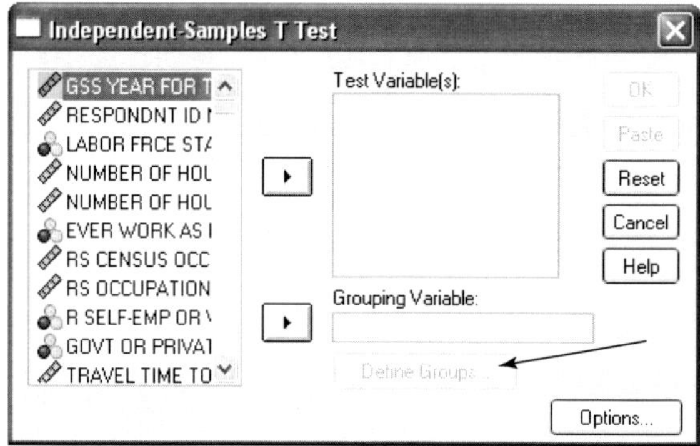

Figure 6.12 SPSS Independent–Samples T Test

SPSS runs the independent-samples t-test comparing the number of hours of TV watched by those under 60 years old with those who are 60 and older, and it produces the printout shown in **Table 6.6**. The first portion of the printout provides descriptive statistics for the two groups. The second part of the printout contains the Levene's Test for Equality of Variances and two t-tests. Researchers use the Levene's test to determine which t-test to use; the null hypothesis for the Levene's test is that the variances are equal. If the significance for

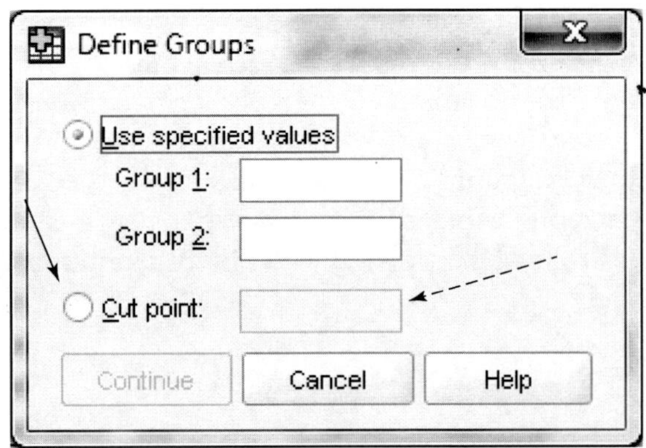

Figure 6.13 SPSS Define Groups within the Independent-Samples T Test

the Levene's test is greater than .05, we fail to reject the null hypothesis and use the Equal Variances Assumed test. However, if the Levene's test significance level is .05 or less, then the Equal Variances not Assumed test is used because the null hypothesis that the variances are equal is rejected. The solid arrow in **Table 6.6** is pointed at the significance level for the Levene's test. Because the significance level of .087 is greater than .05, Shannon and Taylor fail to reject the null hypothesis of the Levene's test that the variances of the two groups are equal in the population. Therefore, Shannon and Taylor use the Equal Variances Assumed test.

Table 6.6 *SPSS Printout of the Independent-Samples T-Test Comparing the Number of Hour of TV Watched Daily by Those under 60 years Old with Those Age 60 and Older*

Group Statistics

Age of Respondent		N	Mean	Std. Deviation	Std. Error Mean
Hours Per Day Watching TV	> = 60	328	3.64	2.632	.145
	< 60	988	2.76	2.642	.084

Independent Samples Test

		Levene's Test for Equality of Variances		t-test for Equality of Means					95% Confidence Interval of the Difference	
		F	Sig.	t	df	Sig. (2-tailed)	Mean Difference	Std. Error Difference	Lower	Upper
Hours Per Day Watching TV	Equal variances assumed	2.926	.087	5.202	1314	.000	.875	.168	.545	1.205
	Equal variances not assumed			5.212	561.543	.000	.875	.168	.545	1.205

The statistics that Taylor and Shannon want to interpret are those for Equal Variance Assumed, found on the top line in **Table 6.6,** concentrating on the t value, the degrees of freedom, and the significance level that give the exact probability at which the null hypothesis can be rejected (the dashed arrows pointing at these statistics). Taylor and Shannon identify that $t_{(1314)} = 5.202$ and conclude that the statistical test was significant, since p is so small that the SPSS program prints .000. Because p is less than α levels of .05, .01, and .001, they reject the null hypothesis. Based on the independent-samples t-test, Taylor and Shannon conclude that those under 60 years old and those 60 and older probably differ in the amount of TV watched daily.

Exercises and Notes

Hypotheses

- Hypotheses always come in pairs.

  - ____________________________________

  - ____________________________________

- Hypotheses are always about ____________________.
  - Only contain symbols of ____________________________.
  - Never have symbols of ____________________________.

Steps in Hypotheses Testing

1. ____________________________________
2. ____________________________________
3. ____________________________________
4. ____________________________________
5. ____________________________________

Types of Errors in Testing Hypotheses

	Decision Based on Sample Data	
	Reject H_0	Fail to Reject H_0
H_0 is true.		
H_0 is False		

Test Statistics and α Levels

- **Test Statistic—**____.

- **Degrees of Freedom** (*df*)**—**____.

- **Type I error, Level of Significance, and α Level—**________________________.

Test of a Single Sample Mean

1. A law professor wants to demonstrate how accurate eye-witness testimony often is. He shows 15 subjects a video in which a robbery scene is suddenly inserted and asks the subjects to describe the robber, including his age, which was 29. The subjects' average judgment of the robber's age was 25.8 years with a variance of 10.03. Test to see if the subjects accurately judged the robber's age.

   $\alpha = .01$

   H_0:

   H_1:

Test Statistic for a Single Mean

$$t = \frac{\overline{X} - \mu}{S_{\overline{X}}}$$

Estimated Standard Error

$$S_{\overline{X}} = \frac{s}{\sqrt{N}}$$

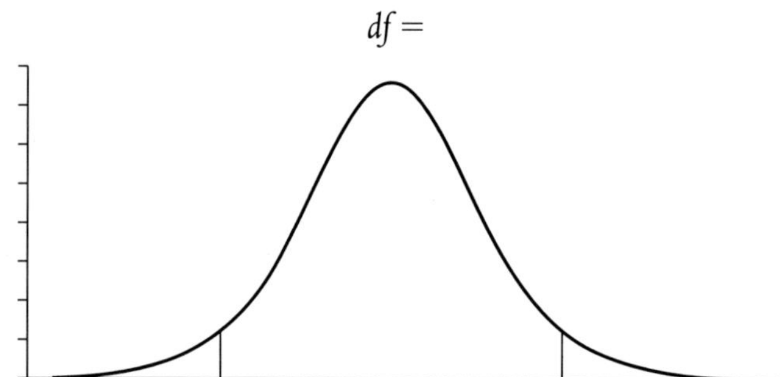

$$df =$$

Decision about H_0 *and Conclusion:* ____

____

____

____

____

Test of the Difference between Two Independent Sample Means

2. Professors were debating whether students performed differently in traditional face-to-face classes as opposed to on-line classes. One of the professors was teaching two sections of the same course, one in the classroom and the other on-line. He decided to do a statistical test to determine if the mean grade in the classroom section differed from the mean grade for the on-line section.

	Classroom	On-Line
Mean	79.63	72.91
S^2	131.42	241.82
N	25	23

Test with $\alpha = .05$.

H_0:

H_1:

Assumed Equal Population Variances

$$t = \frac{(\overline{X}_1 - \overline{X}_2) - (\mu_1 - \mu_2)}{S_{\overline{X}_1 - \overline{X}_2}}$$

$$S_{\overline{X}_1 - \overline{X}_2} = \sqrt{\frac{[(N]_1 - 1)S_1^2 + [(N]_2 - 1)S_2^2}{N_1 + N_2 - 2}} \; \sqrt{\frac{1}{N_1} + \frac{1}{N}}$$

OR

Assumed Unequal Population Variances

$$S_{\overline{X}_1 - \overline{X}_2} = \sqrt{\frac{S_1^2}{N_1} + \frac{S_2^2}{N_2}}$$

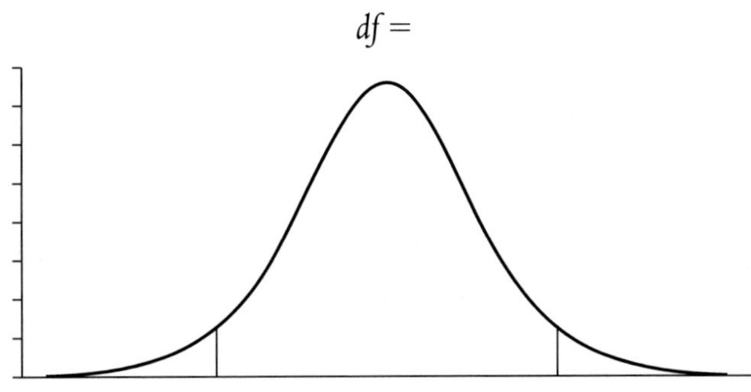

$$df =$$

Decision about H$_0$ and Conclusion: ____

____

____

____

____

Test of the Difference between Dependent or Paired Sample Means

3. A physician is concerned that a new drug may affect her patients' sleep. Before prescribing the medication, she asks her patients how many hours they slept the previous night. After a month of her patients' taking the medication, the doctor schedules a follow-up appointment with each patient and asks them how many hours they slept the previous night. Test to see if the new medication had an effect of the number of hours that patients slept.

H_0:

H_1:

Let $\alpha = .01$.

$$t = \frac{\bar{X}_1 - \bar{X}_2}{S_{\bar{D}}} \qquad S_{\bar{D}} = \frac{S_D}{\sqrt{N}} \qquad S_D = \sqrt{\frac{\Sigma D^2 - \dfrac{(\Sigma D)^2}{N}}{N-1}}$$

Subject	Before	After	D	D^2
1	8	8		
2	8	5		
3	7	5		
4	6	8		
5	7	7		
6	6	6		
7	7	6		
8	9	9		
9	6	7		
10	7	6		
11	7	7		
12	6	4		

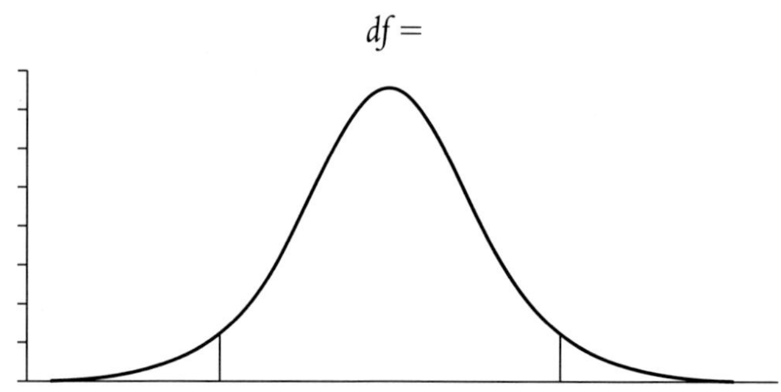

Decision about H_0 and Conclusion: _______________________________________

____

____

____

____

One-Tailed vs. Two-Tailed Tests

- *t-tests* (of a single sample mean, two independent samples, two dependent samples, a correlation, regression coefficient) can be _____________________.

- **Two-tailed** test allow for rejection of the H_0 in _____________.

- Hypotheses that specify a direction require one-tailed tests that allow for the rejection of the H_0 in

  _____________________.

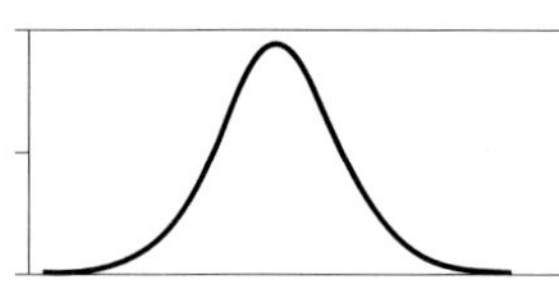 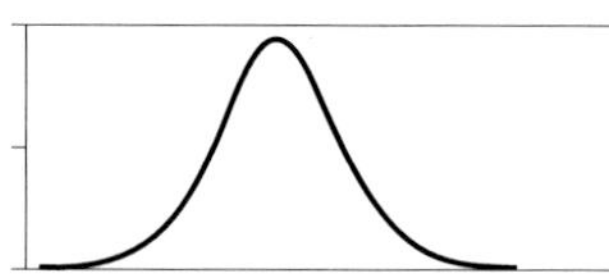

4. Suppose the professor thought subjects would underestimate the robber's age. State the hypotheses.

   H_0:

   H_1:

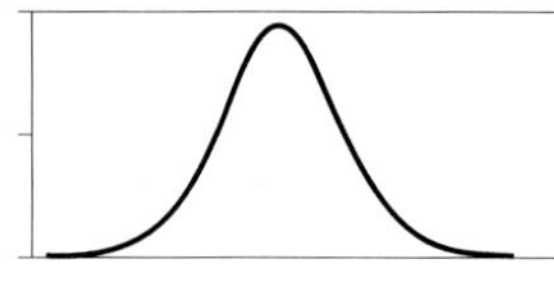 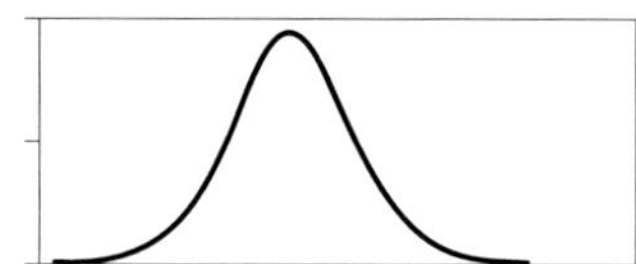

Decision about H_0 and Conclusion: _______________________________________

____

____

____

____

5. In the test of two independent means, suppose the professor suspects that students performed better in face-to-face classes as opposed to on-line classes. State the hypotheses.

   H_0:

   H_1:

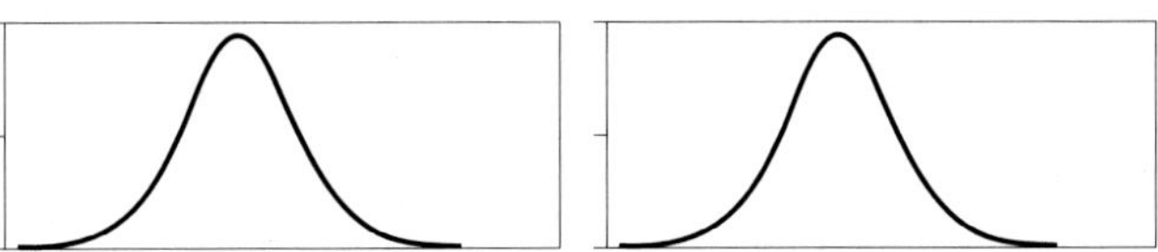

Decision about H_0 and Conclusion: ______________________________________

____

____

____

____

6. How would the hypotheses be stated if a physician thought that a new drug may cause insomnia? State the hypotheses.

   H_0:

   H_1:

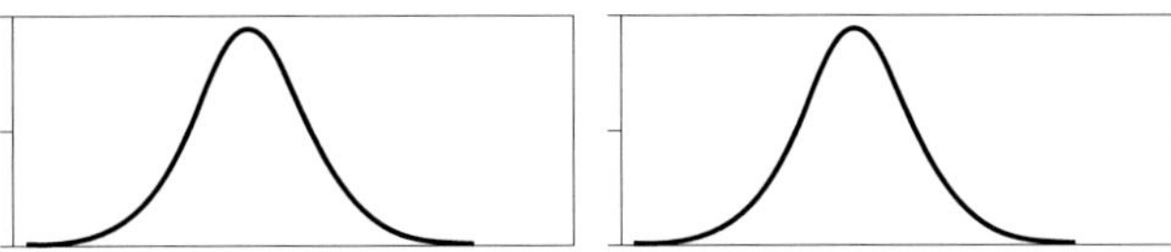

Decision about H_0 and Conclusion: ______________________________________

____

____

____

____

7. A professor believes that students do better in smaller classes. She compares two randomly chosen sections of the same course. Was she correct in assuming that students do better (*i.e.,* earn higher scores) in smaller classes than they do in larger classes? Let $\alpha = .05$.

	Small Class	Large Class
Average Score	79	74
S^2	121	218
N	30	150

H_0:

H_1:

Calculate the test statistic:

$$df =$$

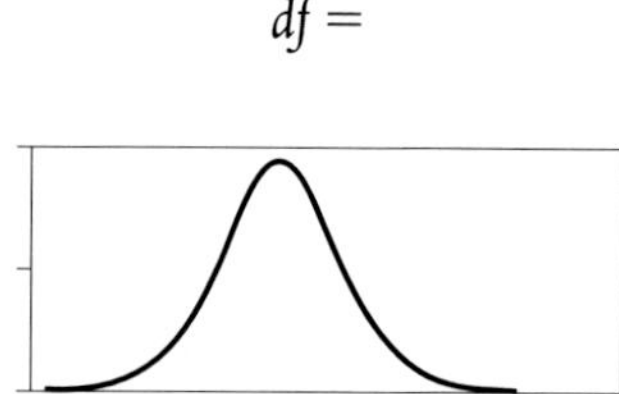

Decision about H$_0$ and Conclusion: ____

____

____

____

____

Name: ____   Date: ________________

Learning Check

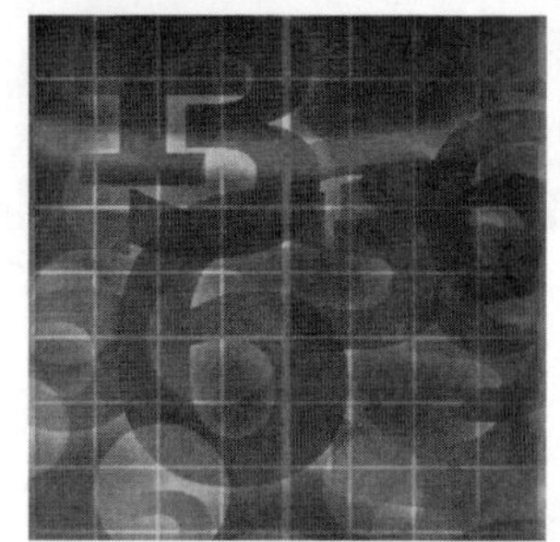

1. A financial-aid officer is concerned about the increasing student loan debt that students at her school
    are incurring. She knows that the national average for students at comparable universities is $28, 654.
    She would like to know if graduates of her institution have a similar level of indebtedness. She draws
    a random sample of students that have graduated from her university. The average amount of student
    loans for a sample of 30 recent graduates was $31, 214, with a standard deviation of $15,130. Do a
    statistical test to determine if graduates at her university incur student loan debt similar to the national
    average. Let $\alpha = .05$.

2. You are worried that your friends have become addicted to texting. You challenge them to record the
    number of texts they send and receive for a month. Eight friends average 3534 texts for the month,
    with a standard deviation of 343. You find that the national average for college students is 3290 text
    messages. Do a statistical test with $\alpha = .05$ to see if your friends text more than the national average.

3. A professor has a theory that eating chocolate before an exam affects test scores. She wants to test this
    as a hypothesis to see if it is true. She randomly divides a class into two groups: one that gets chocolate
    before the test and one that doesn't. Then she gives the exam. Do a statistical test to determine if she was
    correct in assuming that eating chocolate before an exam affects test scores. Test with $\alpha = .05$.

Chocolate Group	No-Chocolate group
81	79
82	60
85	87
55	98
90	95
100	81

4. A music lover believes that she will have to pay more to go to concerts on the East or West Coasts
    compared to the Midwest. She took a sample of 15 artists and compared their ticket prices for the East
    and West Coasts versus the Midwest. Do a statistical test to determine if she was correct in assuming
    that she would pay more for concert tickets on the East and West Coasts compared to the Midwest.
    Test with $\alpha = .01$.

	East and West Coasts	Midwest
Mean	$80.87	$70.60
S	29.97	30.55
N	15	15

5. The professor in question 3 was determined to see if chocolate has a higher value than a snack food.
    She argues that chocolate is the perfect diet food. She contends that eating chocolate reduces the
    appetite so people who eat it lose weight overall. She identifies ten people who want to lose weight
    and puts them on her chocolate diet. After two months, she weighs them again. Do a statistical test to
    determine if her subjects have lost a significant amount of weight. Let $\alpha = .05$.

Start Weight X_1	End Weight X_2
150	152
155	143
165	150
150	156
175	165
210	190
225	215
224	225
158	150
192	160
$\Sigma X_1 = 1804$	$\Sigma X_2 = 1706$

6. A sport sociologist wanted to study the effect of steroids on baseball. Part of his study involved comparing the average number of homeruns hit by the top 20 homerun hitters during the height of the steroid era compared with the number of homeruns hit after Major League Baseball started testing for performance-enhancing drugs, including steroids. He found that at the height of the steroid era, the average number hit by the 20 top hitters was 41.7 homeruns, while after performance-enhancing drug testing occurred, the average number by the 20 top hitters was 34.4 homeruns. The sum of the difference is 146, while the sum of the difference squared between the two time periods was 1128. Test to see if the top 20 homerun hitters hit fewer homeruns in the time period after performance-enhancing drug testing occurred. Let $\alpha = .05$.

7. Researchers wanted to examine the effect that viewing violent images have on aggression. Twenty-five subjects were administered a test at Time 1 that included an aggression scale (0–40). The mean score on the aggression scale was 18.24 at Time 1. The subjects then were asked to watch a series of movies that contained graphic violence over a period of two weeks. The subjects were given the test again. The mean score on the aggression scale at Time 2 was 21.80. The sum of the difference between Time 1 and Time 2 was - 89, while the sum of the difference squared between the two time periods was 603. Test to see if subjects' test scores increased after watching the graphic violence. Let $\alpha = .05$.

8. Use SPSS and the GSS data to test this research question: Did respondents work a significantly different number of hours last week (HRS1) other than 40 hours? State the hypotheses. Before analyzing the data, be sure that missing values have been properly defined. When reporting your results, include the t value, df, and the p value. Be sure to include whether you reject or fail to reject the null hypothesis and what you can then conclude.

9. Use SPSS and the GSS data to test this research question: Do the average white and black workers differ in the number of hours worked last week? In other words, does the race of respondents (RACE) affect the number of hours worked last week (HRS1)? State the hypotheses. Before analyzing the data, be sure that missing values have been properly defined. Indicate whether you are reporting results in which equal variances are assumed or not assumed, and why. Your results should include the t value, df, and the p value. Be sure to include whether you reject or fail to reject the null hypothesis, and what you can then conclude.

Chapter Seven

Chi-Square Tests

In the previous chapter, the *t* tests we performed were examples of **parametric tests** that involve estimating population parameters, such as the mean, from sample data. However, the use of parametric tests requires that certain conditions be met. Among these conditions are that the variables be measured at the interval/ratio level and be normally distributed. These requirements add to the power of the test. You may remember from our discussion of Type II error in Chapter 6 that the probability of a Type II error (or beta) equals 1–power. Therefore, the more powerful the test, the more likely we will be able to reject the null hypothesis when it is false.

If we only have nominal and ordinal data and we cannot assume that the data are normally distributed, we can still test hypotheses using **nonparametric tests** of significance, since these tests do not require interval/ratio-level data nor that we make assumptions about the distribution. In this chapter, we will concentrate on the nonparametric test known as the chi-square (pronounced ki-square) to test the significance of nominal and ordinal level data. Later, in Chapter 9, we will explore another nonparametric test, Spearman's correlation, which uses ordinal data.

Chi-Square Test for the Goodness of Fit

In Chapter 2, we used frequency distributions and cross-tabs to organize data. Once the data were organized into tables, we could compare categories of the data using absolute frequency or percentages. However, we could not indicate that there was a statistical significant difference between categories. The chi-square test for the goodness of fit allows us to determine whether a frequency distribution from sample data significantly differs from an even distribution or any other hypothesized distribution. Let's look at two examples: one in which the null hypothesis predicts an even distribution in the population and another with a null hypothesis that expects a distribution in which the cases are not predicted to be evenly divided over the categories of the variable.

Example 1: H_0 Predicts an Even Distribution

Suppose we have the intended academic major of a sample of 100 incoming students, as shown in **Table 7.1**, and we want to determine if their reported majors significantly differ from an even distribution. We need to apply the hypothesis testing steps described in Chapter 6 to the chi-square goodness of fit test used here.

Table 7.1 *The Intended Major of 100 Incoming Students*

Intended Major	f
Science	35
Social Sciences or Humanities	40
Professional	25
Total	100

Step One: State null and alternative hypotheses. Since the null hypothesis indicated no difference, our null hypothesis reflects an even distribution among the students' majors. The alternative hypothesis, on the other hand, indicates a divergence from an even distribution.

H_0: Incoming students are equally divided in their choice of major.

H_1: Incoming students are not equally divided in their choice of major.

Step Two: Determine the significance or alpha level that will be used in the statistical test. The same levels of significance (.05, .01, or .001) are used for the chi-square test. For this test, let's use .05.

Step Three: Calculate the statistical test. Since we are interested in determining if incoming students' reported majors significantly differ from an even distribution, we need to calculate a one-way chi-square test statistic. The chi-square formula is:

$$x^2 = \Sigma \frac{(f_o - f_e)^2}{f_e}$$

The x^2 formula compares the observed frequencies from the data (f_o) with the frequencies expected (f_e) if the null hypothesis is true. Therefore, step 1 in calculating x^2, is to calculate the expected frequencies. In the goodness of fit test concerning an equal distribution, the expected frequencies (f_e) are determined by:

$$f_e = \frac{N}{k}$$

N = the total number of observations

k = the number of categories in the variable

In this case, $N = 100$ and $k = 3$. Therefore, the expected frequency (f_e) equals 33.3, since:

$$f_e = \frac{100}{3} = 33.3$$

Table 7.2 shows the observed frequency data (f_o) for intended major, along with the calculations needed to determine the x^2 value. The expected frequencies (f_e) reflect the expected even distribution if the null hypothesis is true; in this case, each category has an expected frequency of 33.3. In the next column (step 2), we see the difference between the observed and expected frequencies ($f_o - f_e$) for each category of the variable. In step 3, we square the difference between the observed and expected frequencies ($f_o - f_e$)^2 for each category of the variable. The last column (step 4) of **Table 7.2** shows the quotient from dividing the difference between the observed and expected frequencies squared ($f_o - f_e$)^2 by the expected frequency (f_e). Lastly, in step 5, we sum the three quotients in the last column. Therefore, our calculated x^2 equals 3.51.

Table 7.2 *Summary Table of the Calculation of x^2 for Intended Majors*

Major	f_o	Step 1 f_e	Step 2 $f_o - f_e$	Step 3 $(f_o - f_e)^2$	Step 4 $\dfrac{(f_o - f_e)^2}{f_e}$
Science	35	33.3	1.7	2.89	.09
Social Studies or Humanities	40	33.3	6.7	44.89	1.35
Professional	25	33.3	-8.3	68.89	2.07
Total	100	99.9*		**Step 5**	3.51

** Due to rounding error*

Step Four: Make a decision regarding the null hypothesis. To find the critical value of χ^2, we need the number of degrees of freedom and the significance level (from step 2, $\alpha = .05$). For a χ^2 goodness of fit test, the degrees of freedom equal the number of categories (referred to as k) minus 1. Therefore, the formula for the degrees of freedom is $k - 1$ or, in this case, $3 - 1 = 2$. We use the chi-square table provided in Appendix A and find the critical value of χ^2 of 5.991 associated with an alpha level of .05 and 2 degrees of freedom. **Figure 7.1** shows the critical value of χ^2 of 5.991, which determines a critical region that designates whether the null hypothesis can be rejected. Since the calculated χ^2 of 3.51 is less than the critical χ^2 of 5.991, the calculated χ^2 of 3.51 falls in the area of the curve in which the null hypothesis is not rejected, as **Figure 7.1** shows. Thus, we fail to reject the null hypothesis.

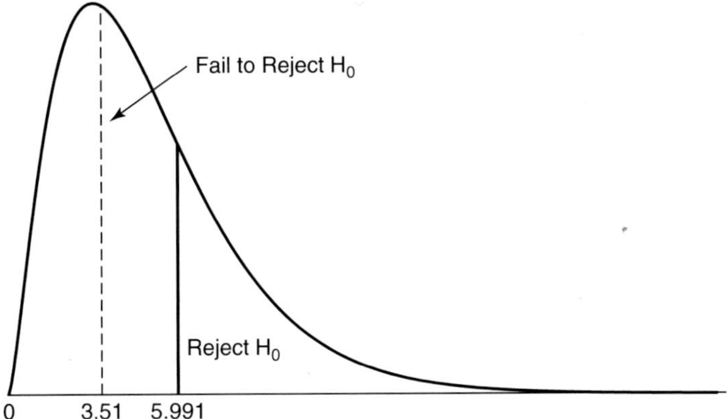

Figure 7.1 The χ^2 Curve with df $= 2$ and $\alpha = .05$.

Step Five: Interpret the results. Since we failed to reject the null hypothesis, we did not have enough evidence in the sample to refute the null hypothesis. The difference between the sample frequency distribution and evenly distributed frequencies is not statistically significant. Therefore, we conclude that all incoming students are equally divided in their choice of major. However, just as we discussed when we did t tests, we could be wrong. Incoming students in the population may vary in their choice of major. Despite correctly conducting the statistical test, we cannot be 100% certain of our results. In this case, we may have still committed a Type II error—failing to reject a false null hypothesis.

Example 2: H_0 Predicts an Alternative to an Even Distribution

If we had not hypothesized an even distribution but instead an alternative distribution, such as 25% in both science and professional majors and 50% in social sciences and humanities, our hypotheses would change. Thus, the hypotheses stated in *step 1* of hypothesis testing would be:

> H_0: Incoming students would reflect the following distribution: 25% in science, 50% in social sciences and humanities, and 25% professional majors.

> H_1: Incoming students would not reflect the following distribution: 25% in science, 50% in social sciences and humanities, and 25% professional majors.

Our level of significance need not change. Therefore, step 2 of hypothesis testing would remain the same with an α of .05. However, in step 3 of hypothesis testing, when we calculate the test statistics, changes would occur. Since our null hypothesis does not reflect an even distribution, our expected frequencies change. For each category of the variable, we take the total number of cases (N) times the expected or hypothesized proportion (p) for each category:

$$f_e = N \times p$$

Remember that proportion equals the percentage divided by 100 ($p = \% \div 100$). Thus, 25% is equivalent to .25. **Table 7.3** shows the calculated expected frequencies based on the hypothesized values in step 1 of the calculations of the chi square. Since the expected frequencies have changed, the calculations in the remaining steps also change. However, the process in each of those steps remains the same. As you can see, the calculated χ^2 is 6.00.

Table 7.3 *Summary Table of the Calculation of x^2 for Intended Majors*

Major	f_0	Step 1 f_e	Step 2 $f_o - f_e$	Step 3 $(f_o - f_e)^2$	Step 4 $\dfrac{(f_o - f_e)^2}{f_e}$
Science	35	25	10	100	4
Social Sciences or Humanities	40	50	−10	100	2
Professional	25	25	0	0	0
Total	100	100		**Step 5**	6.00

In step 4 of hypothesis testing, we make a decision about the null hypothesis. Since the alpha level ($\alpha = .05$) and the degrees of freedom ($df = 2$) do not change, the critical value of χ^2 remains 5.991. However, our calculated χ^2 is different; it is now 6.00. **Figure 7.2** shows that the calculated χ^2 is greater than the critical value of χ^2. Therefore, we reject the null hypothesis.

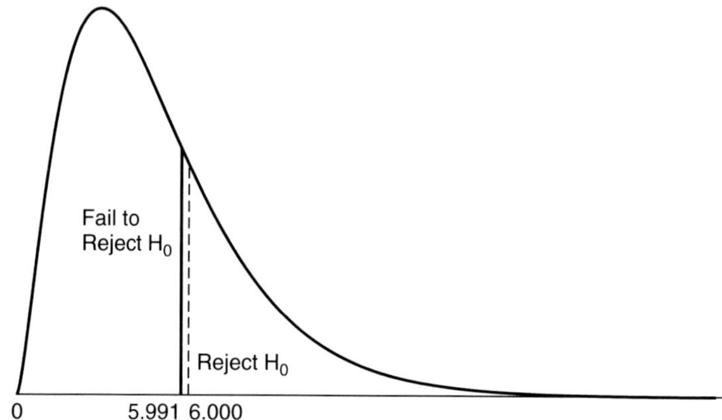

Figure 7.2 The χ^2 Curve with df $= 2$ and $\alpha = .05$.

In step 5 of hypothesis testing, we make inferences about the population based on the statistical test. Because we rejected our null hypothesis, we would conclude that the incoming students deviate from the hypothesized distribution. More specifically, we expect that the incoming students' majors would not be 25% in science, 50% in the social sciences or humanities, or 25% in a professional major.

In conclusion, we follow virtually the same procedures when testing if the categories in a frequency distribution are hypothesized to be evenly distributed or there is some other hypothesized distribution. As we saw, the hypotheses change, and the way that we calculate the expected frequencies differ. Although other differences occur due to these changes in hypotheses and expected frequencies, the procedures are the same. Therefore, the only real differences between the tests in which the null hypothesis predicts an even distribution or some other hypothesized distribution are the stating of the hypotheses and the way we calculate the expect frequencies.

Chi-Square Test for Independence

The chi-square goodness of fit test allows us to do statistical tests on frequency distributions to determine whether a frequency distribution from sample data significantly differs from an even distribution or any other hypothesized distribution. The chi-square test for independence, on the other hand, lets us perform statistical tests on cross-tabs. In Chapter 2, we calculated column or row percentages to compare more easily categories of a variable that have a different absolute number. For example, suppose we are interested in comparing how male and female incoming students vary by intended major, as shown in **Table 7.4.** Converting the absolute frequencies in the table to column percentages, we see that among the sample of 100 incoming students, men were more likely than women to be science and professional majors, 44.7% compared to 26.4% and 27.7% compared to 22.6%, respectively. Women entering college were much more likely than their male counterparts to plan to major in the social sciences or humanities. Although it appears that men and women entering college differ in their intended majors, we cannot generalize the difference that appears in this sample to all incoming students—the population—without doing a statistical test. Therefore, we will do a chi-square test for independence to determine whether the distribution of intended majors does not depend on the sex of the incoming students.

Table 7.4 *Cross-tab of Intended Major by Sex*

	Sex		
Intended Major	Male	Female	
Science	21 (44.7%)	14 (26.4%)	35 (35.0%)
Social Sciences and Humanities	13 (27.7%)	27 (50.9%)	40 (40.0%)
Professional	13 (27.7%)	12 (22.6%)	25 (25.0%)
Total	47 (100.1%)*	53 (99.9%)*	100 (100.0%)

** Due to rounding*

Step One: State null and alternative hypotheses. The null hypothesis in a chi-square test for independence indicates that the distribution of a variable does not depend on the categories of another variable. The alternative hypothesis, on the other hand, indicates that the distribution of a variable does depend on the categories of another variable.

H_0: The distribution of intended majors does not depend on the sex of the incoming students.

H_1: The distribution of intended majors does depend on the sex of the incoming students.

Step Two: Determine the significance or alpha level that will be used in the statistical test. The same levels of significance (.05, .01, or .001) are used for the chi square. For this test, let's use .01.

Step Three: Calculate the statistical test The chi-square test for independence uses the same formula as the chi-square goodness of fit test. As in the one-way chi-square tests, step 1 in calculating χ^2 is to determine the expected frequencies. However, the formula for expected frequencies (f_e) differs. To determine the expected frequency (f_e) for each cell, we multiply the corresponding row marginal by the column marginal for the cell and then divide by the total number of cases:

$$f_e = \frac{(row\ marginal\ total) \times (column\ marginal\ total)}{N}$$

Step 1 of calculating the chi square is to compute the expected frequencies.

$$f_e 11 = \frac{(35) \times (47)}{100} = 16.45 \qquad f_e 12 = \frac{(35) \times (53)}{100} = 18.55$$

$$f_e 21 = \frac{(40) \times (47)}{100} = 18.80 \qquad f_e 22 = \frac{(40) \times (53)}{100} = 21.20$$

$$f_e 31 = \frac{(25) \times (47)}{100} = 11.75 \qquad f_e 32 = \frac{(25) \times (53)}{100} = 13.25$$

Now that we have the expected frequency (f_e), we can use the formula to calculate the chi-square statistic:

$$x^2 = \Sigma \frac{(f_o - f_e)^2}{f_e}$$

Table 7.5 shows the calculations for the chi-square statistic. In step 2 in the calculation, we subtract the expected frequencies (f_e) from the observed frequencies (f_o) from **Table 7.3** for each cell. Remember from Chapter 2 that f_{11} refers to the frequency in the cell that is in the first row and first column of the table, while f_{21} refers to the frequency in the cell that is in the second row and first column of the table, and so on. In step 3, we square the difference between the observed frequencies (f_o) and expected frequencies (f_e) for each cell of the cross-tab table. In step 4, the square of the difference between the observed frequencies (f_o) and expected frequencies (f_e) that we calculated in step 3 is divided by the expected frequencies (f_e) for each cell. Lastly, in step 5, we sum the calculations from step 4 and find that the calculated chi-square statistic is 6.01.

Table 7.5 Summary Table of the Calculation of x^2 for Intended Majors by Sex

	Step 1	Step 2	Step 3	Step 4
	f_e	$f_o - f_e$	$(f_o - f_e)^2$	$\dfrac{(f_o - f_e)^2}{f_e}$
f_{11}	16.45	$21 - 16.45 = 4.55$	20.70	1.26
f_{21}	18.80	$13 - 18.80 = -5.80$	33.64	1.79
f_{31}	11.75	$13 - 11.75 = 1.25$	1.56	.13
f_{12}	18.55	$14 - 18.55 = -4.55$	20.70	1.12
f_{22}	21.20	$27 - 21.20 = 5.80$	33.64	1.59
f_{32}	13.25	$12 - 13.25 = -1.25$	1.56	.12
			Step 5	6.01

Step Four: Make a decision regarding the null hypothesis.   To find the critical value of the chi square, we need the number of degrees of freedom and the significance level (from step 2, $\alpha = .01$). The degrees of freedom represent the amount of unique information available for the statistical test. This is sometimes an abstract concept to students. However, the use of an example for the chi-square test for independence often makes the concept of the degrees of freedom clearer. Suppose a professor teaching an advanced undergraduate course with 9 juniors and 14 seniors asks her students if they preferred tests or papers to assess their knowledge of the material in the course. Eleven students preferred a test, while 12 favored writing papers. These frequencies are presented in the marginal distributions of **Table 7.6**, with the cell frequencies missing. We only know that the maximum number for each cell cannot exceed the smallest marginal total. In a 2 × 2 table, once we determine a value for one of the cells, all other values become fixed.

Table 7.6 Marginal Distributions of Means of Assessment by Class Rank

	Means of Assessment		
Class Rank	Tests	Papers	Total
Juniors			9
Seniors			14
Total	11	12	23

For example, if four juniors prefer to write papers, then we know that eight seniors (12 − 4 = 8) also want to write papers, while five juniors (9 − 4 = 5) and six seniors (14 − 8 = 6) opted for tests. **Table 7.7** shows the value that we chose in the larger, bold type and the calculated values in regular typeface. As you can see, only one cell is free to vary; since once we know one cell's value, we can calculate the others. Therefore, in a two × two table, we have one degree of freedom.

Table 7.7 *Marginal Distributions of Means of Assessment by Class Rank*

Class Rank	Means of Assessment		Total
	Tests	Papers	
Juniors	5	**4**	9
Seniors	6	8	14
Total	11	12	23

Many cross-tabulations, however, are not two × two tables. For example, what if some of the students wanted a combination of tests and papers? **Table 7.8** shows the marginal distributions of those data. This time, if three juniors prefer to do papers, we cannot determine how many juniors want tests or both papers and tests. The only cell that becomes fixed is seniors who prefer papers (9 − 3 = 6). Another cell must be determined before we can calculate the other cell values. For example, if six seniors favor having both tests and papers, we can determine all the cell values.

Table 7.8 *Marginal Distributions of Means of Assessment by Class Rank*

Class Rank	Means of Assessment			Total
	Tests	Papers	Both	
Juniors				9
Seniors				14
Total	6	9	8	23

Table 7.9 shows the values that we chose in the larger, bold type and the calculated values in regular typeface. Once two values are determined, the other cell values can be calculated. Therefore, in two × three and three × two tables, there are two degrees of freedom.

Table 7.9 *Means of Assessment by Class Rank*

Class Rank	Means of Assessment			Total
	Tests	Papers	Both	
Juniors	4	**3**	2	9
Seniors	2	6	6	14
Total	6	9	8	23

While this exercise is illustrative, it can be cumbersome for larger tables. The following formula allows us to calculate the degrees of freedom for the chi square by multiplying the number of row minus 1 by the number of columns minus 1:

$$df = (r - 1)(c - 1)$$

Now that you understand how to calculate degrees of freedom, let's return to our problem that examined intended major by sex. **Table 7.3** has three rows and two columns. Therefore, its degrees of freedom are:

$$df = (3 - 1)(2 - 1)$$

$$df = 2 \times 1 = 2$$

With two degrees of freedom and an alpha of .01, we can find the critical value of the chi square from the chi-square table provided in Appendix A. **Figure 7.3** shows the critical value of the chi square of 9.210, along with our calculated value of 6.01. Thus, we fail to reject the null hypothesis.

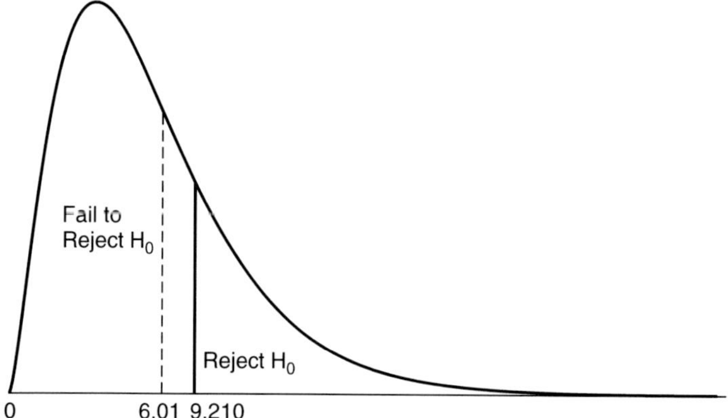

Figure 7.3 The χ^2 Curve with df = 2 and α = 01.

Step Five: Interpret the results. We failed to reject the null hypothesis; the sample data did not provide enough evidence to refute the null hypothesis. It appears that the distribution of intended majors does not depend on the sex of the incoming students. In other words, incoming men and women do not tend to differ with regard to intended majors.

Chi-Square Tests in SPSS

The chi-square tests covered in this chapter can also be done with SPSS. First, let's look at a chi-square goodness of fit test in which the null hypothesis indicates that the cases will be evenly distributed across the categories of the variable, in this case, marital status. Therefore, the hypotheses are:

H_0: Respondents' are equally divided in their marital status.

H_1: Respondents' are not equally divided in their marital status.

For SPSS to calculate the chi square, you need to go to the pull-down menu for Analyze and choose Nonparameteric Tests, followed by Chi Square, as illustrated in **Figure 7.4.**

The Chi Square Test dialogue box below should open after you click on chi square. Find the variable (in this case Marital Status) in the left-hand box. Next, click on the arrow between the two boxes to move the variable into the right-hand box labeled Test Variable List.

Since the null hypothesis is that the cases are evenly divided over the categories of the variable, under the Expected Values, leave the default choice of All Categories Equal checked. Once a variable is in the right-hand box labeled Test Variable List, you can click OK. The following printout would be produced for the chi-square goodness of fit test for an equal distribution for marital status.

The printout shows the observed frequencies (Observed N) for each category of variable, along with the expected frequencies (Expected N). Note that the total number of cases is evenly distributed across the categories of the variable. SPSS has calculated the chi square (x^2 = 1295.930), the degrees of freedom (df = 4), and the precise probability that the null hypothesis is true (Asymp. Sig. = .000). This level of significance indicates that

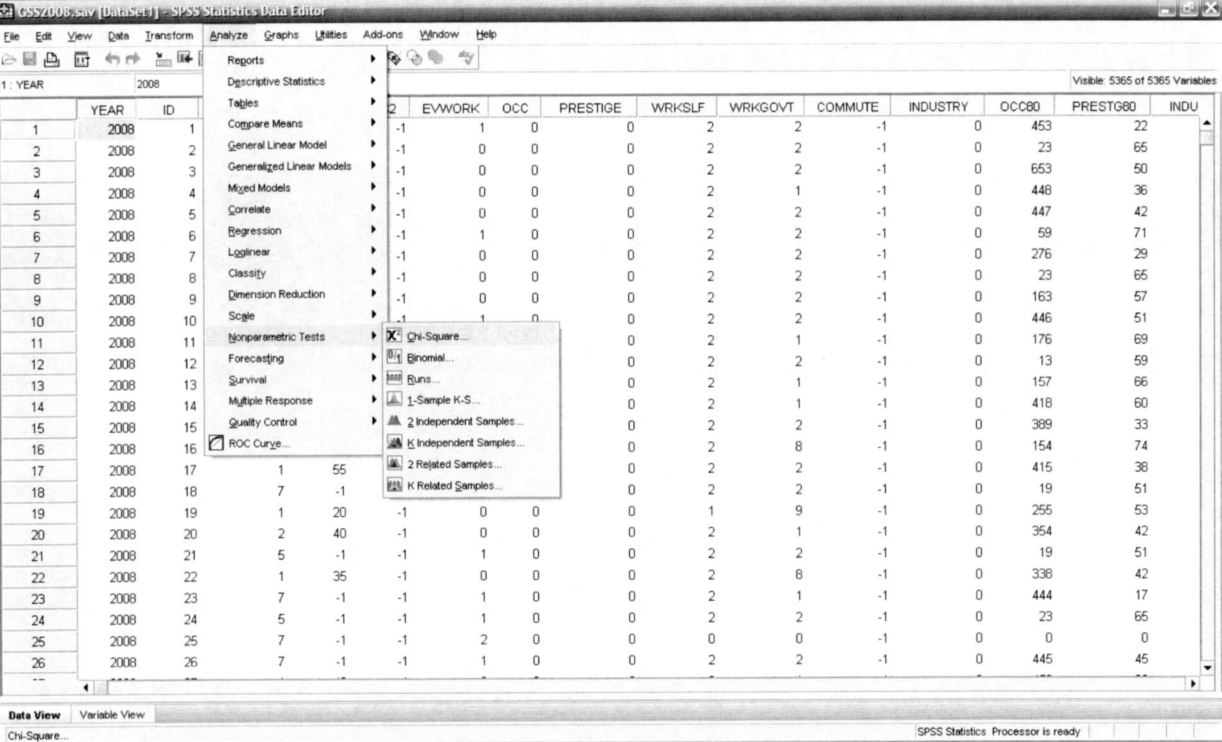

Figure 7.4 SPSS Menu Display of Chi-Square Goodness of Fit.

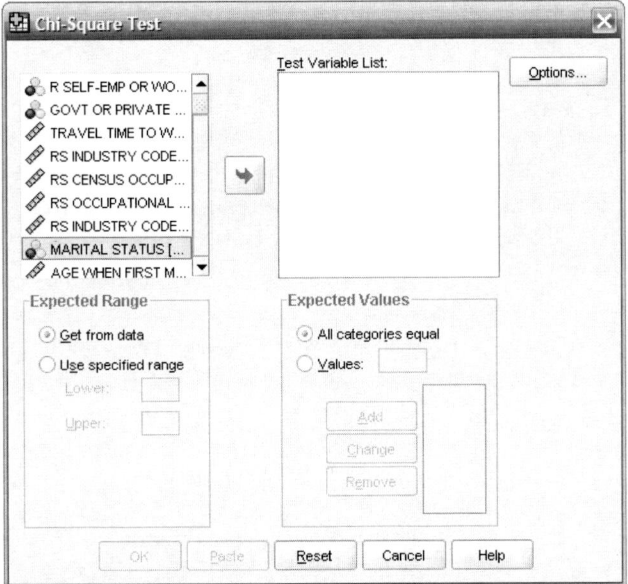

Figure 7.5 Chi-Square Goodness of Fit Dialogue Box.

the probability is very small. The computer program prints only three decimal places, in this case, three zeros. Therefore, this probability does not have a number other than zero until at least the fourth decimal place. Since we reject the null hypothesis, if $p < \alpha$ and we use .05, .01, and .001 alpha levels, we could reject the null hypothesis at each of those levels. Therefore, we can conclude that respondents do not appear to be equally divided among the various marital statuses.

Now we will look at an example in which the null hypothesis does not expect the categories to be equally distributed. Suppose we want to test to see if 15% of Americans have less than a high school diploma as their

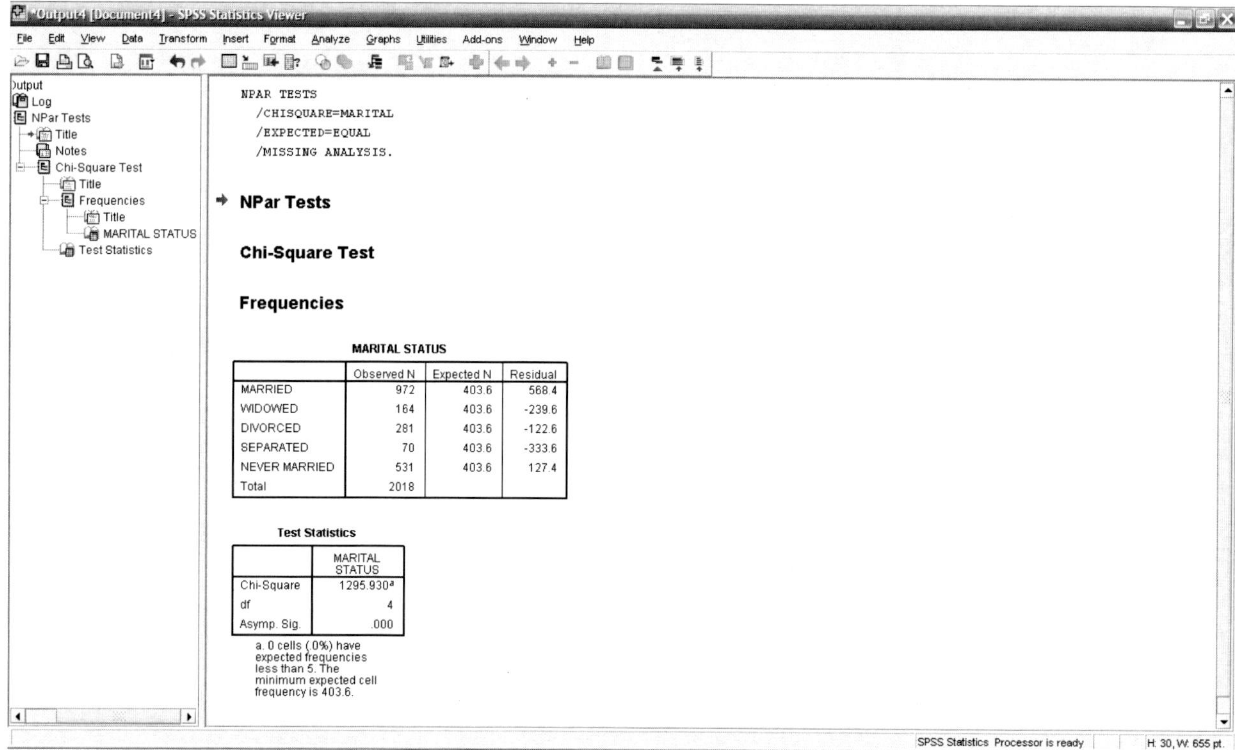

MARITAL STATUS

	Observed N	Expected N	Residual
MARRIED	972	403.6	568.4
WIDOWED	164	403.6	-239.6
DIVORCED	281	403.6	-122.6
SEPARATED	70	403.6	-333.6
NEVER MARRIED	531	403.6	127.4
Total	2018		

Test Statistics

	MARITAL STATUS
Chi-Square	1295.930[a]
df	4
Asymp. Sig.	.000

a. 0 cells (.0%) have expected frequencies less than 5. The minimum expected cell frequency is 403.6.

Figure 7.6 Printout of the Chi-Square Goodness of Fit Test for Marital Status.

highest educational degree, with 50% having a high school diploma, 10% with a junior college degree, 15% with a bachelor's degree, and 10% with a graduate degree. Therefore, our hypotheses would be:

H_0: Americans' highest educational degrees would reflect the following distribution: 15% less than high school, 50% high school, 10% junior college, 15% bachelor's degree, and 10% graduate degree.

H_1: Americans' highest educational degrees would not reflect the following distribution: 15% less than high school, 50% high school, 10% junior college, 15% bachelor's degree, and 10% graduate degree.

After going to the pull-down mean for Analyze and choosing Nonparameteric Tests, followed by Chi Square, the Chi Square Test dialogue box should open, as shown in **Figure 7.7.** Next, find the variable (R's Highest Degree) in the left-hand box and click to the right-hand box, labeled Test Variable List, by highlighting the variable and then clicking on the arrow located between the boxes. Under Expected Values, choose the Value button. Then, in the order the categories appear in the variable, add each category's hypothesized value in the box and click on Add. The SPSS display presented here shows that four out five of the values already have been added. The last value has been entered into the Values box. After Add is clicked, the 10 in the Values box moves down with the other hypothesized values that have been added. Then click OK, and SPSS will calculate the chi square.

SPSS produces the printout shown in **Table 7.10** We can see that the Expected N values are not equal; SPSS used the percentages that we entered for each category and calculated the Expected N values for each category of the variable. SPSS also has given us the calculated chi-square value ($\chi^2 = 13.556$), the degrees of freedom ($df = 4$), and the precise probability that the null hypothesis is true (Asymp. Sig. = .009). Since we reject the null hypothesis if $p < x$ and we use .05, .01, and .001 alpha levels, we could reject the null hypothesis at the .05 and .01 levels. We fail to reject the null hypothesis at the .001 level because .009 (p) is not greater than α of .001(α). In this case, researchers typically report that $p < .01$. Because we can reject the null hypothesis with $p < .01$, we can conclude that it appears that 15% of Americans do not have less than a high school degree, that 50% do not have high school degrees, that 10% do not have junior college degrees, that 15% do not have bachelor degree's, and that 10% do not have graduate degrees.

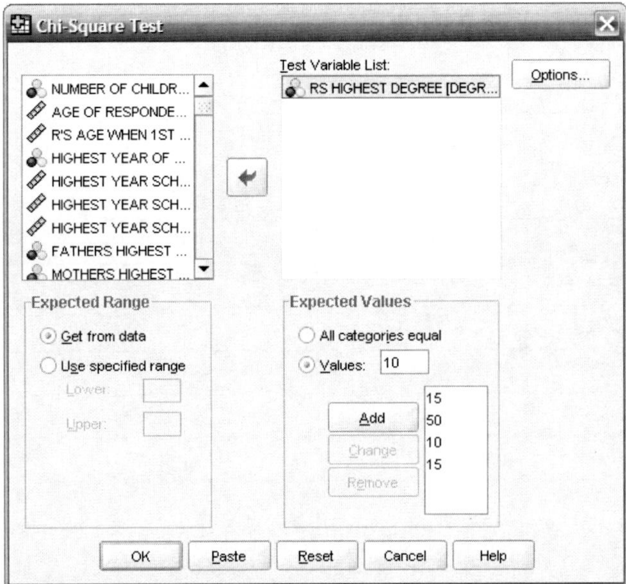

Figure 7.7 Chi-Square Goodness of Fit Dialogue Box.

Table 7.10 *Printout of the Chi-Square Goodness of Fit Test for Respondents' Highest Degree*

	Observed N	Expected N	Residual
LT HIGH SCHOOL	297	303.3	-6.3
HIGH SCHOOL	1003	1011.0	-8.0
JUNIOR COLLEGE	173	202.2	-29.2
BACHELOR	355	303.3	51.7
GRADUATE	194	202.2	-8.2
Total	2022		

Test Statistics	
	RS HIGHEST DEGREE
Chi-Square	13.556[a]
df	4
Asymp. Sig.	.009

a. 0 cells (.0%) have expected frequencies less than 5. The minimum expected cell frequency is 202.2.

Our last example shows how we would conduct a chi-square test for independence. The two variables that we will test for independence are people's views on our country's spending on parks and recreation by respondents' sex. Therefore, our hypotheses are:

H_0: The distribution of people's views on our country's spending on parks and recreation does not depend on their sex.

H_1: The distribution of people's views on our country's spending on parks and recreation does depend on their sex.

When we run frequencies on these two variables, shown in **Table 7.11,** we find that for the variable, Parks and Recreation, those who responded that they did not know (labeled as DK) were not defined as missing values.

However, those who provided no answer (NA) are defined as missing. Before we calculate the chi-square statistics, we need the "don't knows" also to be defined as missing.

Table 7.11 *Printout of the Frequencies of Parks and Recreations*

		Frequency	Percent	Valid Percent	Cumulative Percent
Valid	TOO LITTLE	617	30.5	30.5	30.5
	ABOUT RIGHT	1252	61.9	61.9	92.4
	TOO MUCH	108	5.3	5.3	97.8
	DK	45	2.2	2.2	100.0
	Total	2022	100.0	100.0	
Missing	NA	1	.0		
Total		2023	100.0		

To define a value as missing, we need to switch to Variable View. If you look at the bottom left-hand corner of **Figure 7.8,** there are two display buttons: Data View and Variable View. If the SPSS display does not look like the one below, click on Variable View. Once you have the Variable View, scroll down until you find the variable People's Views on our Country's Spending on Parks and Recreation (NATPARK). Click on the Values column for that variable.

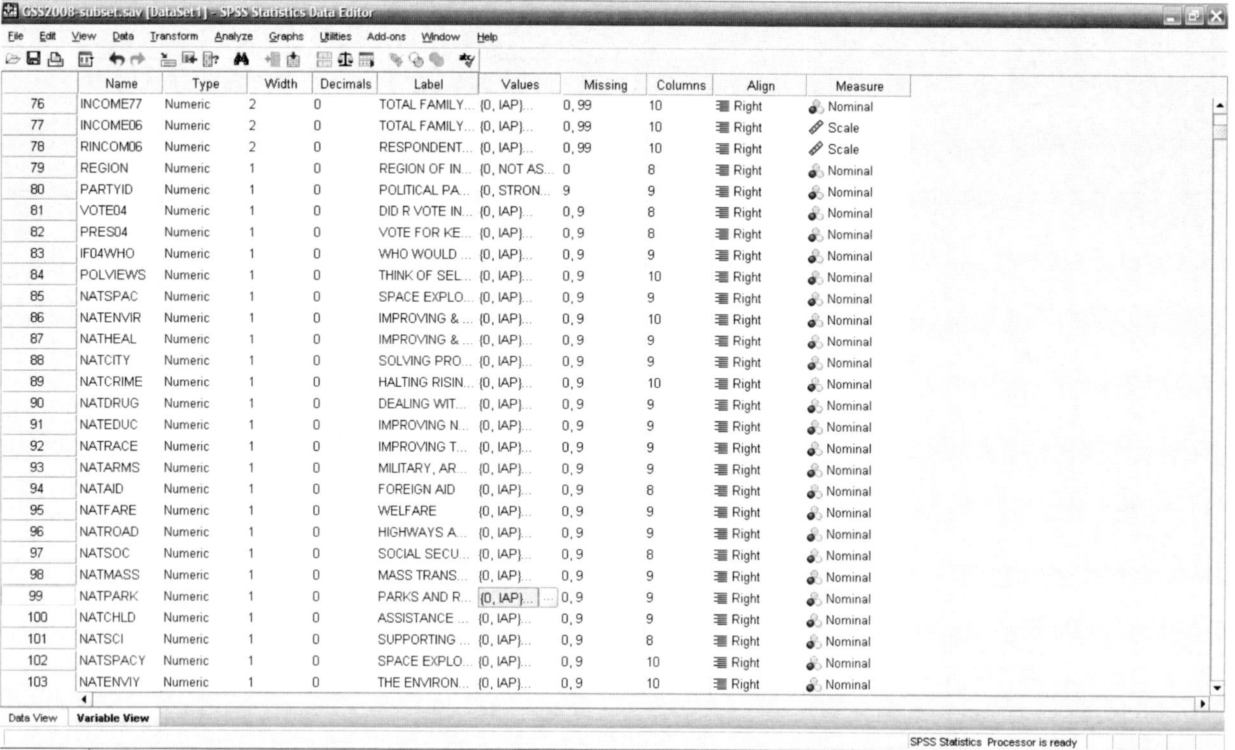

Figure 7.8 SPSS Variable View Display.

A small box with an ellipsis opens. When you click on the ellipsis box, the dialogue box shown in **Figure 7.9** opens and shows you how each response was coded numerically. You can see that DK was coded as 8.

Therefore, 8 must be included as a missing value. The column next to the Values column is labeled as Missing, meaning missing values. If you click on the Missing column for the Parks and Recreation variable (NATPARK), another ellipsis box opens. Once you click on the ellipsis box, the dialogue box in **Figure 7.10** opens. You can see that 0 for Inappropriate and 9 for No Answer have already been coded as missing, with an empty third box. You need to type an 8 in the third box and then click OK. Run another frequency check on the Parks and Recreation variable (NATPARK). DK should now be listed as missing, along with NA.

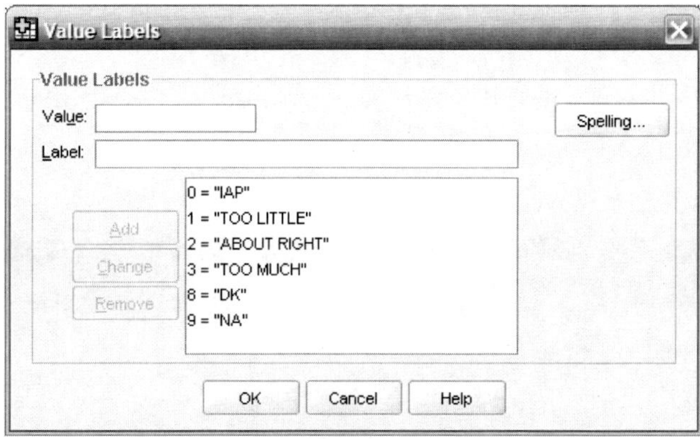

Figure 7.9 Value Labels Dialogue Box for Parks and Recreation.

Figure 7.10 Missing Values Dialogue
Box for Parks and Recreation.

With the missing values defined, we can now calculate the chi square. Because the chi-square test for independence performs statistical tests on cross-tabs, we start the analysis by going to Analysis ⇒ Descriptive Statistics ⇒ Cross-tabs.

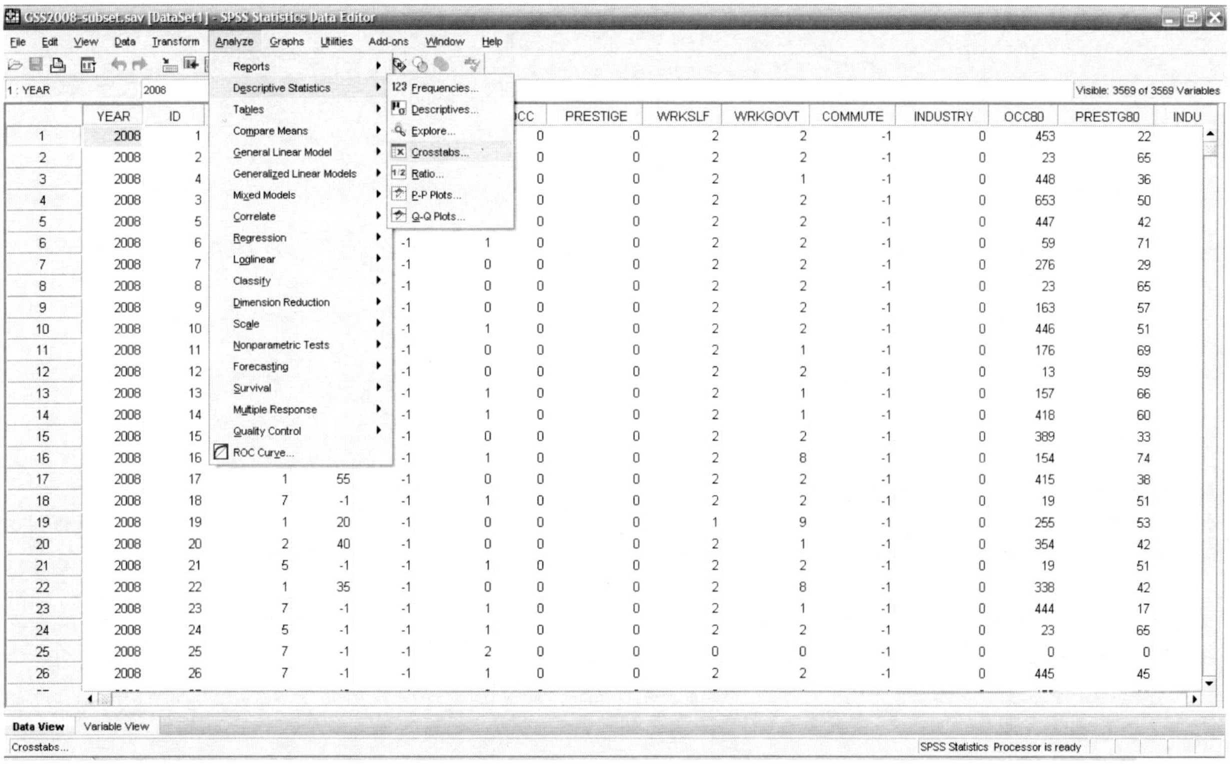

Figure 7.11 SPSS Menu Display for Crosstabs.

After going to Analysis ⇒ Descriptive Statistics ⇒ Cross-tabs, the dialogue box shown in **Figure 7.12** should open. Next, find the variables (Respondent's Sex and Parks and Recreation) in the left-hand box and click them to the right-hand boxes, labeled Row(s) and Column(s), by highlighting each variable and then clicking on the arrows located between the boxes.

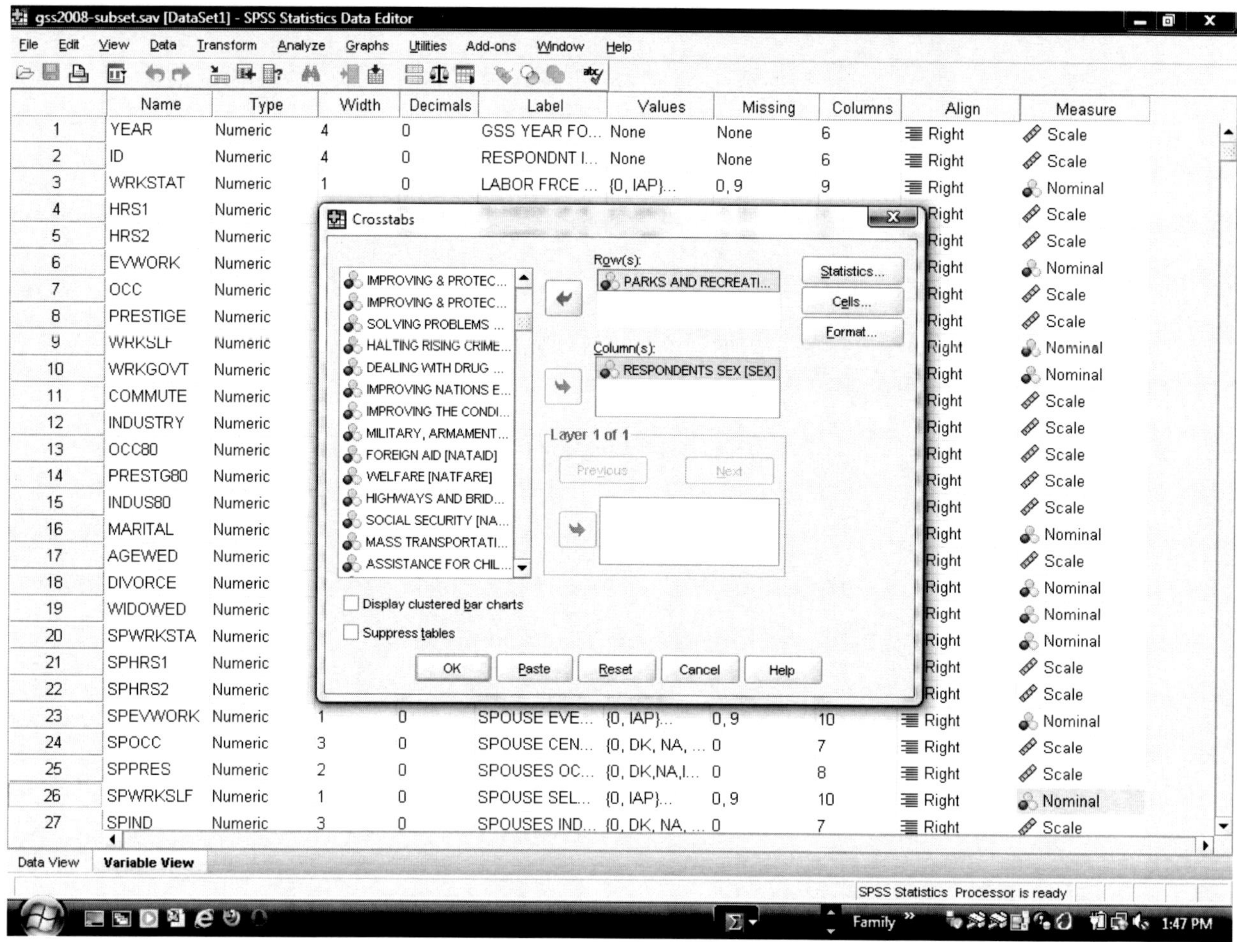

Figure 7.12 Crosstab Dialogue Box with Park and Recreation and Respondent's Sex.

To the right of the Row(s) and Column(s) boxes, there are three buttons. Click on the one labeled Statistics. The Crosstab: Statistics dialogue box should open, as shown in **Figure 7.13.** Click on Chi Square, as shown. Then click Continue. The Crosstab: Statistics dialogue box should close, allowing you to click on OK from the Crosstab dialogue box.

SPSS will run the analysis. It will produce a cross-tab table and a table with the Chi-square test. We are interested in the analysis labeled the Pearson Chi-Square. SPSS calculates the Chi Square, in this case $x^2 = .372$, and is printed it in the column labeled Value. The degrees of freedom are in the next column ($df = 2$). The last column, labeled Asymp. Sig. (2-sided), provides us with the exact probability that the null hypothesis is true ($p = .830$). Because the probability is greater than any of the acceptable alpha levels (.05, .01, and .001), we fail to reject the null hypothesis. Therefore, we can conclude that the distribution of people's views on our country's spending on parks and recreation does not appear to depend on their sex. In other words, men and women's views on our country's spending on parks and recreations do not seem to differ.

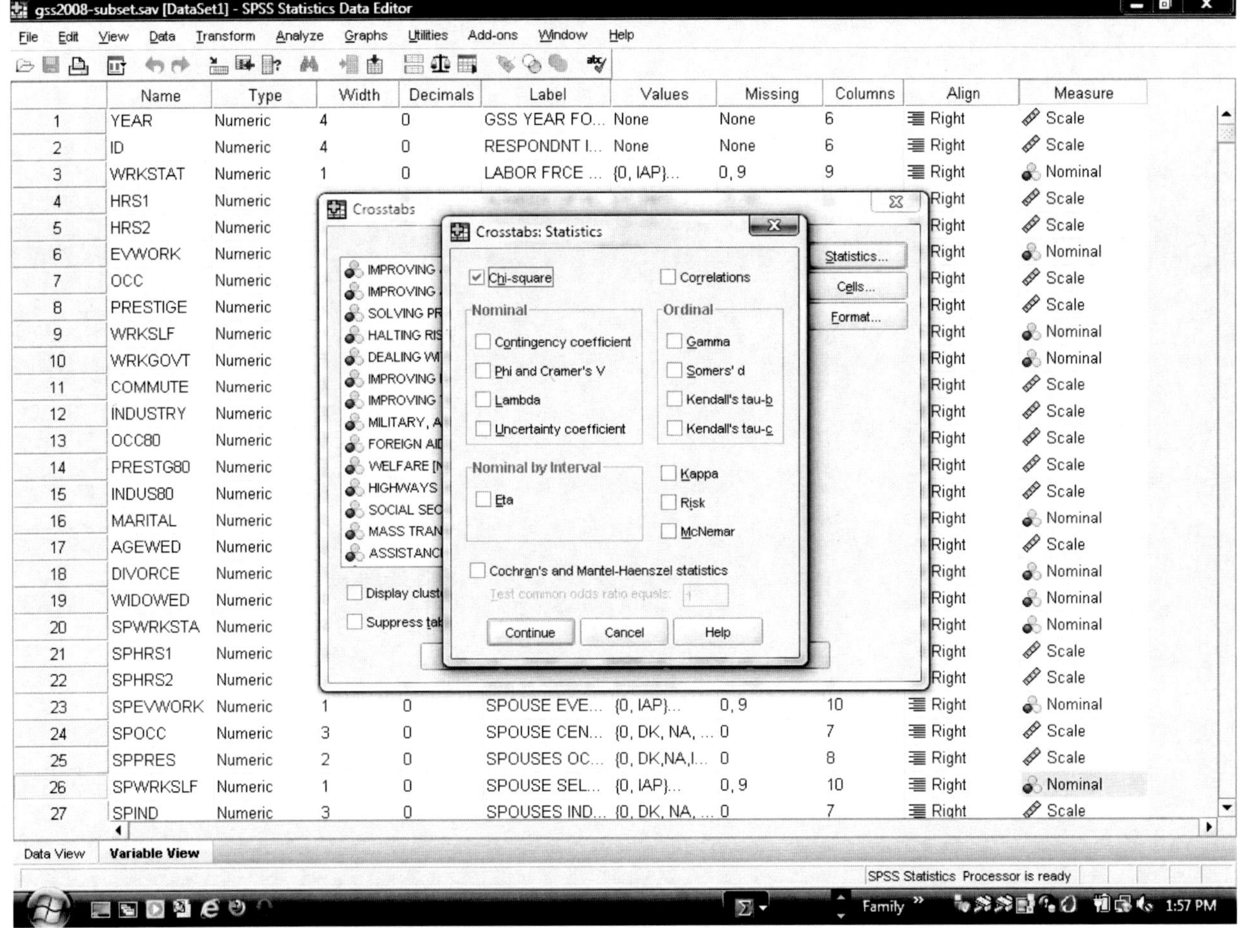

Figure 7.13 Crosstab Statistics Dialogue Box.

Table 7.12 *Printout of Chi-Square Tests*

	Value	df	Asymp. Sig. (2-sided)
Pearson Chi-Square	.372[a]	2	.830
Likelihood Ratio	.372	2	.830
Linear-by-Linear Association	.363	1	.547
N of Valid Cases	1977		

a. 0 cells (.0%) have an expected count less than 5. The minimum expected count is 49.93.

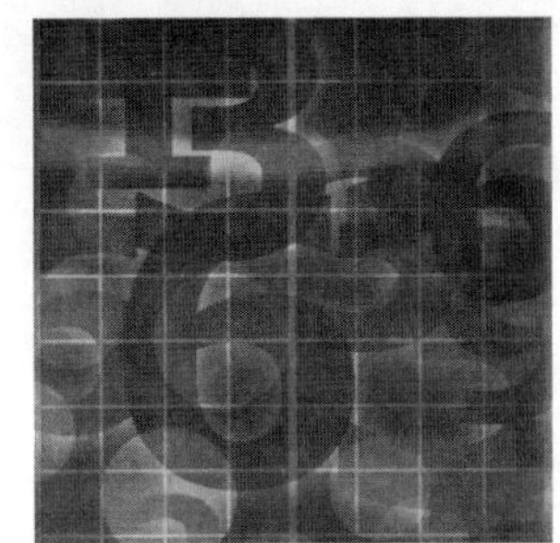

Exercises and Notes

- Chi square is a _____________________ test of significance.

- _____________ level data may be tested.

- _______________________________ are assumed.

- Hypotheses do not indicate a __________________.

Chi-Square Test for the Goodness of Fit

The χ^2 test for goodness of fit determines whether the _____________________________ for a sample differs significantly from an even distribution or any other hypothesized distribution.

Even Distribution Example

You are doing an internship with a local city official who wishes to determine residents' views on a proposed change in garbage collection. You randomly sample 180 residents and ask their opinion on the issue. Do a statistical test to determine whether the residents are equally divided on the issue.

Proposed Change in Garbage Collection	f
Agree	62
Somewhat Agree	49
Somewhat Disagree	38
Disagree	31

H_0:

H_1:

Let $\chi = .05$.

$$x^2 = \Sigma \frac{(f_o - f_e)^2}{f_e} \qquad\qquad f_e = \frac{N}{k}$$

$f_o =$ $N =$

$f_e =$ $k =$

Proposed Change	f	Step 1 f_e	Step 2 $f_o - f_e$	Step 3 $(f_o - f_e)^2$	Step 4 $\dfrac{(f_o - f_e)^2}{f_e}$
Agree	62				
Somewhat Agree	49				
Somewhat Disagree	38				
Disagree	31				
	100			Step 5	

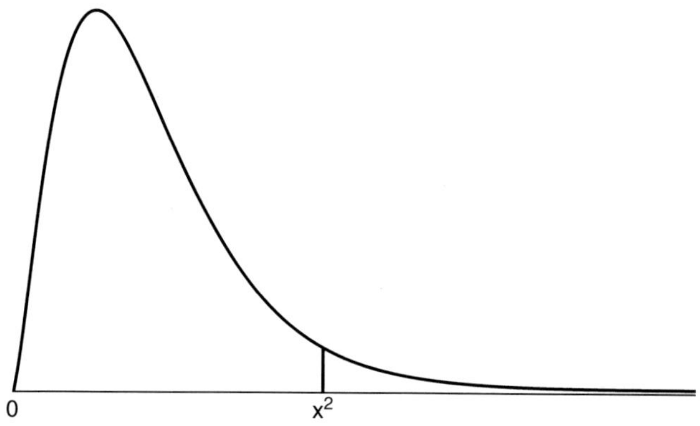

____

____

____

____

Uneven Distribution Example

Instead of hypothesizing an even distribution, suppose the city official expects residents' views on a proposed change in garbage collection to break down as follows: 35% to agree, 30% to somewhat agree, 20% to somewhat disagree, and 15% to disagree. Do a statistical test to see if the data support the city official's expectation.

H_0:

H_1:

$$x^2 = \Sigma \frac{(f_o - f_e)^2}{f_e}$$

$$f_e = N \times p$$

Proposed Change	f_0	Step 1 f_e	Step 2 $f_0 - f_e$	Step 3 $(f_0 - f_e)^2$	Step 4 $\dfrac{(f_0 - f_e)^2}{f_e}$
Agree	62				
Somewhat Agree	49				
Somewhat Disagree	38				
Disagree	31				
	100			**Step 5**	

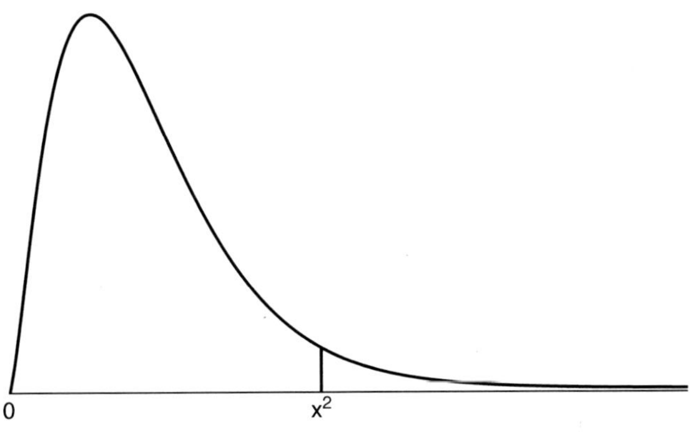

____

____

____

Chi-Square Test for Independence

The x^2 test for independence determines whether the distribution of ____

________________________.

The student government asks you to determine whether the distribution by students' views about a proposed change in campus food service is related to their class rank.

Proposed Food Service Change	Class Rank			
	Freshmen	Sophomores	Juniors	Seniors
Support	37	28	14	11
Oppose	14	13	21	22

H_0:

H_1:

Let $\alpha = .05$

$$x^2 = \Sigma \frac{(f_o - f_e)^2}{f_e}$$

$$f_e = \frac{(row\ marginal\ total) \times (column\ marginal\ total)}{N}$$

	f_o	Step 1 f_e	Step 2 $f_o - f_e$	Step 3 $(f_o - f_e)^2$	Step 4 $\frac{(f_o - f_e)^2}{f_e}$
f_{11}					
f_{21}					
f_{31}					
f_{12}					
f_{22}					
f_{32}				Step 5	

Degrees of Freedom

$$df =$$

Marginal Distributions of Means of Assessment by Class Rank

Class Rank	Means of Assessment		
	Tests	Papers	Total
Juniors			9
Seniors			14
Total	11	12	23

Marginal Distributions of Means of Assessment by Class Rank

Class Rank	Means of Assessment			
	Tests	Papers	Both	Total
Juniors				9
Seniors				14
Total	6	9	8	23

How many degrees of freedom?

$$df =$$

$$CV\ \chi^2 =$$

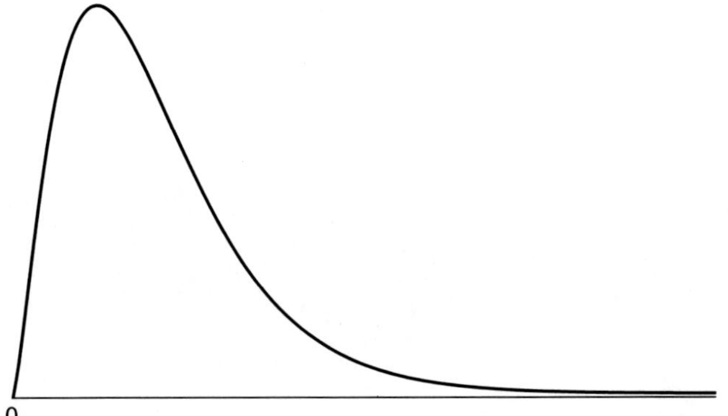

How many degrees of freedom?

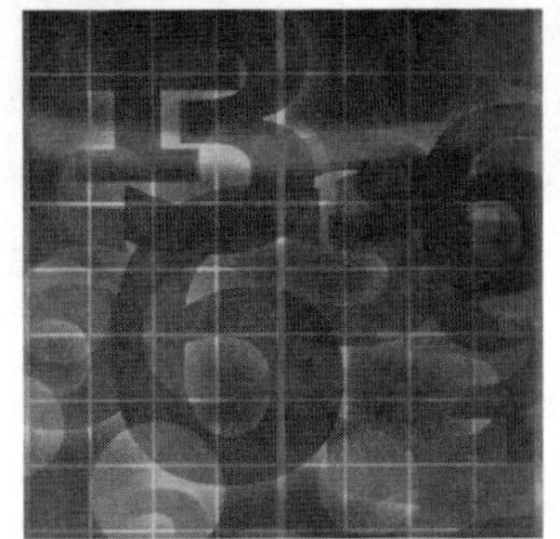

Learning Check

1. The manager of an on-line movie-rental business wonders if the rentals are equally divided among the top six movie genres. Below is a random sample of 300 movie rentals. Do a statistical test to answer his question. Let $\alpha = .05$.

Genre	f
Action	47
Animation	51
Comedy	63
Drama	59
Horror/Thriller	44
Romance	36

2. Two statistics students working at a local ice-cream shop notice that they sold more vanilla ice cream than any other flavor. They hypothesize that 40% of sales were vanilla, followed by chocolate at 25%, cookies and cream at 20%, cookie dough at 10%, and mint at 5%. They take a random sample of 200 orders to test their hypothesized distribution. What did they find? Let $\alpha = .05$. Would your conclusion change if $\alpha = .01$?

Ice-Cream Flavor	f
Vanilla	98
Chocolate	44
Cookies and Cream	28
Cookie Dough	17
Mint	13

3. Do a statistical test to determine whether the distribution by marital status depends on respondents' race/ethnicity. Let $\alpha = .05$.

Race/ Ethnicity

Marital Status	Whites	Blacks	Latinos	Total
Married	57	7	12	
Not Married	40	13	11	
	97	20	23	

4. You want to determine if Americans are equally likely to live in the same city, live in the same state but different city, or live in a different state from the one they lived in when they were 16 years old. Using the GSS 2008 data set and SPSS, test to see if respondents are evenly distributed across the categories of the variable Geographic Mobility Since Age 16 (MOBILE16).

5. You want to determine whether the distribution of respondents' views on gun permits laws depends on the respondents' sex. Using the GSS 2008 data set and SPSS, test the null hypothesis using the variables Favor or Oppose Gun Permits (GUNLAW) and Respondents Sex (SEX). Note: before doing the statistical test, be sure that all missing values are defined as missing.

Chapter Eight

ANOVA

Chapter 7 introduced you to nonparametric measures and tests, such as the χ^2 (chi-square) test, that can be used as analytical tools with this type of data. Analysis of variance (ANOVA) is another of these nonparametric measures. While χ^2 allows us to test the significance of patterns of distribution of a sample, ANOVA allows us to determine whether there is more variation between groups than there is within groups.

It helps to compare ANOVA to the t test for clarification. For example, suppose that, as researchers, we are interested in whether males or females score higher on an exam. The easiest way to test this difference is to compute a t-test, which determines whether or not there are significant differences between the two groups, males and females. Thus, if we have a sample of 500 and our t is 5.96, then we know that males and females are significantly different ($p = .000$) in how well they do on the test.

What if we wanted to test the differences in quality of life by sex? We ask our sample: How satisfied are you with your quality of life? We have the following categories for quality of life:

1. Very dissatisfied
2. Somewhat dissatisfied
3. Somewhat satisfied
4. Very satisfied

Since we have coded our variable in an ordinal scale, we could use the mathematical formula for the t-test and get an answer. However, to get reliable results with a t-test, we would have to construct the following groups:

Very dissatisfied males

Very dissatisfied females

Somewhat dissatisfied males

Somewhat dissatisfied females

Somewhat satisfied males

Somewhat satisfied females

Very satisfied males

Very satisfied females

Once the groups are constructed, we would need to run a series of t-tests to determine whether or not there are significant differences in satisfaction level by sex. However, if we do the analysis this way, not only are we doing a large amount of work, but also, we are far more likely to introduce Type I error. The likelihood of Type I error increases when you use multiple equations on one hypothesis. Also, how would we interpret the data if we got only one significant t-test on the categories proposed above? For example, what if there is a significant difference only between very satisfied females and very dissatisfied males?

Rather than use a series of t-tests, we can perform ANOVA. ANOVA provides an F score, which measures whether or not the differences within the group are significantly different from those between the groups. In other words, in our example above: Do men vary more from each other than they do from women?

For this example we will use the following hypothesis:

Marital status is an important predictor for number of hours spent watching TV.

In this example, marital status is ranked Married, Divorced, Widowed, or Single. TVHOURS (How many hours a day to you spend watching TV?) is an interval-level measure.

The formula for ANOVA is:

$$F = \frac{MS_{between}}{MS_{within}}$$

In this equation, MS stands for the mean of squares. This is a very complex, though not difficult, equation, which has to be broken into several steps.

Consider these data:

Marital	TVhours	Marital	TVhours	Marital	TVhours	Marital	TVhours
Married	4	Widowed	0	Divorced	4	Single	5
Married	3	Widowed	3	Divorced	2	Single	4
Married	3	Widowed	4	Divorced	1	Single	1
Married	3	Widowed	1	Divorced	6	Single	1
Married	2	Widowed	2	Divorced	2	Single	2
Married	2	Widowed	3	Divorced	4	Single	4
Married	4	Widowed	16	Divorced	1	Single	4
Married	1	Widowed	4	Divorced	2	Single	2
Married	2	Widowed	3	Divorced	1	Single	5
Married	2	Widowed	4	Divorced	2	Single	3

We want to know whether or not there is a difference in how much time people spend watching TV if we compare them by marital status. The first step is to find the sum of squares for the entire sample (SS_{total}).

The formula for SS_{total} is:

$$SS_{total} = \Sigma(X - \bar{X}_{total})^2$$

x = any raw score

$\bar{x}_{total}$ = the total mean for all groups combined

Now that we've defined our terms, we can begin to set up our equation.

Step 1: Find the differences for the entire group. You'll recognize this step, as it is the same step that we've done in other equations. We find the mean for the entire sample and then subtract that from every value of X (hours spent watching TV).

4	− 3.1	=	0.95
3	− 3.1	=	−0.05
3	− 3.1	=	−0.05
3	− 3.1	=	−0.05
2	− 3.1	=	−1.05
2	− 3.1	=	−1.05
4	− 3.1	=	0.95
1	− 3.1	=	−2.05
2	− 3.1	=	−1.05
2	− 3.1	=	−1.05
0	− 3.1	=	−3.05
3	− 3.1	=	−0.05
4	− 3.1	=	0.95
1	− 3.1	=	−2.05
2	− 3.1	=	−1.05
3	− 3.1	=	−0.05
16	− 3.1	=	12.95
4	− 3.1	=	0.95
3	− 3.1	=	−0.05
4	− 3.1	=	0.95

4	− 3.1	=	0.9
2	− 3.1	=	−1.1
1	− 3.1	=	−2.1
6	− 3.1	=	2.9
2	− 3.1	=	−1.1
4	− 3.1	=	0.9
1	− 3.1	=	−2.1
2	− 3.1	=	−1.1
1	− 3.1	=	−2.1
2	− 3.1	=	−1.1
5	− 3.1	=	1.9
4	− 3.1	=	0.9
1	− 3.1	=	−2.1
1	− 3.1	=	−2.1
2	− 3.1	=	−1.1
4	− 3.1	=	0.9
4	− 3.1	=	0.9
2	− 3.1	=	−1.1
5	− 3.1	=	1.9
3	− 3.1	=	−0.1

Step 2: Square the differences.

0.95^2	=	0.90
-0.05^2	=	0.00
-0.05^2	=	0.00
-0.05^2	=	0.00
-1.05^2	=	1.10
-1.05^2	=	1.10
0.95^2	=	0.90
-2.05^2	=	4.20
-1.05^2	=	1.10
-1.05^2	=	1.10

0.9^2	=	0.81
-1.1^2	=	1.21
-2.1^2	=	4.41
2.9^2	=	8.41
-1.1^2	=	1.21
0.9^2	=	0.81
-2.1^2	=	4.41
-1.1^2	=	1.21
-2.1^2	=	4.41
-1.1^2	=	1.21

-3.05^2	$=$	9.30
-0.05^2	$=$	0.00
0.95^2	$=$	0.90
-2.05^2	$=$	4.20
-1.05^2	$=$	1.10
-0.05^2	$=$	0.00
12.95^2	$=$	167.70
0.95^2	$=$	0.90
-0.05^2	$=$	0.00
0.95^2	$=$	0.90

1.9^2	$=$	3.61
0.9^2	$=$	0.81
-2.1^2	$=$	4.41
-2.1^2	$=$	4.41
-1.1^2	$=$	1.21
0.9^2	$=$	0.81
0.9^2	$=$	0.81
-1.1^2	$=$	1.21
1.9^2	$=$	3.61
-0.1^2	$=$	0.01

Step 3: Sum the squared differences.

$$SS_{total} = 244$$

The next step is to find the "within" group sum of squares. This is the sum of squares when we calculate the average for each group separately.

The formula is:

$$SS_{within} = \Sigma(X - \bar{X}_{group})^2$$

It is easier to set this up in a table by group. We have divided the data into four tables by marital status. Then we find the average score for each group.

Married: $\bar{X} = 26$

Widowed: $\bar{X} = 4$

Married		
x	$x - \bar{x}$	$x - \bar{x}^2$
4	1.4	1.96
3	0.4	0.16
3	0.4	0.16
3	0.4	0.16
2	-0.6	0.36
2	-0.6	0.36
4	1.4	1.96
1	-1.6	2.56
2	-0.6	0.36
2	-0.6	0.36

Widowed		
x	$x - \bar{x}$	$x - \bar{x}^2$
0	-4	16
3	-1	1
4	0	0
1	-3	9
2	-2	4
3	-1	1
16	12	144
4	0	0
3	-1	1
4	0	0

Divorced: $\bar{X} = 2.5$ Single: $\bar{X} = 3.1$

Divorced		
x	$x - \bar{x}$	$x - \bar{x}^2$
4	1.5	2.25
2	−0.5	0.25
1	−1.5	2.25
6	3.5	12.25
2	−0.5	0.25
4	1.5	2.25
1	−1.5	2.25
2	−0.5	0.25
1	−1.5	2.25
2	−0.5	0.25

Single		
x	$x - \bar{x}$	$x - \bar{x}^2$
5	1.9	3.61
4	0.9	0.81
1	−2.1	4.41
1	−2.1	4.41
2	−1.1	1.21
4	0.9	0.81
4	0.9	0.81
2	−1.1	1.21
5	1.9	3.61
3	−0.1	0.01

Step 4: Sum the squared differences for the entire sample.

$$SS_{\text{within}} = 229.8$$

We now need to find the sum of squares between the groups. The formula for this is:

$$SS_{between} = \Sigma N_{group} (\bar{X}_{group} - \bar{X}_{total})^2$$

In this equation:

N_{group} = *the number of scores in any group*

$\bar{X}_{group}$ = *the mean of any group*

$\bar{X}_{total}$ = *the mean of all groups combined*

As with most complicated equations, it is easier to set these data up in a table. This equation is simplified, since all groups have an N of 10. Typically, data are not this neatly distributed and the N will vary by group.

	Married	Widowed	Divorced	Single
N_{group}	10	10	10	10
$\bar{X}_{group}$	2.6	4	2.5	3.1
$\bar{X}_{total}$	3.1	3.1	3.1	3.1

Now we can plug the number into our equation.

$$SS_{between} = 10(2.6 - 3.1)^2 + 10(4 - 3.1)^2 + 10(2.5 - 3.1)^2 + 10(3.1 - 3.1)^2$$
$$SS_{between} = 14.2$$

Check your math.

$$SS_{between} + SS_{within} = SS_{total}$$

$$14.2 + 229.8 = 244$$

The next step is to calculate the F score to see if there is a significant difference between the sums of squares. To do this, we have to calculate the mean squares (MS). There are two formulas for this:

$$MS_{between} = \frac{SS_{between}}{df_{between}} \qquad MS_{within} = \frac{SS_{within}}{df_{within}}$$

The only thing that we don't have is the df for the two equations. We need to use the following calculations to find the F score:

$$df_{between} = k-1$$

$k = $ the number of groups

$$df_{between} = 4 - 1$$

$$MS_{between} = \frac{14.2}{3} = 4.7$$

$$df_{within} = N_{total} - k$$

$$df_{within} = 40 - 4 = 36$$

$$MS_{within} = \frac{229.8}{36} = 6.4$$

Now we can calculate our F score:

$$F = \frac{MS_{between}}{MS_{within}} = \frac{4.7}{6.4} = .73$$

We determine the significance of F by finding our values on the critical values of the F table provided in the Appendix (Appendix A). We find our degrees of freedom for the denominator (within) in the row and determine which value for the numerator (between) intersects in the column. So, we need an F of 4.51 for significance ($\alpha = .05$). Given that our F is much lower, we must reject the hypothesis that there is a significant difference in hours spent watching TV by marital status. This means that there is more variation within the groups than there is between groups.

Using SPSS to Compute Anova

Now that you understand the math needed to compute an ANOVA, we will use SPSS to test the hypothesis: Marital status is likely to influence how often someone prays. Make sure that you have SPSS open, and open your data file (GSS 2008). Click on the Analyze screen and then on the Compare Means screen. Select OneWay ANOVA.

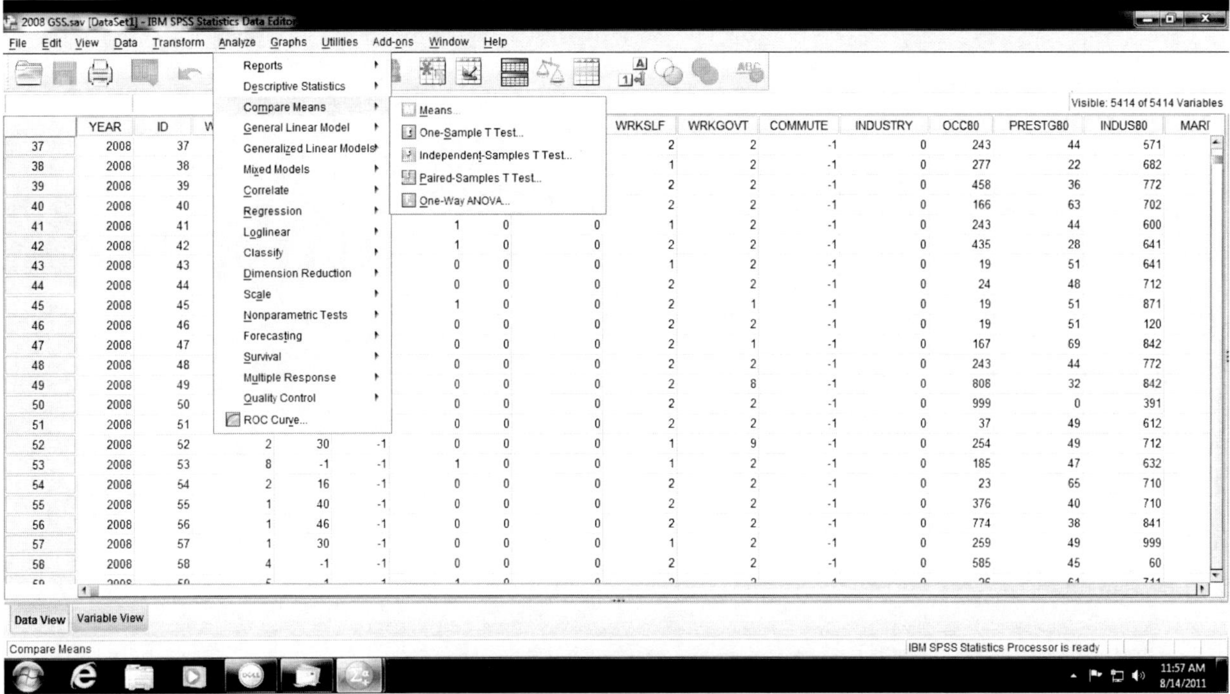

This will bring up the data window. We will test the relationship between how often people pray and marital status. Move PRAYFREQ into the Dependent list and MARITAL into the Factor list. Select OK to run the analysis.

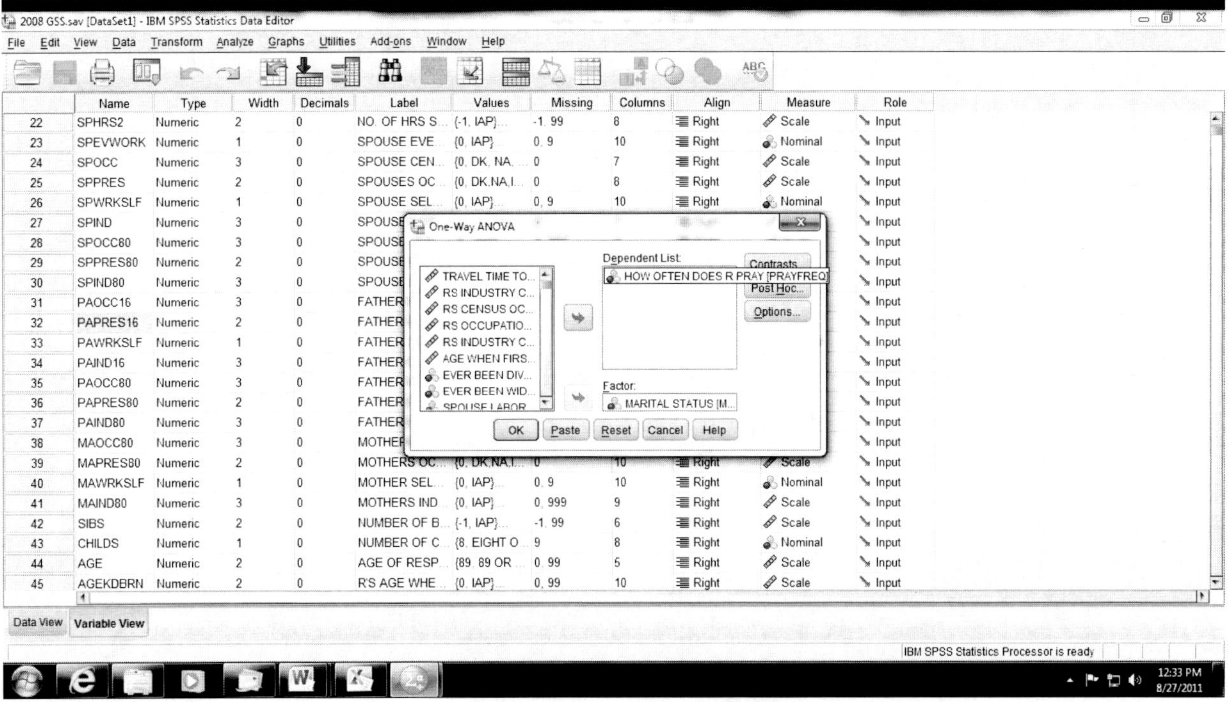

This will open the output window and give you the following results.

Table 8.1 Oneway ANOVA

How Often Respondents Pray					
	Sum of Squares	**df**	**Mean Square**	**F**	**Sig.**
Between Groups	1314.208	4	328.552	7.140	.000
Within Groups	62024.871	1348	46.013		
Total	63339.079	1352			

As you can see, we have an *F* of 7.140 and a significance level of .000. This means that we can say that there is a significant relationship between marital status and how often people pray. In other words, there is more variation between the groups than there is within the groups.

In order to accurately interpret this statistic, we need understand how the data are distributed. For this, we need to run a cross-tab. (If you don't remember how to run a cross-tab, refer to Chapter 2.) The cross-tab gives us the following data:

	Marital Status				
Frequency of Prayer	**Married**	**Widowed**	**Divorced**	**Separated**	**Never Married**
Never	8.5%	7.1%	7.4%	7.8%	16.4%
Less Than Once a Year	2.8%	0.0%	1.1%	0.0%	2.3%
About Once or Twice a Year	1.3%	0.8%	4.7%	0.0%	2.0%
Several Times a Year	3.4%	2.4%	4.7%	2.0%	9.5%
About Once aMonth	2.2%	4.0%	1.1%	2.0%	4.6%
2−3 Times a Month	4.4%	4.0%	6.3%	5.9%	4.0%
Nearly Every Week	1.3%	1.6%	2.6%	0.0%	3.5%
Every Week	5.0%	2.4%	4.2%	3.9%	6.1%
Several Times a Week	12.1%	4.0%	10.5%	13.7%	11.5%
Once a Day	28.6%	32.5%	24.7%	27.5%	20.5%
Several Times a Day	29.9%	40.5%	32.6%	35.3%	19.6%
Dk	0.6%	0.8%	0.0%	2.0%	0.0%
Total	100.0%	100.0%	100.0%	100.0%	100.0%

SPSS provides frequencies with output. It is useful to convert these to percentages for discussion and presentation, as we have done here. When we review the distribution, we can note that single people are not only less likely to pray frequently, but also,they are more likely to say that they never pray.

The last step is to interpret these data within the social context. We would need to look at social factors and religious practices. We might argue that religious attendance is less popular today than it was for previous generations, and that this contributes to the lack of prayer. We should also consider age; singles are more likely to be younger than married or once-married groups. What this *F* score does not allow us to determine is which group has a higher probability of praying. For that, we would need to run other tests. This analysis also does not allow us to control for other factors that are likely important, such as gender and age. For example, are older married women more likely to pray than younger married men?

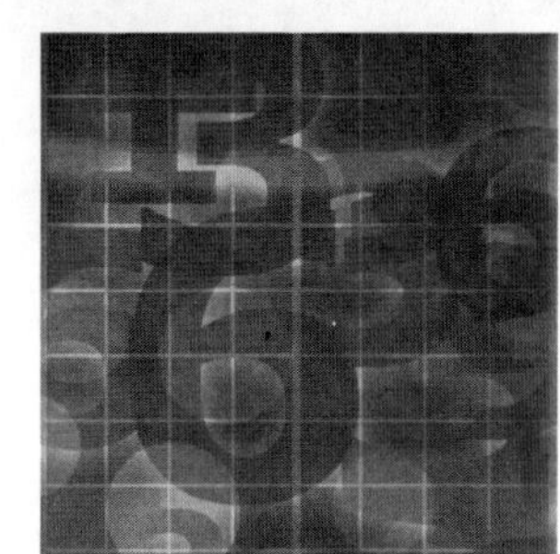

Exercise and Notes

Compute an ANOVA using the following data:

Hours	SEX
45	1
35	1
55	1
40	1
0	2
35	1
25	2
45	2
60	2
40	2

Sex: 1 = Male
Sex: 2 = Female

Hours: Number of hours worked last week

Is the *F* significant?

Name: ____   Date: __________________

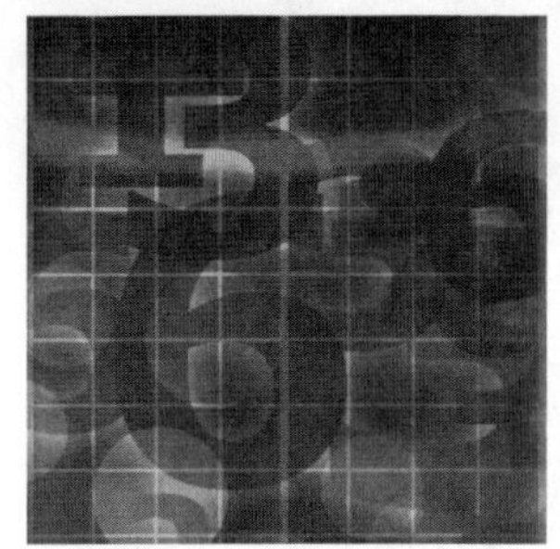

Learning Check

Use SPSS and calculate an ANOVA for WRKSTAT and SEX. Use the data to answer the following questions.

1. What is the *F* between the variables?

2. What is the significance of the *F*?

3. Write a paragraph (200 words or more) discussing the sociological implications of this analysis.

4. Use the following data to compute an ANOVA. Discuss the relationship between race and years of education.

Black	Hispanic	Asian	White
16	12	12	12
12	10	16	16
10	10	11	12
12	16	12	12
22	11	16	22
11	12	22	16
12	10	22	21
21	10	12	12
16	15	21	11
16	12	16	10

## Chapter Nine

Correlation

Another mechanism for testing a hypothesis is to use a correlation. A correlation shows the strength and direction of the relationship between two variables. The **correlation coefficient** is a number that we use to determine whether the independent variable and the dependent variable are related to one another. In other words, does a change in the independent variable correlate with a change in the dependent variable? The correlation coefficient ranges between -1 and $+1$.

There are two types of correlation coefficients: Pearson's r and Spearman's r. There are a few basic facts to remember about these coefficients. First, Pearson's r is used with parametric data and Spearman's r is used with nonparametric data. Other than this distinction, we interpret the coefficients in the same way.

Interpreting the Coefficient:

The **direction** of a coefficient refers to the sign that accompanies the r (negative or positive). When a correlation is negative (for example, $r = -.345$), we say that the variables move in opposite directions. This means that as the value of X increases, the value of Y decreases. Alternatively, a negative correlation means that as the value of X decreases, the value of Y increases. Let's say that we are interested in the relationship between the cost of cell phones and the use of cell phones. If we look back to the 1980s, when cell phones were first introduced, we will see that it was expensive to purchase a cell phone and the use of minutes was also very expensive. Thus, very few people owned or used cell phones then. This would mean a negative correlation: high cost (X) = low use (Y). Examining cell phone usage in 2011 shows us that the cost of cell phones has come down considerably. In the 1980s, a person might have spent \$800 on a phone that he or she carried around in a shoulder bag and might have had only 30 minutes per month to use the phone. The cell-phone bill might have been several hundred dollars a month. In contrast, in 2011, a person is likely to get a free cell phone, have free cell phone-to-cell phone minute plans, and spend between \$50 and \$100 a month. Despite this change in costs, we still have a negative correlation; as cost decreases (X) usage increases (Y).

As you likely have guessed, a **positive** correlation (for example, $r = .345$) is the opposite. It means that our variables change in the same direction. Suppose we wanted to look at the relationship between education and income. We would expect to see a positive correlation: As education increases, income also increases, or alternatively, as education decreases, income also decreases. With a positive correlation, an increase in X will correlate with an increase in Y. A decrease in X will correlate with a decrease in Y.

Direction does not tell us if a relationship is strong or significant. It only tells us if the variables change in the same or opposite directions. To fully understand the correlation, we also need to interpret its strength and significance. **Strength,** refers to the absolute value of the correlation. Thus, an r of $-.345$ is identical to an r of $.345$ in strength. But what do we consider strong? Disciplines have some variation in what is an acceptable strength for r. In sociology, we use the following guidelines:

No correlation $= 0$
Weak $= > 0$ but $< .30$
Moderate $= 3.0 - < .69$

Strong $= > .69$ but $< .95$

Multicollinear $= \geq .95$

We also need to understand the **coefficient of determination** or the r^2 to fully understand the strength of a correlation. Let's look again at the example of cell phone cost and usage. Suppose that we compute our correlation for these values and determine that our $r = -.545$. This tells us that our variables are strongly related, meaning they share a lot of covariance. In order to determine how much of their covariance is shared, we take the r^2, which for this correlation is .297. We can then convert this r^2 to a percentage and find that the cost of cell phones and cell phone usage share roughly 30% of their covariation.

A correlation can be weak and significant or strong and insignificant. In order to determine **significance,** we use a t score for the Pearson's coefficient and a z score for the Spearman's coefficient. We will learn new formulas for t and z later in this chapter. This t will be interpreted in the same way as previous t scores, where $t \geq 1.96$ corresponds to $\alpha = .05$ and $t \geq 2.58$ corresponds to $\alpha = .01$. If we have a weak but significant correlation, this means that while there may only be a small amount of covariation between X and Y, we can be confident that this covariation is not a result of our sample. When we have a nonsignificant r, then, regardless of the strength, we cannot say that the correlation is not a result of a characteristic of our sample.

The last term to discuss is **multicollinearity.** Multicollinearity means that the variables have a strong linear relationship. A perfect relationship, -1 or $+1$, typically only occurs when we correlate a variable with itself. Basically, when variables are multicollinear, they are measuring some element of the same variable. For example, if we run a correlation between "sex" (male $= 1$, female $= 2$) and "Have you ever been pregnant? ($0 = $ no, $1 = $ yes), we would expect a multicollinear relationship since only women can get pregnant. We would not have a perfect relationship because not every woman has been pregnant. Thus, while female sex doesn't measure pregnancy, "Have you ever been pregnant?" does measure a quality of the female of sex.

Spearman's Correlation Coefficient

As mentioned earlier, Spearman's correlation is the one that we use when we have nonparametric data. The formula for Spearman's correlation is:

$$r_s = 1 - \frac{6\Sigma D^2}{N(N^2 - 1)}$$

We will test this hypothesis: The highest degree that a person earns is strongly related to the highest degree that his or her mother earned. To do this, we will use the following data.

X Respondents' Highest Degree	Y Respondents' Mothers' Highest Degree
1	1
4	1
1	3
1	1
1	1
3	1
1	3
1	1
1	0
0	0

0 = Less Than High School
1 = High School
2 = Junior College
3 = Bachelor's Degree
4 = Graduate Degree

The first step is to calculate the D^2, which we have done before. This is the difference between X and Y. Again, it is easiest to set this up in a table. We'll begin with the numerator in our problem: $6\Sigma D^2$.

X	Y	D $(X - Y)$	D^2
1	1	0	0
4	1	3	9
1	3	−2	4
1	1	0	0
1	1	0	0
3	1	2	4
1	3	−2	4
1	1	0	0
1	0	1	1
0	0	0	0

Now we can sum D^2 and calculate r_s: $\Sigma D^2 = 22$.

$$r_s = 1 - \frac{6(22)}{10(10^2 - 1)}$$

$$r_s = 1 - \frac{6(22)}{10(10^2 - 1)}$$

$$r_s = 1 - \frac{132}{990}$$

$$r_s = 1 - .133$$

$$r_s = .866$$

Our correlation is positive and strong. Based on this correlation, we are able to say that as a person's mother's education increases, his or her education is also likely to increase. We can also take the r^2 and say that 75% of the variation in a person's highest degree covaries with his or her mother's highest degree. Now we need to determine the significance of our correlation. To do this we need to calculate our z score.

$$z = r(\sqrt{n - 1})$$

$$z = .866(\sqrt{9})$$

$$z = 2.6$$

Our z means that our findings are significant with an $\alpha = .01$.

Interpretation

Now that we have our correlation coefficient and our coefficient of determination, we can discuss sociological reasons for why a mother's education strongly and significantly covaries with her children's education. The most obvious answer perhaps is that the mother is typically the primary caregiver. This means that she is more likely to spend time with a child doing homework, more time in the classroom and on field trips, and more time interacting with the teacher to facilitate her child's learning. The more educated a parent is, the more likely that the parent will encourage their children to seek higher education. Also, more-educated parents will likely have an easier time providing assistance to their children, which may result in higher academic achievement for these children. While we can say that these variables are strongly related, we can't use these data to make a causal argument. While a mother's education is related, clearly there are other things that will influence her child's education level, such as the father's education, access to education, and the child's individual motivations.

Using SPSS

We have run a correlation using a very small sample. Now we can use SPSS to test the relationship between these variables in a larger sample. Make sure you have your data file (GSS 2008.sav) open. From the Analysis menu, select Correlation and then Bivariate. You will use the same Analysis tab for both Spearman's and Pearson's r. For now, check Spearman's and move DEGREE and MADEG into the Variable box, and then click OK.

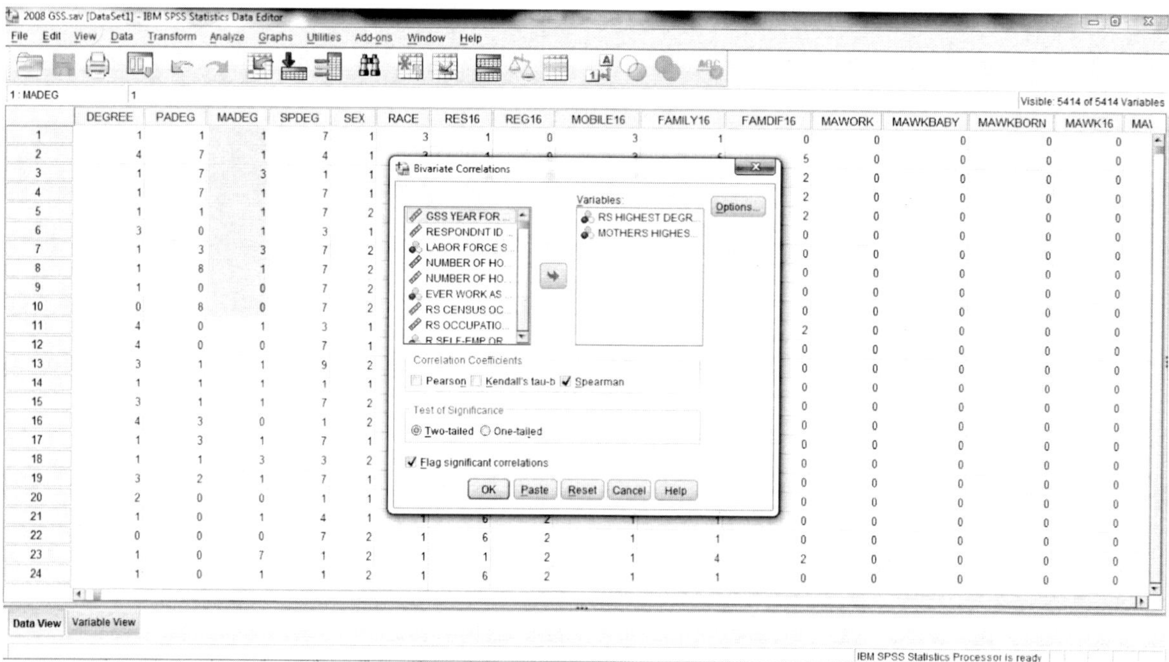

This is the output from SPSS. For now, we will use this table, but at the end of the chapter we will look at the proper formatting of a correlation matrix for presentation or publication. The first thing to take note of is that your variable names appear in both the columns and in the rows. When a variable intersects with itself in this table, the correlation is a perfect 1. To determine the correlation between our variables, we find the number that occurs where our variables intersect. As you can see, our r_s = .285. You will also notice that there are two asterisks (**) beside our correlation. The legend of our table reads "**Correlation is significant at the 0.01 level (2-tailed)." SPSS marks significant correlations for us. This means that our correlation has an α = .01 or 99% confidence. When the correlations are significant at the 95% level, they are marked with one asterisk (*) and the legend reflects this as well. The actual significance level for each correlation is listed below the correlation. Finally, the N tells us the number of cases that were used in this calculation.

This gives us the following results:

Correlations			RS Highest Degree	Mothers Highest Degree
Spearman's rho	RS Highest Degree	Correlation Coefficient	1.000	.285**
		Sig. (2-tailed)	.	.000
		N	2023	1939
	Mothers Highest Degree	Correlation Coefficient	.285**	1.000
		Sig. (2-tailed)	.000	.
		N	1939	1939

*** Correlation is significant at the 0.01 level (2-tailed).*

You can see that more data went into the respondents' highest degree (2023) than into the mothers' highest degree (1939). This merely means that we did not have the mother's highest degree for all of our respondents. By default, SPSS only includes data in the correlation calculation when there are data for both variables in the data set.

Pearson's Correlation Coefficient

Pearson's formula is used when we have parametric data. That formula is:

$$r = \frac{\Sigma(x - \bar{x})(y - \bar{y})}{\sqrt{\Sigma(x - \bar{x})^2 \Sigma(y - \bar{y})^2}}$$

Another way of writing this is:

$$r = \frac{SP}{\sqrt{SS_x SS_y}}$$

Again, while this looks difficult, it is merely complex. We already know that *SS* is the abbreviation for the Sum of Squares. *SP* is the abbreviation for the Sum of Products. As with most of our complex equations, the easiest way to work through these problems is to organize the data into a table.

For this example, we will test the hypothesis: A person's education is likely to be strongly and significantly correlated with his or her father's education. In our previous example, we used Spearman's correlation because we were analyzing ordinal, or nonparametric data. We used the highest degree earned as our variable, coding for the respondents and the mothers. For this example, we will use the number of years of school completed for both the respondent and his or her father. This is an interval measure, or parametric.

X Respondents' Education (years)	Y Fathers' Education (years)
17	6
19	8
18	12
16	12
16	14
18	16
15	16
13	12
16	12
14	4

First we need to find the average years of education for both the respondents and their fathers.

$$\bar{x} = 16.2$$
$$\bar{y} = 11.2$$

Now we can set up a table for our calculations.

X EDUC	Y PAEDUC	$(x - \bar{x})$	$(y - \bar{y})$	$(x - \bar{x})(y - \bar{y})^2$ SP	$(x - \bar{x})^2$ SS_x	$(y - \bar{y})^2$ SS_y
17	6	0.8	-5.2	-4.16	0.64	27.04
19	8	2.8	-3.2	-8.96	7.84	10.24
18	12	1.8	0.8	1.44	3.24	0.64
16	12	-0.2	0.8	-0.16	0.04	0.64
16	14	-0.2	2.8	-0.56	0.04	7.84
18	16	1.8	4.8	8.64	3.24	23.04
15	16	-1.2	4.8	-5.76	1.44	23.04
13	12	-3.2	0.8	-2.56	10.24	0.64
16	12	-0.2	0.8	-0.16	0.04	0.64
14	4	-2.2	-7.2	15.84	4.84	51.84
$\bar{x} = 16.2$	$\bar{y} = 11.2$			SP = 3.6	$SS_x = 31.6$	$SS_y = 145.6$

Now we have the data that we need for our calculations.

$$r = \frac{SP}{\sqrt{SS_x SS_y}}$$

$$r = \frac{3.6}{\sqrt{(31.6)(145.6)}}$$

$$r = \frac{3.6}{67.8}$$

$$r = .05$$

Next, we calculate our t to determine the significance level. That formula is:

$$t = \frac{r\sqrt{N-2}}{\sqrt{1-r^2}}$$

$$t = \frac{.05\sqrt{10-2}}{\sqrt{1-.0025}}$$

$$t = \frac{.141}{.998}$$

$$t = .141$$

The weakness of this correlation and its accompanying t is a function of the sample size. Let's see what happens when we test this hypothesis in SPSS using these variables from the GSS 2008.

Make sure you have the data file open. Follow the same steps to open the Correlation Analysis window that you used for Spearman's. This time, move EDUC and PAEDUC into the Variable window and make sure that Pearson's is checked before you click OK.

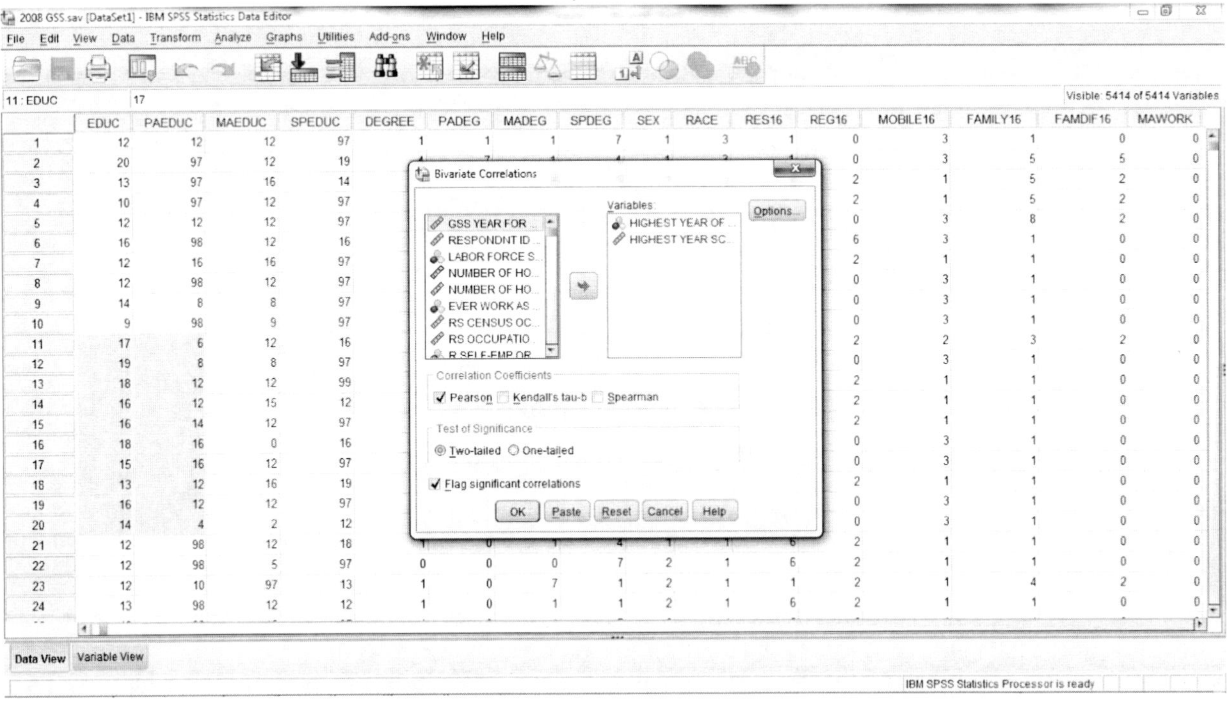

This gives us the following results.

Correlations		Highest Year of School Completed	Highest Year School Completed, Father
Highest Year of School Completed	Pearson Correlation	1	−.135**
	Sig. (2-tailed)		.000
	N	2021	1663
Highest Year School Completed, Father	Pearson Correlation	−.135**	1
	Sig. (2-tailed)	.000	
	N	1663	1664

***Correlation is significant at the 0.01 level (2-tailed).*

We read this table exactly as we read the Spearman's table. The sample size makes all the difference when analyzing data. As you can see, everything about our r has changed; direction, strength, and significance. We have a negative correlation here, which means that, lower levels for fathers' education are correlated with higher levels of respondents' education. Our r is weak, but it is significant at the 99% level. Our correlation of determination is .18, which means that fathers' education accounts for 18% of the variation in respondents' education. While this is not a large amount of covariation, we can be confident that the relationship exists and is not a result of our sample.

Interpretation

It is likely that fathers who are not educated encourage their children to get an education so that they will have better occupational and employment opportunities. However, it is also likely that all fathers, educated or not, encourage their children to get an education. Unlike mothers, who are likely to be the primary caregivers and actively engaged in the educational process, fathers are likely to influence education though encouragement and positive or negative reinforcement. The likelihood for all fathers to encourage education and the more passive role in the education process may result in the lower, though still significant, covariation between a father's education and his child's education.

Formatting a Correlation Matrix

As mentioned above, SPSS does not provide presentation-ready correlation tables. Below are two examples of correlations tables that have been formatted correctly. You will note that these examples have more than two variables. Read these tables in the same way as the examples used above.

In both tables, the data are identical. The only difference is whether we list the correlations above or below the diagonal line of 1's. Either way is correct and is a matter of personal preference. The identifiers in the rows are the variable labels from the GSS. The identifiers in the columns are the variable names from the GSS. Listing the variables like this accomplishes two things. First, it makes the data manageable and easier to fit on a page. Second, it informs the reader in a glance not only of the variables used should they want to re-run the data independently, but also, what the variables are actually measuring. CHILDS, for example, could mean anything. Our table lets the reader easily make sense of the variable label Number of Children and find the variable in GSS.

Table 9.1 *Example 1*

	EDUC	PAEDUC	HRS1	Marital	Childs	Age
Highest Year of School Completed	1					
Highest Year School Completed, Father	−.135**	1				
Number of Hours Worked Last Week	.010	.001	1			
Marital Status	−.050*	.006	−.069*	1		
Number of Children	−.121**	.072**	.029	−.391**	1	
Age of Respondent	.014	.146**	−.028	−.394**	.386**	1

**Correlation is significant at the 0.01 level (2-tailed).*
Correlation is significant at the 0.05 level (2-tailed).

Table 9.2 *Example 2*

	EDUC	PAEDUC	HRS1	Marital	Childs	Age
Highest Year of School Completed	1	−.135**	.010	−.050*	−.121**	.014
Highest Year School Completed, Father		1	.001	.006	.072**	.146**
Number of Hours Worked Last Week			1	−.069*	.029	−.028
Marital Status				1	−.391**	−.394**
Number of Children					1	.386**
Age of Respondent						1

**Correlation is significant at the 0.01 level (2-tailed).*
Correlation is significant at the 0.05 level (2-tailed).

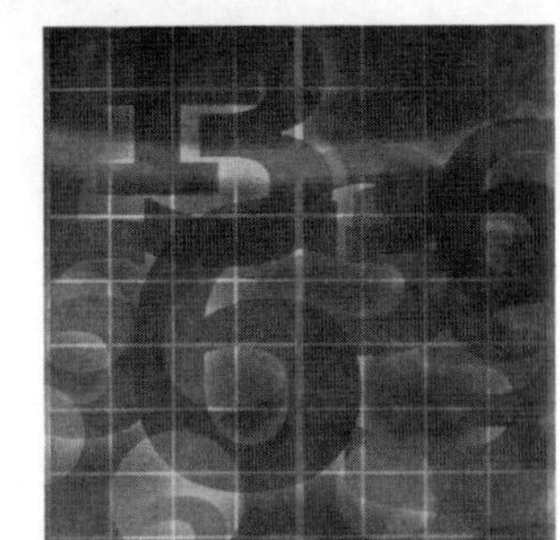

Exercises and Notes

Use the following data tables to calculate correlations and their significance levels. Make sure to use the correct formulas.

Problem 1:

Level of Education	Political Affiliation
2	1
9	2
4	3
5	4
8	5
1	6
7	7
3	8
10	9
6	10

1 = Highest Level of Education

1 = Most Conservative

Problem 2:

Education	Income
12	25,000
13	30,000
15	42,000
16	50,000
11	17,000
10	15,000
12	22,000
12	21,000

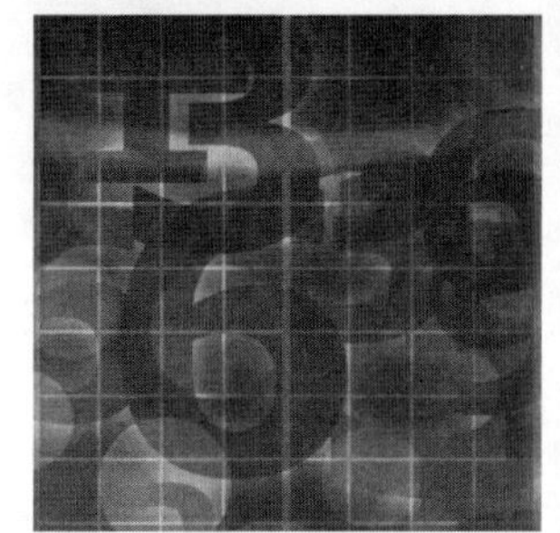

Learning Check

Use this correlation matrix to answer the following questions:

	EDUC	PAEDUC	HRS1	Marital	Childs	Age
Highest Year of School Completed	1	−.135**	.010	−.050*	−.121**	.014
Highest Year School Completed, Father		1	.001	.006	.072**	.146**
Number of Hours Worked Last Week			1	−.069*	.029	−.028
Marital Status				1	−.391**	−.394**
Number of Children					1	.386**
Age of Respondent						1

***Correlation is significant at the 0.01 level (2-tailed).*
**Correlation is significant at the 0.05 level (2-tailed).*

1. What is the correlation between the number of children a person has and his or her age?

2. What is the percentage of covariability between the number of children a person has and his or her age?

3. Write a paragraph discussing the sociological interpretation of this relationship. Be sure to discuss the direction, strength, and significance of the relationship.

Chapter Ten

Bivariate Regression

We have explored several different ways of examining the relationship between two or more variables. Depending on the level of measurement of the variables, different statistical techniques were used. In Chapter 2, you learned that cross-tabulations of nominal or ordinal data show us the relationship of variables by presenting the cross-classified frequencies or percentage of two or more variables. Later, in Chapter 7, you learned how to conduct a test of significance on a cross-tab, and that the chi-square test of independence allows us to determine whether the distribution of a variable depends on the distribution of another variable. Other techniques that we covered allow us to examine the relationship between variables if at least one of the variables is measured at the interval-ratio level. When we are interested in comparing two groups in terms of an interval-ratio measure, such as hours worked, we would do a t-test of the difference between two independent sample means (see Chapter 6). More specially, if we want to compare blacks and whites regarding the number of hours worked, we would do a t-test of the difference between two independent sample means. However, if we wish to compare whites, blacks, and Latinos in terms of hours worked, as you learned in Chapter 8, a t-test would not be appropriate but ANOVA would be.

In the last chapter, on correlation, Pearson's r provided us with a statistic that measures the linear relationship between two interval-ratio variables. In this chapter, you will learn that the correlation coefficient, Pearson's r, shares some similarities with the regression coefficient, b. For example, both use the sum of the products (SP) and sum of squares of X (SS_X) in their calculations and both statistics provide us with information about the strength and direction of the relationship. However, the correlation coefficient, Pearson's r, and the regression coefficient, b, are not the same. Differences between the two statistics will be pointed out throughout this chapter. Also, b is just one part of regression analysis that will be explored in this chapter.

The Regression Model

In Chapter 9, you learned how scatterplots could inform us about the relationship between variables. They can tell us if the relationship is linear, if it is a positive or negative relationship, and, if the relationship is linear, can give us an idea about the strength of the linear relationship. The more closely the data points form a straight line, the stronger the relationship will be. However, we have to calculate a correlation (r) to actually measure the strength of the relationship, and yet a correlation never exactly indicates where the line passes through the data. Regression analysis fits a straight line through the data. A form of regression analysis, known as ordinary least squares (OLS), minimizes the square distance between each data point and the straight line. We will discuss this point later, but let's first address the regression model equation, which in bivariate regression takes the form of:

$$Y = a + b X + e$$

In the equation, X and Y represent the two variables in the analysis, with X denoting the independent variable and Y the dependent variable. When calculating the correlation, we also had an X and Y variable. However, mathematically, it did not matter which variable was X and which was Y, since the correlation between, for example, education (X) and hourly pay (Y) equals the correlation between hourly pay (Y) and education (X). The same cannot be said for regression analysis because the regression equation explains or predicts the dependent variable based on the independent variable(s). Therefore, in regression analysis, it is crucial to correctly identify the independent and dependent variables.

As you learned in Chapter 1, the independent variable is often referred to as the cause, and the dependent variable is often referred to as the effect. There are three conditions necessary to establish causation that can help us try to identify the independent and dependent variables. First, there needs to be a relationship between the variables, that is, a correlation. The second condition is time ordering. This means that the independent variable must come before the dependent variable. For example, researchers might use the number of siblings to predict the number of children that people might have rather than expecting the number of children that people have to affect the number of siblings they would have. The number of siblings is the independent variable because one's parents typically complete their childbearing before their offspring do. Lastly, to establish causation, we need to eliminate any other potential factors that might affect the relationship between the independent and dependent variables. The relationship between the two variables may be spurious, not causal. If the relationship is **spurious,** it can be explained by a third variable. Since our focus here is bivariate regression, we will discuss this third condition more in the next chapter, on multiple regression.

Besides X and Y, there are other components of the regression equation. The a is referred to as the constant or the Y-intercept. It is known as the Y-intercept because that is the point at which the regression line crosses the Y-axis. The b is called the slope because it indicates how much tilt the regression line has. If the independent variable has no effect on the dependent variable, b will equal zero, and the regression line will have no tilt; it will be parallel to the X-axis. The last component of the regression equation is e, which stands for the error term, or residual. Since the regression line does not go through all the data points, estimating the data involves some error. The error term, or residual, captures the effect of omitted variables that also affect the dependent variable or the error in measuring variables included in the analysis.

Let's look at an example to see how we would calculate and interpret these elements of regression. Suppose we are interested in determining whether additional education is a good predictor of hourly pay among people with at least a high school education. For most workers, their years of education are completed before they earn their current pay level. Since we suspect that post-secondary education affects hourly pay, the years of post-secondary education are the independent variable and hourly pay is the dependent variable. **Table 10.1**

Table 10.1 *Years of Post-Secondary Education and Hourly Pay*

Years of Post-Secondary Education	Hourly Pay
1	10.00
0	9.00
1	16.50
2	14.50
3	16.00
4	26.00
0	18.50
2	23.00
5	19.00
6	37.00
$\Sigma X = 24$	$\Sigma Y = 189.50$
Average = 2.40	Average = 18.95

contains the data on the years of post-secondary education and hourly pay of 10 workers with at least a high school education.

Figure 10.1 is a scatterplot of the data on years of post-secondary education and hourly pay. As you learned in the previous chapter, we can discern that the relationship between the two variables is positive and appears to be fairly strong. If we visualize a straight line placed through the data, none of the data points would fall too far from the line. While we can guess where the line should go, regression analysis gives us the ability to accurately draw the line. However, before we can determine where to draw the regression line, we need to know how to calculate a and b.

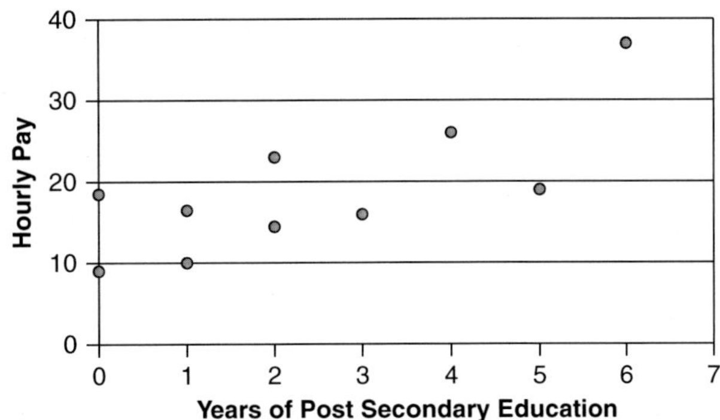

Figure 10.1 Scatterplot of Hourly Pay and Years of Post Secondary Education.

We will begin by calculating b, the slope. As previously mentioned, Pearson's r, which we calculated in the previous chapter, and b are mathematically very similar. **Table 10.2** shows the equations for Pearson's r and b. The numerator of both r and b are the sum of the products (SP). Although the denominators differ, the sum of the square of X (SS_X), which is the sole component of the denominator of b, is also a part of the denominator of r. Therefore, components that are used in calculating r will also be used to compute b.

Table 10.2 *Comparing r and b*

Correlation Coefficient r	Regression Coefficient b
$r = \dfrac{SP}{\sqrt{SS_X\,SS_Y}}$	$b = \dfrac{SP}{SS_X}$
$SP = \Sigma(X_i - \bar{X})(Y_i - \bar{Y})$	$SP = \Sigma(X_i - \bar{X})(Y_i - \bar{Y})$
$SS_x = \Sigma(X_i - \bar{X})^2$ $SS_y = \Sigma(Y_i - \bar{Y})^2$	$SS = \Sigma(X_i - \bar{X})^2$

Table 10.3 shows the calculations needed to compute the sum of the products (SP), the sum of the squares of X (SS_X), and the sum of the squares of Y (SS_Y). While the first two components are used in calculating b, all three are used to compute r, as shown in **Table 10.2**. To calculate the SP, we need to subtract $\bar{X}$ from each value of X and $\bar{Y}$ from each value of Y. Then for each case, the difference of each X is multiplied by each corresponding difference of Y. The final step is to sum the products, as shown in **Table 10.3**. Therefore, the sum of the products (SP) equals:

$$SP = \Sigma(X - \bar{X})(Y - \bar{Y})$$

$$SP = 115.70$$

To calculate the sum of the squares of X (SS_X), we take the difference between each value of X and $\bar{X}$ and square each difference. As seen in **Table 10.3,** the sum of the squares of X (SS_X) equals:

$$SS_X = \Sigma(X_i - \bar{X})^2$$
$$SS_X = 38.40$$

Although we do not need the sum of the squares of Y to calculate b, the sum of the squares of Y is used to calculate another statistic that is a part of a regression analysis. Also, we can use the sum of the squares of Y to calculate r and compare it to b. The sum of the squares of Y (SS_Y) follows the same procedures for the sum of the squares of X (SS_X), but for the Y variable. The sum of the squares of Y (SS_Y) equals:

$$SS_Y = \Sigma(Y_i - \bar{Y})^2$$
$$SS_Y = 605.725$$

Table 10.3 Worksheet for r and b for Education and income

Education	Pay	$(X - \bar{X})$	$(Y - \bar{Y})$	$(X - \bar{X})(Y - \bar{Y})$	$(X - \bar{X})^2$	$(Y - \bar{Y})^2$
1	10	−1.4	−8.95	12.53	1.96	80.1025
0	9	−2.4	−9.95	23.88	5.76	99.0025
1	16.5	−1.4	−2.45	3.43	1.96	6.0025
2	14.5	−0.4	−4.45	1.78	0.16	19.8025
3	16	0.6	−2.95	−1.77	0.36	8.7025
4	26	1.6	7.05	11.28	2.56	49.7025
0	18.5	−2.4	−0.45	1.08	5.76	0.2025
2	23	−0.4	4.05	−1.62	0.16	16.4025
5	19	2.6	0.05	0.13	6.76	0.0025
6	37	3.6	18.05	64.98	12.96	325.8025
$\Sigma X = 240$	$Y = 189.5$			$SP = 115.70$	$SS_X = 38.40$	$SS_Y = 605.73$
$\bar{X} = 2.40$	$\bar{Y} = 189.5$					

As shown in **Table 10.2,** we can use SP, SS_X, and SS_Y to calculate r and b. As you learned in the previous chapter, the correlation between education and hourly pay equals the SP divided by the square root of the product of the SS_X and SS_Y. The slope, or b, in regression equals the SP divided by the SS_X:

$$r = \frac{115.70}{\sqrt{38.40 \times 605.73}} = .759 \qquad b = \frac{115.70}{38.40} = 3.013$$

The correlation (r) of .759 indicates a positive relationship between years of post-secondary education and hourly pay. Just like the correlation, the sign of b indicates whether the relationship is positive or negative. In fact, the SP, the numerator of both r and b, determines the sign of r and b. If SP is negative, r and b are negative, while if SP is positive, r and b are positive. Therefore, given a b of 3.013, we know that the relationship between years of post-secondary education and hourly pay is positive.

However, when interpreting the strength of the relationship, there are differences between r and b. Unlike the r, the b is not constrained to the range of values between -1.0 and $+1.0$ because it is based on the metric of the dependent variable. The b indicates how much Y changes given a one-unit change in X. When interpreting b, we should use the metric in which the variables were measured. In this case, education was measured in years of education after high school and hourly pay was measured in terms of dollars and cents. Therefore, the b of 3.013 indicates that for each additional year of education beyond high school, hourly pay increases by $3.01.

Another element in the regression equation is a, the Y-intercept or constant. To calculate a, we use the following equation:

$$a = \overline{Y} - b\overline{X}$$
$$a = 18.95 - 3.013 \times 2.40$$
$$a = 11.719$$

The a indicates the point at which the regression line crosses the Y-axis. In other words, the value of a tells us what that the value of Y is when X is zero. Therefore, our value of a indicates that the typical worker with only a high school degree (zero years of post-secondary education) is expected to make $11.72 per hour.

You may be wondering why years of post-secondary education was used instead of just years of education. Years of education could have been used and would have produced the same effect that education has on income. Although the b would have been the same, the a would have been different because the a would indicate how much someone who has zero education makes per hour. This would mean that we would be looking at measuring education in an entirely different way: A person with zero years of post-secondary education (or 12 years of education) would be expected to make $11.72 per hour, while someone with no years of formal education (actual zero years of education) would be expected to make $- $24.44 per hour. Such a measurement would make no sense. Negative Y-intercepts often have nonsensical interpretations, as in this case. Therefore, we chose to measure education in terms of years of post-secondary education, not years of education.

Remember that a person's education would not change, but only the way that we measured it. Looking at metric change another way, a person could be 5 feet, 4 inches tall, or 5.33 feet tall, or 64 inches tall, or 162.56 centimeters tall. The values change; however, the person did not shrink in size or grow taller. Only the metric in which we measured the variable changed. In practice, researchers do not typically worry about a negative Y-intercept. If a negative Y-intercept is nonsensical, it is not interpreted. The Y-intercept, positive or negative, often is not discussed in research results.

Predicting Values of the Dependent Variable

The regression analysis allows us to predict values of the dependent variable based on knowledge of the independent variable(s). In this case, we are interested in how much money we expect workers to make per hour given how many years of education they have beyond high school. If we did not know anything about a worker's education (or any other factors that might affect a person's income), the best predictor of a worker's hourly pay is the mean hourly pay ($\overline{Y}$). Thus, regression analysis allows us to improve our prediction of the dependent variable and therefore reduces the error in our prediction.

We are making a distinction between how much the respondents *actually* make per hour and how much we *expect* them to make per hour given how many years of education they have. Since we often only have sample data, we can use regression to predict what a typical worker with a certain level of education is expected to make. The equation used to predict values of the dependent variable is slightly different from the regression equation that we have been discussing. In the predictive equation, Y is replaced by the predictive Y, symbolized as $\hat{Y}$ and referred to as Y hat. In addition, the e (the error term or residual) is dropped from the equation.

Regression Equation	Predictive Equation
$Y = a + bX + e$	$\hat{Y} = a + bX$

The predictive equation of the effect of post-secondary education on hourly pay with the calculated values of a and b is:

$$\hat{Y} = 11.719 + 3.013X$$

If we wanted to know how much a worker with two years of post-secondary education is expected to make, we plug in 2 for X with the values we previously calculated for a and b. We could also compare their income with what it would be when they have a college degree (assuming they graduate in four years). We would modify the equation by using 4 in the equation, instead of 2.

$$\hat{Y} = 11.719 + (3.013 \times 2)$$
$$\hat{Y} = 17.745$$

AND

$$\hat{Y} = 11.719 + (3.013 \times 4)$$
$$\hat{Y} = 23.771$$

Therefore, we would expect that a person with two years of education beyond high school would make \$17.75 an hour, compared to someone with a college degree, who would be expected to make \$23.77 per hour.

Mathematically, we can plug in any value of X and compute a predictive hourly pay level. However, statistically, it is not advisable to include values outside the scope of your data for multiple reasons. First, you cannot assume that a linear relationship will continue. For example, given the data we have, we should not try to predict the hourly pay of someone with a PhD because this is beyond the scope of the data. Those with PhDs typically have at least five years of education beyond a bachelor's degree (usually, 4 years of post-secondary education). Therefore, we would be making predictions about someone with nine or more years of post-secondary education, while the highest year of post-secondary education in the data is six years. We should not try to predict the pay of workers with more than six years of post-secondary education because we do not have any data to suggest that the relationship will continue to be linear; it could become nonlinear at the extremes.

Another reason why we should not use values of X outside the scope of our data is that nonsensical values could be predicted. If the dependent variable has a floor or a ceiling, impossible values could be predicted. As we saw in our discussion of the Y-intercept, a negative a can lead to a nonsensical interpretation. In addition, if we had measured education in terms of years of education, instead of years of post-secondary education, and predicted the hourly pay of someone with only eight years of education, we would have obtained the impossible predicted wage of negative 33 cents ($-.33$) as the person's hourly pay. A different variable, grade point average (GPA), typically has a ceiling of 4.0, a perfect A average. If we predicted GPA from the number of hours studied and used a number of hours studied that was higher than reported by any student in the sample, the calculated predicted GPA could be higher than 4.0. Therefore, we should not predict the dependent variable using values of the independent variable that are lower or higher than those that occur in the data set.

Drawing the Regression Line and Prediction Errors

To draw a straight line, we need two points. Therefore, in order to draw the regression line through the data, we need to determine two points that will be on the line. Although any value of $\hat{Y}$ will be on the regression line, two points are typically used. Since the Y-intercept is the point where the line crosses the Y-axis, we know that ($X = 0$, $Y = a$) is one point on the line. Another point that always falls on the regression line is the intersection of the two means ($X = \bar{X}$, $Y = \bar{Y}$). **Figure 10.2** is a scatterplot of the data of years of post-secondary education and hourly pay with the regression line. As you can see, the Y-intercept ($X = 0$, $Y = 11.73$) and the intersection of the means ($X = 2.40$, $Y = 18.95$) are relatively close to each other; these are the first two black plotted points. If these two points are close together, it is difficult to draw the line accurately. In such cases, one should choose a relatively large value of X, calculate the predictive Y, and use those two values (in this case, $X = 6$, $Y = 29.79$, the last black point in the scatterplot) rather than the two means.

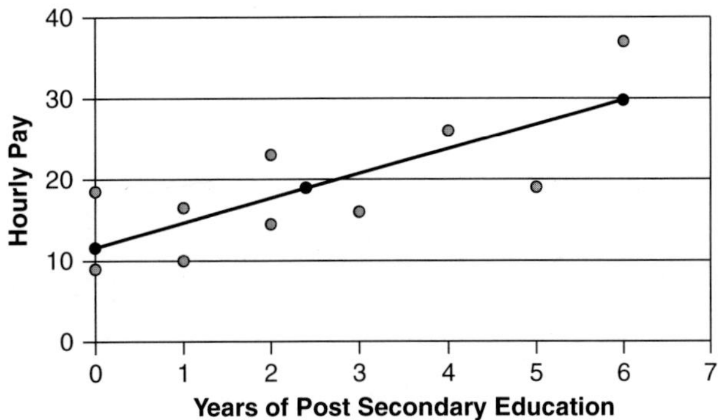

Figure 10.2

Figure 10.2 also shows that the data points (the gray points) do not fall far from the regression line. This is not surprising given the strong correlation ($r = .759$) between years of education and hourly pay. However, the correlation is not perfect (1.00) and the data points do not fall on the regression line. The difference between the data points and the regression line is the error, or residual, term (e). Therefore,

$$e = Y - \hat{Y}$$

The error term is the portion of the value of the dependent variable that cannot be predicted by the regression model equation. We can see that $e = Y - \hat{Y}$ by manipulating the regression equation. First, let's rewrite the equation, so that we are solving for e, not Y. Since $\hat{Y}$ equals $a + b$, we can substitute, $\hat{Y}$ for $a + b$.

Regression equation	$Y - a + bX + e$
Rearrange equation to solve for e	$Y - (a + bX) = e$ or $e = Y - (a + bX)$
Substitute $\hat{Y}$ for $a + bX$	$e = Y - \hat{Y}$

Visually, the error or the residual for each case is the distance each data point is from the regression line. OLS is a method used to estimate the regression coefficients that minimizes the error sum of squares. In other words, collectively, the predictive values of Y ($\hat{Y}$), which fall on the regression line, have the smallest possible distance between them and the actual data points. **Figure 10.3** illustrates for two cases the errors in the regression of hourly pay on education. The two white-filled points indicate the expected hourly pay ($\hat{Y}$) for workers with two and five years of secondary education. If we use the predictive equation to predict hourly pay for those who have two and five years of post-secondary education, we would expect those workers to make $17.74 and $26.78 per hour, respectively.

Predictive equation	$\hat{Y} = a + bX$
Predictive equation with values of a and b	$\hat{Y} = 11.719 + 3.013X$
Predicted pay when $X = 2$	$17.745 = 11.719 + 3.013(2)$
Predicted pay when $X = 5$	$26.784 = 11.719 + 3.013(5)$

However, workers with those years of education in our sample did not actually make $17.75 and $26.78 per hour. We can determine the distance between the first predicted white point ($X = 2$, ($\hat{Y}$) $= 17.74$) and an actual gray data point ($X = 2$, $Y = 23$) and the second predicted white point ($X = 5$, ($\hat{Y}$) $= 26.78$) and the actual gray data point ($X = 5$, $Y = 19$). We calculate the error, or residual (e), by taking the predicted value ($\hat{Y}$) from the actual value of Y.

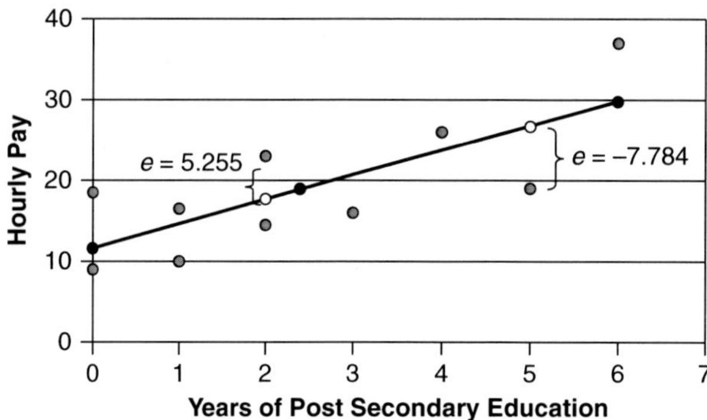

Figure 10.3 Predicting Error in the Regression of Hourly Pay on Years of Post Secondary Education.

$$\text{Error or Residual } (e) = \text{actual } (Y) - \text{predicted } (\hat{Y})$$
$$e = Y - \hat{Y}$$
$$e = 23 - 17.745 = 5.255$$
$$e = 19 - 26.784 = -7.784$$

R^2: The Coefficient of Determination

As we have discussed, OLS fits a regression line that most closely provides a linear representation of the data points in which the degree of error is minimized. There is another statistic in regression that allows us to assess how accurate the fit is. This statistic tells us how well the independent variable(s) explain the dependent variable. Earlier, we indicated that without knowing the number of years of post-secondary education (X), our best estimate of a worker's hourly pay would be the average hourly pay, $\bar{Y}$. Therefore, the error of not knowing X is $Y - \bar{Y}$. You may remember that when calculating Pearson's r, one of the components we needed was the sum of the squares of Y (SS_Y), which equals $\Sigma(Y - \bar{Y})^2$. In regression analysis, since we are interested in predicting Y, the sum of the squares of Y (SS_Y) is also known as the **total sum of square** (SS_{total}), which is divided into two components: the error sum of squares (SS_{error}) and the regression sum of squares (SS_{reg}).

$$SS_{total} = SS_{reg} + SS_{error}$$

Although regression analysis takes into account knowledge of X, social data do not lend themselves to perfect predictions. The error in predicting Y with knowledge of X is equal to the difference between the actual Y's and the predictive Y's, or $\Sigma(Y - \hat{Y})$. The sum of the squared errors with knowledge of X is called the **error sum of squares**:

$$SS_{error} = \Sigma(Y - \hat{Y})^2$$

Without Knowledge of X	With Knowledge of X
$\bar{Y}$	$\hat{Y}$
$SS_{total} = \Sigma(Y - \bar{Y})^2$	$SS_{error} = \Sigma(Y - \hat{Y})^2$

Conversely, the part of the total sum of squares explained by X is called the **regression sum of squares.** The regression sum of squares is the difference between predicting Y without knowledge of X [$\Sigma(Y - \bar{Y})^2$] compared to predicting Y with knowledge of X [$SS_{error} = \Sigma(Y - \hat{Y})^2$]:

$$SS_{reg} = [\Sigma(Y - \bar{Y})^2] - [\Sigma(Y - \hat{Y})^2]$$
$$SS_{reg} = SS_{total} - SS_{error}$$

Researchers typically use the sum of squares to compute a statistic that indicates how much knowing X reduces the error in predicting Y over knowing only $\bar{Y}$. R^2 (read as R-squared) is the proportional reduction in error

(PRE) statistic associated with regression analysis. All PRE statistics have a minimum value of 0 that indicates that there was no reduction in error. If R^2 equals 0 in the regression analysis, knowing X did not improve the prediction of Y over knowing only the mean, $\bar{Y}$. On the other hand, the maximum value of all PRE statistics is 1.00, which means that all error was eliminated, and that knowing X allowed us to perfectly predict Y. The value of a PRE statistic, such as R^2, is that it provides us with the amount of variance in the dependent variable explained by the independent variable(s).

Table 10.4 *Worksheet for Total and Error Sum of Squares*

Educ.	Pay	$Y - \bar{Y}$	$(Y - \bar{Y})^2$	$\hat{Y} = a + bX$	$(Y - \hat{Y})$	$(Y - \hat{Y})^2$
1	10	-8.95	80.1025	14.73902	-4.73902	22.45832
0	9	-9.95	99.0025	11.726	-2.726	7.431076
1	16.5	-2.45	6.0025	14.73902	1.760979	3.101048
2	14.5	-4.45	19.8025	17.75204	-3.25204	10.57578
3	16	-2.95	8.7025	20.76506	-4.76506	22.70582
4	26	7.05	49.7025	23.77808	2.221917	4.936914
0	18.5	-0.45	0.2025	11.726	6.774	45.88708
2	23	4.05	16.4025	17.75204	5.247958	27.54107
5	19	0.05	0.0025	26.7911	-7.7911	60.7013
6	37	18.05	325.8025	29.80413	7.195875	51.78062
$\Sigma X = 240$	$\Sigma Y = 189.5$		SS_Y or $SS_{Total} = 605.73$			$SS_{Error} = 257.12$
$\bar{X} = 2.40$	$\bar{Y} = 18.95$					

R^2 is the portion of the total sum of squares (SS_{total}) that is explained by the independent variables in the regression (SS_{reg}). **Table 10.5** shows how we move from this conceptualization of R^2 to a formula to calculate R^2. Although we do not have the regression sum of squares (SS_{reg}), we do have the total sum of squares (SS_{total}) and the error sum of squares (SS_{error}). We can take the error sum of squares (SS_{error}) from the total sum of squares (SS_{total}) to obtain the regression sum of squares (SS_{reg}). We also can manipulate the equation to arrive at a more easily calculated formula:

$$R^2 = 1 - \frac{SS_{error}}{SS_{total}}$$

Table 10.5 *R^2: From Conceptualization to Calculation*

Conceptualization of R^2	$R^2 = \dfrac{SS_{reg}}{SS_{total}}$
	$R^2 = \dfrac{SS_{tot} - SS_{error}}{SS_{total}}$
	$R^2 = \dfrac{SS_{tot}}{SS_{total}} - \dfrac{SS_{error}}{SS_{total}}$
Calculation of R^2	$R^2 = 1 - \dfrac{SS_{error}}{SS_{total}}$

The calculated R^2 is:

$$R^2 = 1 - \frac{257.12}{605.73}$$

$$R^2 = .576$$

Since R^2 is interpreted as the amount of variance in the dependent variable that is explained by the independent variable(s), our R^2 indicates that 57.6% of the variance in hourly pay is explained by the number of years of post-secondary education. This example has produced a relatively large R^2. As previously mentioned, social data do not lend themselves to perfect predictions. In fact, much lower R^2 values are actually common with social data.

In bivariate regression only, we can square the bivariate correlation coefficient (r) to obtain R^2. Earlier in this chapter, when we calculated b, we also calculated r. The correlation between years of post-secondary education and hourly pay equals .759. If we square .759, we obtain R^2.

Only in Bivariate Regression:

$R^2 = r^2$	$\sqrt{R^2} = r$
$R^2 = .759^2$	$\sqrt{.576} = r$
$R^2 = .576$	$.759 = r$

This is limited to bivariate regression because R^2 takes into account how *all* the independent variables in the regression explain the variance in the dependent variable. Since bivariate regression only has one independent variable contributing to the R^2, squaring the bivariate correlation (r) is equal to the R^2.

Testing the Regression Coefficient

Our analysis of the *sample* data suggests that for every additional year of education beyond high school, hourly pay is expected to increase by $3.01. However, the question remains as to whether workers in the *population* receive higher pay for each year of post-secondary education, that is, is the slope in the population (β) positive? Similar to testing to see if a correlation (r) is significant, as we did in the previous chapter, we will do a *t*-test to determine whether post-secondary education increases workers' hourly pay in the population. Therefore, we follow the five steps of hypothesis testing.

Step 1. State the hypotheses.

H_0: $\beta \leq 0$: Post-secondary education does not have a positive effect on hourly pay.

H_1: $\beta > 0$: Post-secondary education has a positive effect on hourly pay.

Step 2. Determine the significance or alpha level that will be used in the statistical test. For this test, let's use .01.

Step 3. Calculate the statistical test. The *t*-test formula to determine whether the b is significant is similar to the *t*-test formula for a mean, except that instead of dealing with a mean, we have a slope. In the numerator, the population parameter is subtracted from the sample statistic and then divided by the standard error. When testing the slope, β is subtracted from b and then divided by the standard error of the b.

$$t_{N-2} = \frac{b - \beta}{S_b}$$

Earlier, we found that b equals 3.01. However, we do not know what β is. Indeed, we are hoping to generalize our sample slope of 3.01 to the population slope (β). As we learned in earlier chapters, we conduct statistical tests based on the assumption that the null hypothesis is true. Since the null hypothesis indicates that β could be zero, we use zero as the value for β in calculating the *t*-test.

To calculate the standard error of the b that is the denominator of the t-test, we use two the sum of squares that we previously calculated—the SS_{error} and SS_X. The formula for the standard error of the b is:

$$S_b = \sqrt{\frac{SS_{error}/N-2}{SS_X}}$$

If we refer back to **Tables 10.3** and **10.4,** we can find the SS_{error} and the SS_X (257.12 and 38.40, respectively). If we l ook at the raw data in **Tables 10.3** and **10.4,** we see that there are 10 cases ($N = 10$). Therefore:

$$S_b = \sqrt{\frac{257.12/10-2}{38.40}}$$

$$S_b = .915$$

Now that we have the standard error of the b, we can calculate the t:

$$t_{N-2} = \frac{b-\beta}{S_b}$$

$$t_{N-2} = \frac{3.013-0}{.915} = 3.293$$

Step 4. Make a decision regarding the null hypothesis. In order to make a decision about the null hypothesis, we need to determine the critical value of t. The subscript in the above formula for t indicates the formula for the degrees of freedom ($df = N - 2$). Thus, our degrees of freedom (df) equal $N - 2$, or $df = 10 - 2 = 8$. We use the $df = 8$ and alpha level of .01 from Step 2 of hypothesis testing to determine the critical value of t. Because the alternative hypothesis indicates a direction (i.e., post-secondary education is expected to have a *positive* effect on hourly pay), we are doing a one-tailed test. If we look for the intersection of eight degrees of freedom and $\alpha = .01$ on the one-tailed, t-test table, we find that the critical value of t equals 2.896. **Figure 10.4** shows that our calculated t of 3.293 falls in the rejection region. Therefore, we reject the null hypothesis.

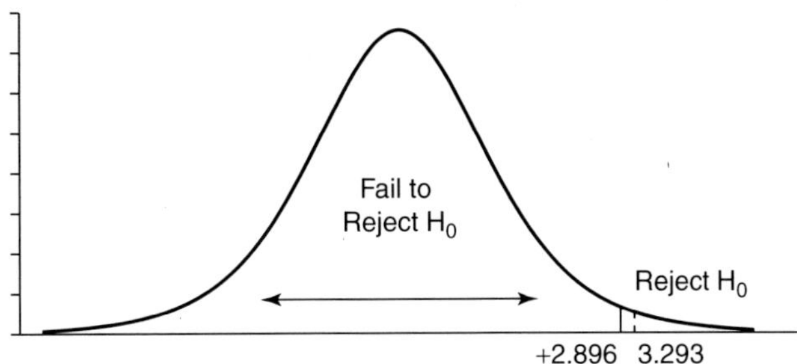

Figure 10.4 The t Curve with df $= 8$ and $\alpha = .01$.

Step 5. Interpret the results. The sample data provided enough evidence to allow us to reject the null hypothesis. We can conclude that years of post-secondary education do have a positive effect on hourly pay in the population. Therefore, years of post-secondary education can be described as having a statistically significant effect on hourly pay. However, as with any other inferential statistical test, we must remember that we cannot be 100% certain. In this case, we still could have committed a Type I Error—rejecting a true null hypothesis.

Standardized Equation

The regression equation that we have been discussing up to this point is known as the unstandardized equation because the regression coefficient (b) is based on the *metric* of the variables in which, in this case, education is measured in *years* of post-secondary education and hourly pay is measured in *dollars and cents*. However, the variables and coefficients can be standardized so that they are no longer in the metric of the variables. Standardization rescales the independent and dependent variables in terms of their standard deviations, as we did in Chapter 5 when we converted raw scores to Z scores. The unstandardized regression coefficient (b) also can be converted to the standardized regression coefficient (β^W), which we will refer to as the beta weight to distinguish it from β—the slope in the population.

We calculate the beta weight (β^W) by multiplying b by the quotient of the standard deviation of the independent variable divided by the standard deviation of the dependent variable:

$$\beta^W = b \times \frac{S_X}{S_Y}$$

Remember that the standard deviation of a variable equals:

$$S = \sqrt{\frac{\Sigma(X - \bar{X})^2}{N - 1}}$$

Since the numerator of the standard deviation is equal to the sum of squares, we can use SS_X and SS_Y from **Table 10.3** to calculate the standard deviations of X and Y.

$$S_X = \sqrt{\frac{38.40}{9}} = 2.066 \qquad S_Y = \sqrt{\frac{605.73}{9}} = 8.204$$

We can now calculate the standardized regression coefficient.

$$\beta^W = 3.013 \times \frac{2.066}{8.204} = .759$$

Besides the variables being converted to Z scores and the b being changed to β^W, the Y-intercept is eliminated in the standardized equation. We can see why the Y-intercept drops out of the equation by manipulating the equation for the Y-intercept. In the unstandardized form:

$$\alpha^W = Z_{\bar{Y}} + \beta^W Z_{\bar{X}}$$

If we replace the variables with their Z scores and a and b with their standardized versions (α^W and β^W, respectively), the equation is:

$$\alpha = Z_Y - \beta^W Z_X$$

Since the mean of a standardized variable equals zero, we replace the standardized means of X and Y ($Z_{\bar{X}}$ and $Z_{\bar{Y}}$, respectively) with zero:

$$\alpha^W = 0 - (\beta^W \times 0)$$

$$\alpha^W = 0$$

Therefore, the term for the Y-intercept is eliminated, and the standardized predictive regression equation with one independent variable is:

$$\hat{Z}_{\bar{Y}} = \beta^W Z_{\bar{X}}$$

You may be wondering why we would want to use the standardized equation. The beta weight β^W can be useful when dealing with more arbitrary units, such as political orientation or religiosity. For example, political orientation measured on a 7-point scale from very conservation to very liberal is much more arbitrary than education measured in years or pay measured in dollars and cents. The interpretation of the unstandardized coefficient (b) for political orientation is not as clear as it is with variables that have clearly defined metrics. In the case of variables with arbitrary units, it may be useful to have the standardized coefficient as well. The β^W is interpreted in standardized units; for one standardized unit change in X, we expect Y to change in term s of the number of standardized units of the β^W. In the case of years of post-secondary education and hourly pay, the β^W equals .759. The value of this coefficient indicates that for one standardized-unit increase in years of post-secondary education, we expect hourly pay to increase by .759 standardized units. Also, beta weights are more useful in some statistical techniques, such as structural equation modeling , which we will discuss in Chapter 14.

Earlier in this chapter, we discussed how squaring the correlation coefficient (r) equals R^2 in only bivariate regression. There is another relationship that also occurs only in bivariate regression: The standardized coefficient, or β^W equals the correlation coefficient (r). If you look back to the point where we calculated r, you will find that it has the same value as β^W: .759. In the next chapter, on multiple regression, we will explain why the beta weight for the effect that post-secondary education has on hourly pay will no longer equal .759 when an additional independent variable is added to the regression equation.

Regression Analysis in SPSS

Most statistical programs, such as SPSS, allow researchers to use computers to perform regression analysis. We will use the same data that we just used to calculate regression statistics to see how SPSS produces those results. Since we are interested in determining whether workers in the *population* receive higher earnings for each year of post-secondary education, we begin by stating the hypotheses that reflect the possible relationships between the variables:

H_0: $\beta \leq 0$: Post-secondary education does not have a positive effect on hourly pay.

H_1: $\beta > 0$: Post-secondary education does have a positive effect on hourly pay.

We begin the data analysis within SPSS by going to the pull-down menu for Analyze, Choose Regression, followed by Linear, as shown below.

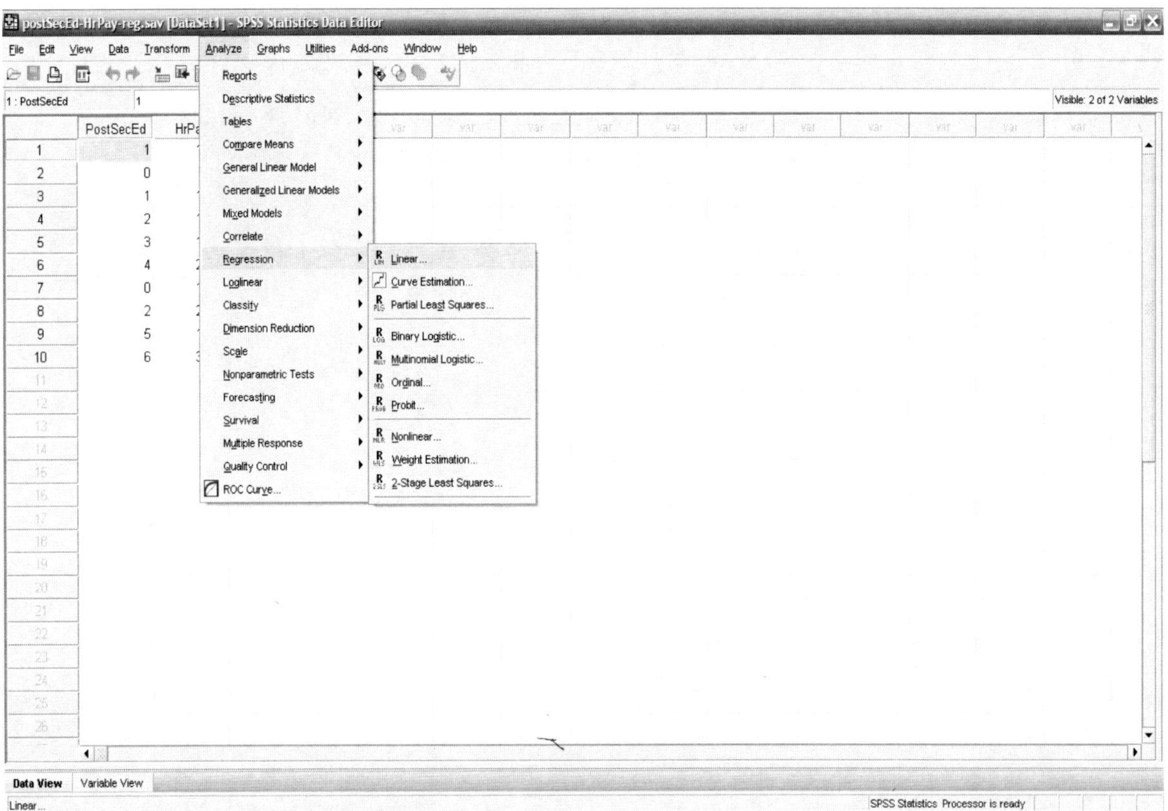

Figure 10.5 SPSS Display of Commands for Linear Regression.

The Linear Regression dialogue box should open after you click on linear. As you see in the dialogue box, you need to identify the independent and dependent variables. Since we expect years of education beyond high school to affect hourly pay, years of post-secondary education is the independent variable and hourly pay is the dependent variable. Find these variables in the left-hand box and click on one of the variables. Next, click on the arrows between the two boxes to move the variable into the appropriate right-hand boxes labeled Dependent or Independent(s). Click on the other variable and move it into the appropriate Dependent or Independent(s) box.

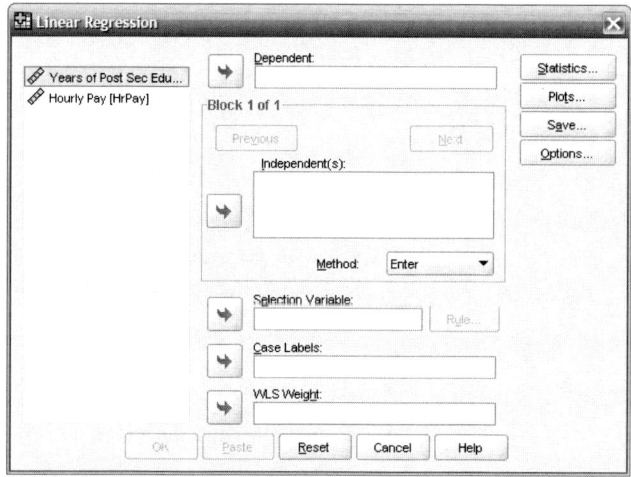

Figure 10.6 SPSS Display for Linear Regression.

Click on Statistics and the dialogue box shown in **Figure 10.7** opens. As you can see, SPSS already has Estimates and Model Fit checked; keep them selected. To obtain the means, standard deviations, and correlation of the variables in the analysis, click on the box for Descriptives. Next, click Continue to close the Linear Regression: Statistics dialogue box. Last, click OK for SPSS to perform the regression analysis.

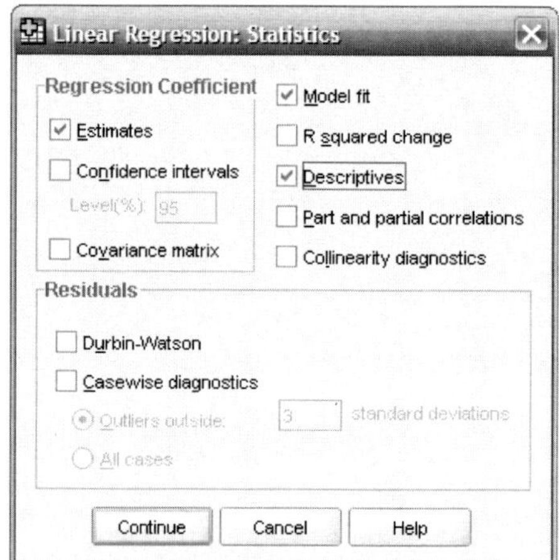

Figure 10.7 SPSS Statistics Display Box within Linear Regression.

SPSS will generate output with several statistics. In an early section of the printout, SPSS produces the Model Summary, which includes the R^2 that we calculated. **Table 10.6** shows that R^2 equals .576. Thus, 57.6% of the variance in hourly pay is explained by post-secondary education. The R in the Model Summary equals the square root of R^2, and because this is a bivariate regression, the R also equals the Pearson's r. As previously discussed, this is only true in bivariate regression.

Table 10.6 SPSS Printout of the Model Summary for the Regression of Hourly Pay on Years of Post-Secondary Education

Model Summary

Model	R	R Square	Adjusted R Square	Std. Error of the Estimate
1	.759[a]	.576	.522	5.66920

[a]*Predictors: (Constant), Years of Post-Secondary Education.*

Another element of the output contains an ANOVA analysis that includes the sum of squares that we calculated in order to compute the R^2. Other elements of ANOVA that were discussed in Chapter 9 are also included in the output.

Table 10.7 SPSS Printout of the ANOVA within the Regression of Hourly Pay on Years of Post-Secondary Education

ANOVA[b]

Model	Sum of Squares	df	Mean Square	F	Sig.
1 Regression	348.607	1	348.607	10.847	.011[a]
Residual	257.118	8	32.140		
Total	605.725	9			

[a]*Predictors: (Constant), Years of Post-Secondary Education.*
[b]*Dependent Variable: Hourly Pay.*

The last section of the printout contains some important elements of regression analysis. The *a* and the *b* are the first numbers in the output. The *a*, which is referred to as the intercept or the constant ,equals 11.719. Underneath the constant, the values for the *b*'s associated with the independent variables are printed. Since we are doing a bivariate regression, only one independent is present: years of post-secondary education. The *b* for years of post-secondary education equals 3.013.

Table 10.8 SPSS Printout of the Coefficients for the Regression of Hourly Pay on Years of Post-Secondary Education

Coefficients[a]

Model	Unstandardized Coefficients		Standardized Coefficients		
	B	Std. Error	Beta	t	Sig.
1 (Constant)	11.719	2.835		4.134	.003
Years of Post-Secondary Education	3.013	.915	.759	3.293	.011

[a]*Dependent Variable: Hourly Pay.*

We can determine if the *b* is significant by looking at the last column of the table. The exact probability at which we can reject the null hypothesis is printed in the last column of the output under "Sig." SPSS automatically performs the test as a two-tailed test, and in this case, with a $p = .011$. However, our research hypothesis indicates a one-tailed test, since we expected years of post-secondary education to have a *positive* effect on hourly pay, and when we calculated the *t*-test by hand, we rejected the null hypothesis at the .01 level with a one-tailed test. To solve this problem, we can convert the two-tailed probability produced by SPSS to a one-tailed probability. In a two-tailed test compared to a one-tailed test, the rejection area is divided into two tails. If $\alpha = .01$, as in this case is a one-tailed test, then the entire .01 is in one tail. However, if it is a two-tailed test, the .01 is divided by 2 and .005 will be in each tail. To convert the level of significance from a two-tailed to a one-tailed test, we follow a similar process; we divide the probability on the printout by 2. Therefore,

$$P_{\text{one-tailed}} = \frac{P_{\text{two-tailed}}}{2}$$

$$P_{\text{one-tailed}} = \frac{.011}{2} = .0055$$

Since the $p = .0055$ for a one-tailed test is less than .01 but greater than .001, the *b* for post-secondary education is significant at the .01 level. We can reject the null hypothesis at the .01 level and conclude that post-secondary education has a *positive* effect on hourly pay. Thus, our calculated results and SPSS results are consistent.

In this chapter, you learned the basic elements of regression analysis, in which there is only one independent variable. In the next chapter, we will discuss how the analysis changes with the inclusion of additional variables. Furthermore, we will explore how nominal variables and nonlinear effects can be studied using regression analysis.

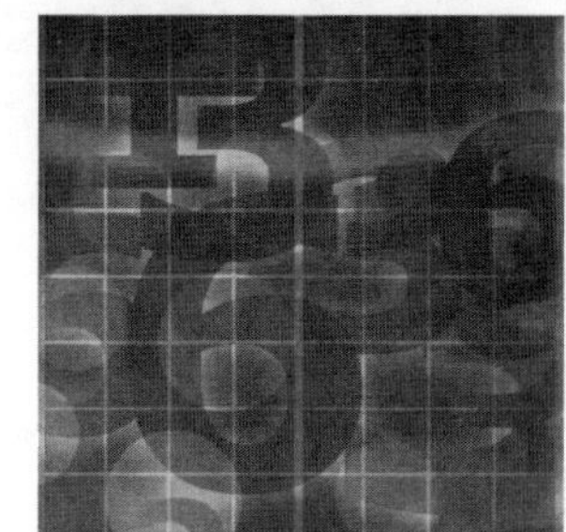

Exercises and Notes

- Regression is a statistical analysis used to assess the relationship between ___________________________.

- Regression is similar to ___________________ because it can tell us about the strength and direction of the relationship.

- However, regression also is used to describe or predict the value of the ___________________ variable on the basis of one or more ___________________ variables.

In bivariate regression, the following equation is used to represent the linear regression model:

$$Y = a + bX + e$$

$Y =$

$a =$

$b =$

$X =$

$e =$

Comparing r and b

Correlation Coefficient r	Regression Coefficient b
$r = \dfrac{}{\sqrt{}}$	$b = \dfrac{}{}$
$SP =$	$SP =$
$SS_X =$	$SS_X =$
$SS_Y =$	

2. A professor wonders if students' grades on the final exam is affected by the number of credit hours that students are enrolled in for that semester. She thinks that the more credit hours a student is taking, the lower his or her exam grades. She randomly samples 10 students to study the possible relationship between enrolled credit hours and final exam grades.

Worksheet for r and b

Hours	Grades	$(X - \bar{X})$	$(Y - \bar{Y})$	$(X - \bar{X})(Y - \bar{Y})$	$(X - \bar{X})^2$	$(Y - \bar{Y})^2$
12	97					
9	76					
15	93					
12	62					
12	83					
18	73					
15	57					
12	84					
16	61					
15	80					

Calculating SP, SS_X, and SS_Y

$SP = \Sigma(X - \bar{X})(Y - \bar{Y})$

$SP =$

$SS_X = \Sigma(X - \bar{X})^2$

$SS_X =$

$SS_Y = \Sigma(Y - \bar{Y})^2$

$SS_Y =$

Calculating r and b

Correlation Coefficient r	Regression Coefficient b
$r = \dfrac{}{\sqrt{}}$	$b = \dfrac{}{}$
$r =$	$b =$

Interpretation of b: ____

____

____

____

____

Calculating *a*

$a = \overline{Y} - b\overline{X}$

$a =$

Interpretation of a: ____

____

____

____

____

____

____

____

Write the predictive equation for exam grades.

____

Predicting Exam Grades

- What grade would you predict for a student who is enrolled for 12 credit hours?

____

____

____

- What grade would you predict for a student who is enrolled for 14 credit hours?

____

____

____

Scatter Plot of Number of Enrolled Credit Hours and Exam Grade

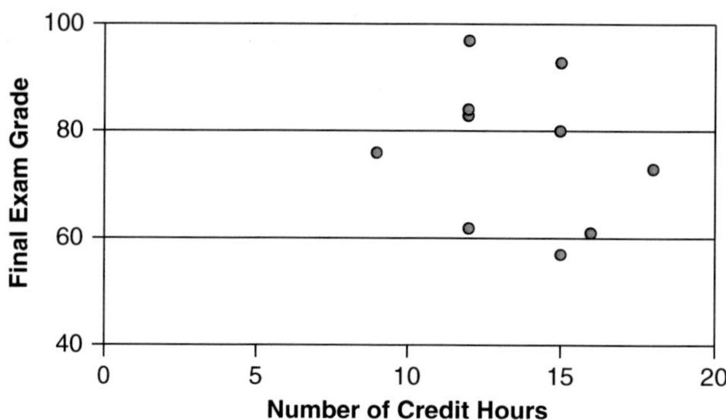

Drawing the Regression Line and Prediction Errors

_______________________ *needed to draw a straight line.*

   1.

   2.

The Difference between Y and $\hat{Y}$

$Y =$

$\hat{Y} =$

The error term, or residual, is the amount that _______________________.

The Error, or Residual, Term

The error term is the portion of the value of the _______________ that cannot be predicted by the _______________.

Since $Y = a + bX + e$ $\hat{Y} = a + bX$

Then $e =$

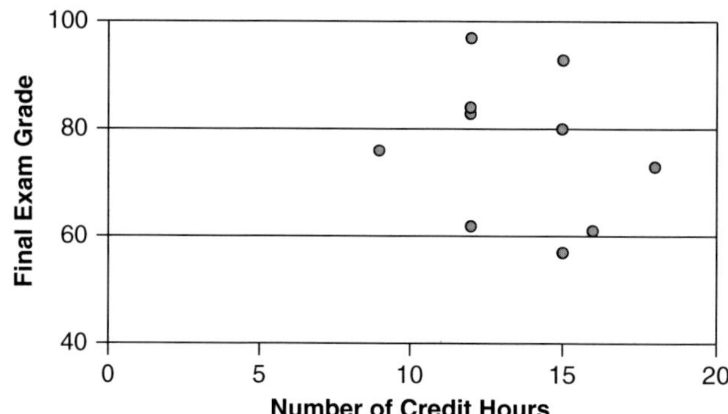

Benefit of Knowing X in Predicting Y

- *Without* knowing the number of credit hours (X), our best estimate of any student's exam grade would be

  _______________________.

- Therefore, the error of *not knowing X* is _______________________.

- The sum of the square of the deviation *without* knowing X is called the _______________________________

  ____.

$SS_{\_} =$

- The error in predicting Y *with* knowledge of X is equal to _____________________.

- The sum of the squares *with* knowledge of X is called the _____________________.

- The sum of squares *explained by* X is called the _____________________.

- The _____________________ is the difference between predicting Y *without* knowledge of X compared to predicting Y *with* knowledge of X.

$SS_{\_} =$

Worksheet for Total and Error Sum of Squares

Hours	Grade	$Y - \bar{Y}$	$(Y - \bar{Y})^2$	$\hat{Y} = a + bX$	$(Y - \hat{Y})$	$(Y - \hat{Y})^2$
12	97					
9	76					
15	93					
12	62					
12	83					
18	73					
15	57					
12	84					
16	61					
15	80					
$\Sigma X =$	$\Sigma Y =$					
$\bar{X} =$	$\bar{Y} =$					

Benefit of Knowing X in Predicting Y

- _____________________ (also known as the coefficient of determination) is a proportional reduction of error (PRE) measure, a measures that compares _____________________ to _____________.

- To calculate _____________________:

Interpretation of _____________________:

____

____

____

Testing Hypotheses about Regression Slopes

Our analysis of the sample data suggests that for every additional ____

____.

However, the question remains as to whether students in the population receive _______________________ *, that is, is the slope in the population (β)_______________________?*

- State the hypotheses.

- Determine the significance level.

- Calculate the test statistic.

- Make a decision regarding the null hypothesis.

- Interpret the results.

____

____

____

____

Standardized Equation

- Standardization rescales the independent and dependent variables in terms of their _______________________.

- One way to standardize involves converting raw scores to _______________________. The regression can then be estimated.

- The unstandardized regression coefficient (b) can be converted to the standardized regression coefficient (β^W) by multiplying _______________________ by _______________________ divided by _______________________:

$\beta^W =$

$S_X =$ $S_Y =$

$\beta^W =$

- In the unstandardized equation:

  $a =$

- In the standardized equation, _______________________ drops out of the equation because:

  - The mean of a standardized variable = _______________________.

  - Therefore, the standardized predictive regression equation for two variables is :

Regression in SPSS

To run a regression, go to Analyze, then Regression, and then Linear.

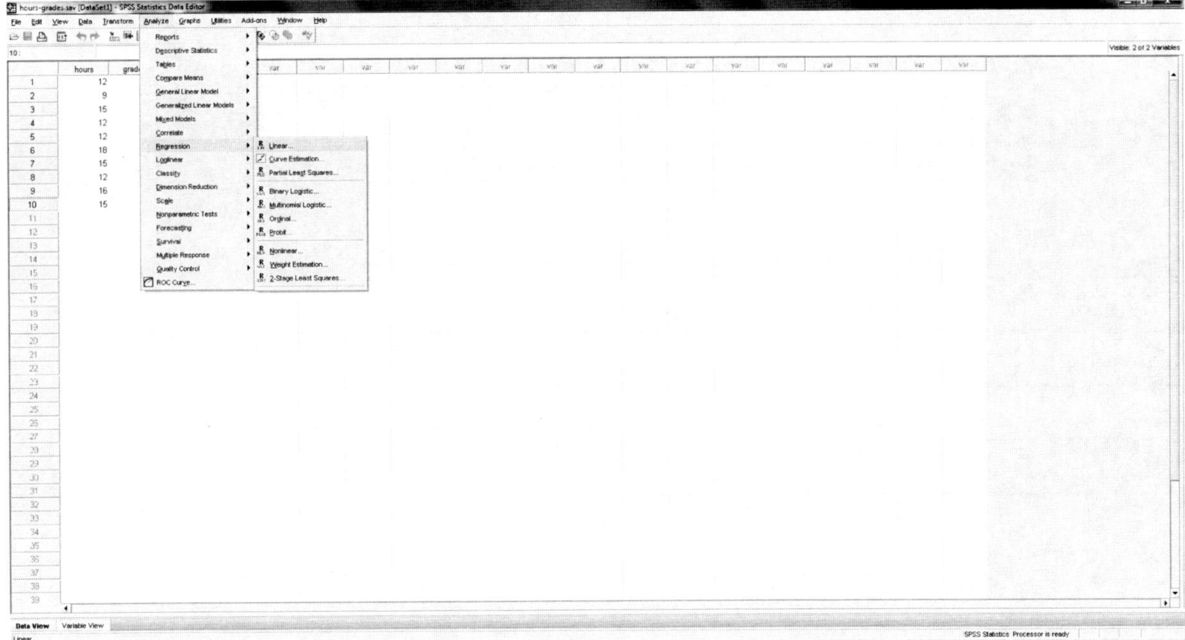

Regression Dialogue Box

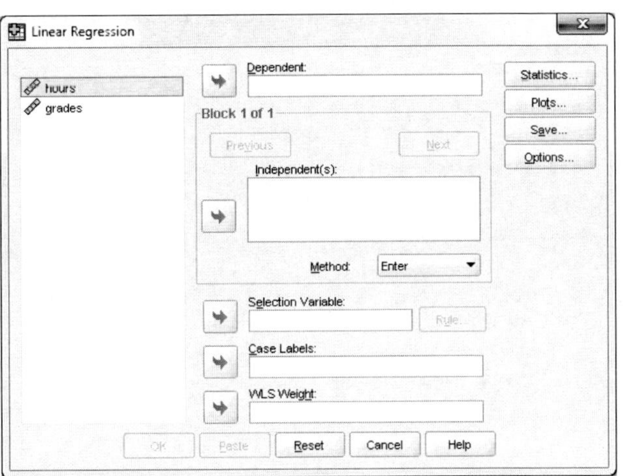

Statistics Dialogue Box within Linear Regression

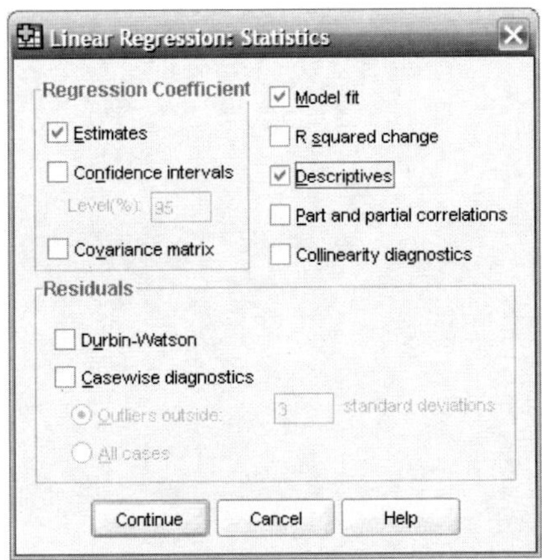

Regression Printout

Model Summary

Model	R	R Square	Adjusted R Square	Std. Error of the Estimate
1	.255[a]	.065	−.052	13.873

[a]*Predictors: (Constant), Hours.*

ANOVA[b]

Model	Sum of Squares	df	Mean Square	F	Sig.
1 Regression	106.708	1	106.708	.554	.478[a]
Residual	1539.692	8	192.462		
Total	1646.400	9			

[a]*Predictors: (Constant), Hours.*
[b]*Dependent Variable: Grades.*

Coefficients[a]

Model	Unstandardized Coefficients		Standardized Coefficients	t	Sig.
	B	Std. Error	Beta		
1 (Constant)	94.385	24.284		3.887	.005
Hours	−1.308	1.756	−.255	−.745	.478

[a]*Dependent Variable: Grades.*

Name: ____   Date: _________________

Learning Check

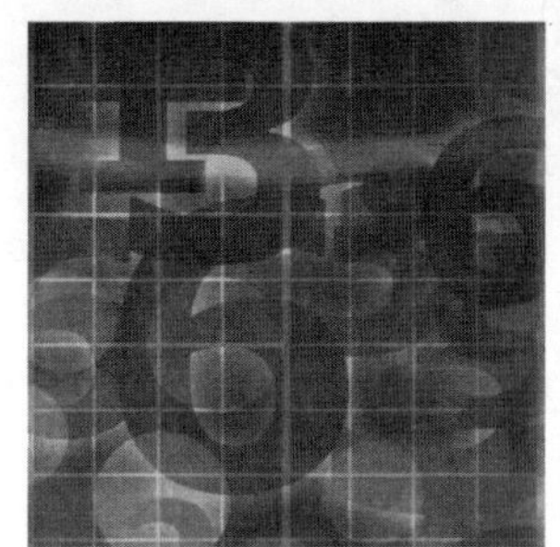

1. A college admission officer wants to determine if the ACT college entrance exam is a good predictor of students' college GPA. She believes that students with higher ACT scores do better in college. Below are the data on the ACT scores and college GPA of 10 college students.

ACT Scores	College GPA
12	2.13
31	3.86
23	2.64
14	1.8
29	3.21
26	3.91
18	2.75
20	3.11
15	2.92
22	1.97
$\Sigma X = 210$	$\Sigma Y = 28.3$
Average = 21.00	Average = 2.83

   a. What are the independent variable and the dependent variable?

   b. State the hypotheses.

   c. Calculate a and b. Interpret a and b.

   d. What GPA would you expect of a student who scored a 25 on the ACT?

   e. How much variance in the dependent variable is explained by the independent variable?

   f. Do a statistical test on the null hypothesis from b.

   g. What is the standardized regression coefficient?

2. The manager of a stadium concession stand wants to use data from his previous two years of home games to help him predict how much coffee he will sell at the different home football games. In the previous years, sales fluctuated with the weather; the manager believes cold weather increases his coffee sales. He wants you to perform a regression using the following data:

Temperature	Coffee Sales
78	32
83	26
74	74
67	92
54	247
51	253
80	27
73	48
48	101
59	104
53	281
56	311

a. What are the independent variable and the dependent variables?

b. State the hypotheses.

c. Calculate a and b. Interpret a and b.

d. How much coffee would you expect to be sold at the concession stand if the temperature was 55 degrees?

e. How much variance in the dependent variable is explained by the independent variable?

f. Do a statistical test on the null hypothesis from b.

g. What is the standardized regression coefficient?

3. A researcher is interested in determining the effect of household size on monthly grocery bills. She expects that, as household size increases, the monthly grocery bill will increase. Given $SP = 6615$, $SS_X = 145.87$, $SS_Y = 1,036,750$, $SS_{error} = 736,762$, mean household size = 3.93, mean grocery bill = 445, and $N = 30$, answer the following questions.

a. What are the independent variable and the dependent variable?

b. State the hypotheses.

c. Calculate a and b. Interpret a and b.

d. What would you expect the monthly grocery bill to be if a household had four people?

e. How much variance in the dependent variable is explained by the independent variable?

f. Do a statistical test on the null hypothesis from b.

g. What is the standardized regression coefficient?

4. A consumer testing agency wants to determine whether a car's mileage per gallon of gas contributes to the car's price. Given $SP = -116200$, $SS_X = 506.1$, $SS_Y = 72{,}400{,}000$, $SS_{error} = 45{,}720{,}000$, mean mpg $= 28.7$, mean price $= 22{,}600$, and $N = 10$, answer the following questions.

    a. What are the independent variable and the dependent variable?

    b. State the hypotheses.

    c. Calculate a and b. Interpret a and b.

    d. What would you expect a car to cost if it gets 31 miles per gallon?

    e. How much variance in the dependent variable is explained by the independent variable?

    f. Do a statistical test on the null hypothesis from b.

    g. What is the standardized regression coefficient?

5. Does the number of siblings you have affect the number of children you want to have? Formulate a hypothesis based on this question. Before doing the statistical analysis, be sure to define the response "As Many As Want," as missing. Regress Ideal Number of Children (CHLDIDEL) on Number of Brother and Sisters (SIBS). Interpret the a, b, and R^2. Does the number of siblings significantly affect people's preferences regarding ideal family size? Be sure to indicate which significance level you use.

6. Are younger people's religious beliefs weaker than older people's beliefs? Formulate a hypothesis based on this question. Before doing the statistical analysis, be sure to check to see if responses such as Inappropriate (IAP), Don't Know (DK), and No Answer (NA) are defined as missing. Regress Strength of Affiliation (RELITEN) on Age of Respondent (AGE). Interpret the a, b, and R^2. Does age positively significantly affect people's religious beliefs? Be sure to indicate which significance level you use.

Chapter Eleven

Multiple Regression

The previous chapter on bivariate regression introduced regression analysis. You learned about the unstandardized and standardized equations, how to test the regression coefficient for significance, and how much variance in the dependent variable is explained by the independent variable. However, our examination of regression analysis was limited. In this chapter, we expand our exploration to regression analysis to include more than one independent variable and delve into some of the unique issues raised by some variables. For example, we will explain how nominal-level variables can be transformed into dummy variables, so that they can be included in a regression analysis. In addition, we will show how independent variables may interact with each other as they affect the dependent variable. We will also examine how an independent variable may have a nonlinear rather than a linear effect on the dependent variable.

Multiple Regression Model

Researchers typically want to look at how multiple factors may affect their dependent variable. On the other hand, researchers do not just include every variable available in a data set as a possible predictor of the dependent variable for several reasons. First, we have to have a theoretical reason for every variable included in the regression model. The researcher formulates a separate research hypothesis for every variable included in the regression analysis. In addition, researchers often adhere to the principle of **parsimony,** which contends that the simplest explanation that can explain the data is the most preferred. In regression analysis, there are practical reasons for adopting the principle of parsimony, especially if the N is small. Each additional independent variable reduces the degrees of freedom, which in turn reduces the likelihood that the null hypothesis will be rejected. Therefore, we must thoughtfully consider the inclusion of additional variables into regression analyses.

Table 11.1 shows how a regression equation changes with the addition of a second independent variable. Therefore, the three-variable equation has two slopes (b's) and two independent variables (X's). In order to distinguish between the different b's and X's in the equation, numbered subscripts are used.

Table 11.1 *Bivariate and Three-Variable Regression Equations*

Type of Equation	Bivariate Equations	Three Variable Equations
Sample regression equation	$Y = a + bX + e$	$Y = a + b_1X_1 + b_2X_2 + e$
Predictive equation	$\hat{Y} = a + bX$	$\hat{Y} = a + b_1X + b_2X_2$
Population predictive equation	$\hat{Y} = \alpha + \beta X$	$\hat{Y} = \alpha + \beta_1X + \beta_2X_2$
Standardized predictive equation	$\hat{Z}_Y = \beta^w Z_X$	$\hat{Z}_Y = \beta_1^w Z_1 + \beta_2^w Z_2$

Suppose you want to know the effect that the number of siblings has on a workers' income. You expect that the fewer siblings a person has, the more advantages his or her parents could have provided that would pay off in higher earnings. However, you also expect education to increase workers' income, regardless of the number of siblings they have. Therefore, you decide to do two regression analyses: a bivariate regression of income on the number of siblings and a multiple regression of income on the number of siblings and education. In the multiple regression, you want to determine the effect that the number of siblings have on workers' income, controlling for their education and the effect of education on workers' income, controlling for the number of siblings. Thus, you would have a bivariate regression and a multiple regression with the following research hypotheses.

In the bivariate regression model:

H_1:   As the number of siblings a worker has decreases, the worker's income increases.

In the multiple regression model:

H_2:   As the number of siblings a worker has decreases, the worker's income will increases, controlling for education.

H_3:   As the number of years of education completed increases, a worker's income will increase, controlling for the number of siblings.

You may have noticed that we did not state the null hypothesis associated with research hypotheses. In the social sciences, researchers often omit the null hypothesis, assuming that the reader understands that there is the underlying null hypothesis. Indeed, researchers may not explicitly state the hypotheses as we just did.

You learned in the previous chapter that the quotient of the sum of the products (*SP*) divided by the sum of squares of X (SS_X) gives us the slope, or *b*. To calculate the *b*'s in a three-variable regression, we use the correlations between the variables and the standard deviations of the variables. The formulas to calculate the two *b*'s are:

$$b_1 = \left(\frac{S_Y}{S_{X_1}}\right) \frac{r_{YX_1} - r_{YX_2} r_{X_1 X_2}}{1 - r^2_{X_1 X_2}} \qquad b_2 = \left(\frac{S_Y}{S_{X_2}}\right) \frac{r_{YX_2} - r_{YX_1} r_{X_1 X_2}}{1 - r^2_{X_1 X_2}}$$

We can use the data on number of siblings, education, and income to calculate the regression coefficients and standard deviation found in **Table 11.2.** First, we need to determine the dependent variable (*Y*). Since we expect the number of siblings and education to have an effect on income, income is the dependent variable. Because the number of siblings is the independent variable in Hypothesis 2 (H_2), the number of siblings will be X_1 and education will be X_2. Therefore, to calculate b_1 and b_2:

$$b_1 = \left(\frac{35262.419}{2.876}\right) \frac{-.153 - (.3923 - .328)}{1 - (-.328)^2} = -335.562$$

$$b_2 = \left(\frac{35262.419}{2.881}\right) \frac{.392 - (-.1533 - .328)}{1 - (-.328)^2} = 4688.068$$

Table 11.2 *Correlation and Standard Deviations of Number of Siblings, Education, and Income (N = 1189)*

	Siblings	Education	Income	Means	Standard Deviations
# of Siblings	1.000			3.47	2.876
Education	−.328***	1.000		13.830	2.881
Income	−.153***	.392***	1.000	40804.46	35262.419

**p < .05 (one-tailed test).*

***p < .01 (one-tailed test).*

****p < .001 (one-tailed test).*

The *b*'s tell the effect the independent variables have on the dependent variable. Therefore, the *b* for the number of siblings indicates that for each additional brother or sister, respondents' income decreased by $335.56, when controlling for years of education. The *b* for education suggests that for each additional year of education, respondents' income increased by $4688.07, when controlling for the number of siblings.

The equation to calculate *a* in a multiple regression is very similar to the equation for *a* in the bivariate case. We simply need to include the additional independent variables and their corresponding *b*'s. Therefore, the calculation of the *a* in the regression of income on the number of siblings and education is:

$$a = \bar{Y} - (b_1\bar{X}_1 + b_2\bar{X}_2)$$

$$a = 40804.46 - [(-335.56 \times 3.47) + (4688.07 \times 13.83)]$$

$$a = -22867.12$$

The unstandardized regression coefficients (the *b*'s) can be standardized by multiplying each by the quotient of the standard deviation of the independent variable divided by the standard deviation of the dependent variable:

$$\beta_j^W = \left(\frac{S_{X_j}}{S_Y}\right)b_j$$

Therefore, the beta weight for the number of siblings is:

$$\beta_1^W = \left(\frac{2.876}{35262.419}\right) - 335.56 = -.027$$

The beta weight for education is:

$$\beta_2^W = \left(\frac{2.881}{35262.419}\right) 4688.07 = .383$$

We have covered a few of the elements of multiple regression. As you may have surmised, the more independent variables that are added to the regression equation, the calculations of the coefficients become more complex. Therefore, computers are typically used. Let's turn our attention to how SPSS produces multiple regression results.

Multiple Regression in SPSS

Instead of obtaining all the regression statistics by hand calculations, we can use SPSS to analyze the data. Not only will SPSS do all the calculations, but also, it will compute the calculations using more decimal places and therefore produce statistics with less rounding error than hand calculations often produce. Whenever we are analyzing data, we need to make sure that the variables are coded appropriately for the analysis. This includes checking to make certain that missing values have been defined. In previous chapters, we discovered how variables are coded by going to the Variable View. You can switch from the Data View to the Variable View by clicking the tab in the lower left-hand corner, as shown in the data file display below. Once your are in Variable View, look for the variables that we want to use in the regression analysis. Education is one of the variables we want to use. If you click on the Values column for the education variable (EDUC—Highest Year of School Completed), the Value Labels box opens and you can see how the variable is coded. As you can see in **Figure 11.1,** the labels assigned correspond to years of schooling completed. Therefore, this variable can be used as a ratio level variable.

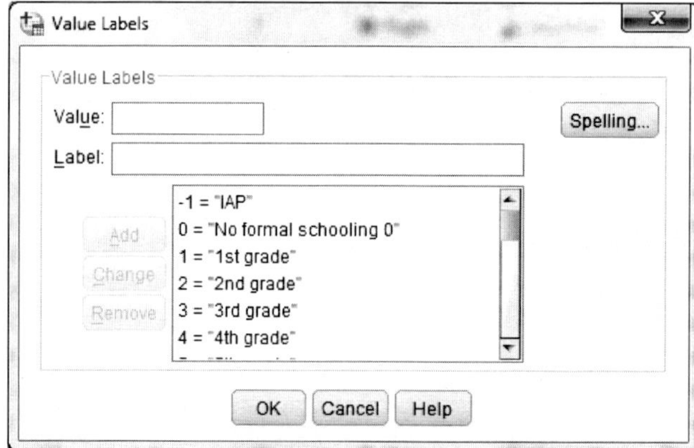

Figure 11.1 Value Labels Dialogue Box for Highest
Year of School Completed (EDUC).

However, you should also check to make sure that those coded as Inappropriate (IAP), Don't Know (DK), or No Answer (NA) are not included in the analysis. It appears that there is an anomaly in the GSS regarding those viewed as being inappropriate (IAP) for the variable Highest Year of School Completed (EDUC). It is appropriate to ask every respondent what is the highest year of school that he or she completed, and yet there are two inappropriate codes (-1 and 97) in the data set, with no cases being coded as inappropriate. Therefore, when using any secondary data, we should never assume that those who were coding the data did everything as we would do it. If we click on the Missing column for the education variable (EDUC—Highest Year of School Completed), we find that 97 and 99 have been defined as missing. However, 98 which is the code used for people who responded "Don't Know," is not defined as missing. If we don't define these data as missing, then the people who responded "Don't Know" would be included in the analysis and would be treated as if they had 98 years of education. You can see how that could be problematic. Therefore, we want to define 98 as missing. If you click on the Missing column, the Missing Values box opens and you can type in the code 98 to define the response "Don't Know" as missing.

The other variables that we are including in our multiple regression are the Number of Brothers and Sisters (SIBS) and Respondent's Income (RINCOM06). Check to see if any missing codes (IAP, DK, and NA) need to be defined as missing and follow the procedure that we used for the Highest Year of School Completed (EDUC) for the Number of Brothers and Sisters (SIBS), if necessary. The Respondent's Income (RINCOM06) and another variable that we will use in future analyses, Total Family Income (INCOME06), are coded as categories of income, not as a ratio level measure. In addition, the categories are not uniform in size, with some low-level categories having income ranges of $1000 (*e.g.*, $3000 to $3999), while some upper-level categories have ranges of $20,000 ($130,000 to $149,999). Therefore, we need to recode the data. To best represent this grouped data as ratio data for regression analyses, we can substitute the midpoint of the class interval. You may remember that to calculate a midpoint, we add the lowest and highest values and divide by 2:

$$m = \frac{lowest\ value\ +\ highest\ value}{2}$$

Table 11.3 is an abbreviated worksheet that we would create to recode Respondent's Income (RINCOM06) and Total Family Income (INCOME06); since both variables are coded the same, if you do them together, you only have to enter the recodes once. The first column has the value label of the category, while the next columns have the old value and the recoded new value, which represents the midpoints of the category of the label. Let's deal with the old values that represent dollar amounts first, then we will address missing values (IAP), even though the first value in the worksheet is a missing value. The first monetary category contains those who made under $1000. Therefore, the lowest possible amount earned would be $1 and the highest would be $999. Therefore, the midpoint would be 500 ($1 + 999 = 1000, \div 2 = 500$). The worksheet shows some of the additional categories' midpoints rounded to the nearest dollar in the New Value column. Now, look at the last monetary category, which represents those who made $150,000 or more. You may be wondering what the highest value is. Unfortunately, we don't know. In the previous category, the size of the interval was 20,000. Since the highest values are more than $170,000, we will make a judgment and use $180,000 as the highest value and calculate a midpoint of $165,000. Remember that **Table 11.3** is an abbreviated worksheet; be sure to calculate the midpoints for the interval associated with the old values between 6 and 23.

Table 11.3 *Abbreviated Recoding Worksheet for Respondents' Income (RINCOM06) and Total Family Income (INCOME06)*

Value Label	Old Value	New Value
IAP	0	
UNDER $1000	1	500
$1000 TO 2999	2	2000
$3000 TO 3999	3	3500
$4000 TO 4999	4	4500
$ 000 TO 5999	5	5500
. . .	. . .	. . .
$130,000 TO 149,999	24	140,000
$150,000 OR OVER	25	165,000
REFUSED	26	
DK	98	
NA	99	

Once you calculate all the midpoints, you can recode the Respondent's Income (RINCOM06) and Total Family Income (INCOME06) by going to Transform ⇒ Recode into Different Variables, as shown in **Figure 11.2.** SPSS then opens the Recode into the Different Variables box (see **Figure 11.3**). In the left-hand box, look for Respondent's Income (RINCOM06) and click it into the middle box, Numeric Variable → Output Variable. Then, in the boxes labeled Name and Label, under the heading Output Variable, type a name (*e.g.,* R$INC) and a label for a new variable (Respondent's Income in Dollars), respectively. Then, click Change. Now, go back to the left-hand box and click Total Family Income (INCOME06) into the middle box, Numeric Variable → Output Variable. Next, in the boxes labeled Name and Label under the heading Output Variable, type a name (*e.g.,* F$INC) and a label for a new variable (Total Family Income in Dollars), followed by clicking Change.

Click on Old and New Values, and the box displayed in **Figure 11.4** opens. **Figure 11.4** shows the first three recodes being entered. First, 1 was typed in the Old Value box; 500 was typed in the New Value box, and Add was clicked. The 1 and 500 moved down from the Old Value and New Value boxes into the Old → New box. Then, the process was repeated for the Old Value of 2 and the New Value of 2000. **Figure 11.4** shows 3 and 3500 entered into the Old Value and New Value boxes. The next step would be to click Add. The process would continue until all the codes are entered. If we wanted to continue to differentiate the various missing codes, we would have to enter old and new variables for each of the missing values for Inappropriate, Refused, Don't Know, and No Answer, and then go to the Missing column under Variable View to define those codes as missing. However, if we just want those cases treated as missing and do not care to differentiate those who, for example, responded Don't Know from those with No Answer, we would not enter the codes associated with Inappropriate, Refused, Don't Know, and No Answer into our recode. SPSS will define any cases not recoded with old and new values as system missing. However, it is important to check your work; if you miss entering a needed recode, SPSS will include it in the system missing category.

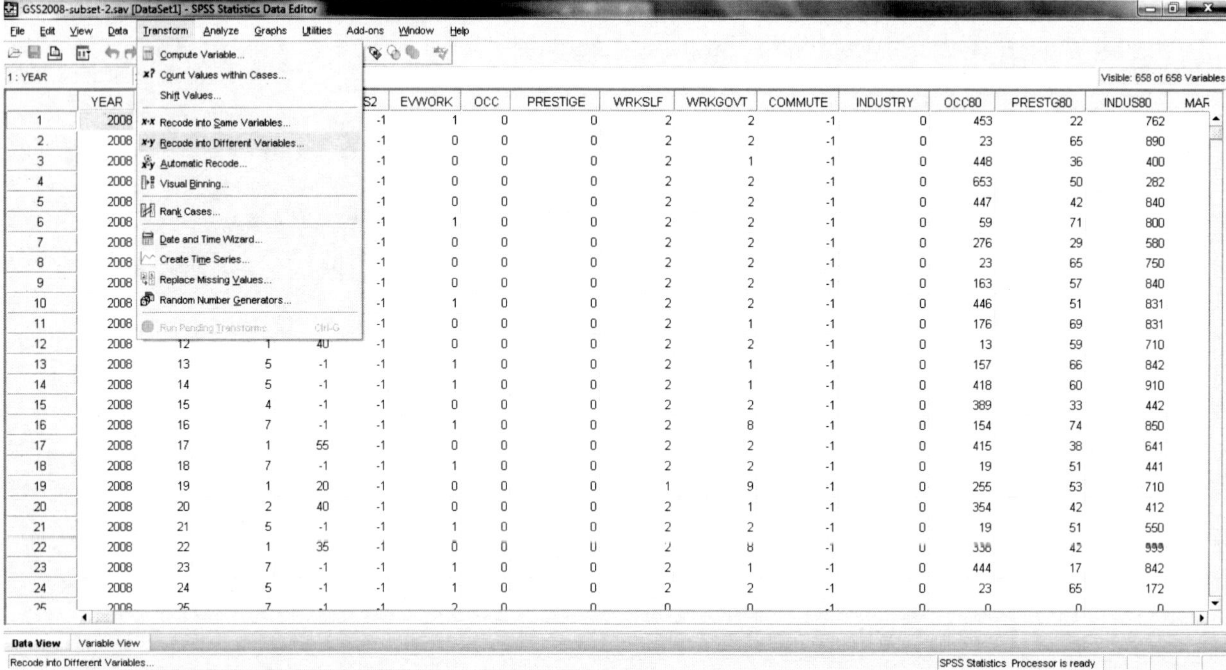

Figure 11.2 SPSS Display of Commands for a Recode into Different Variables.

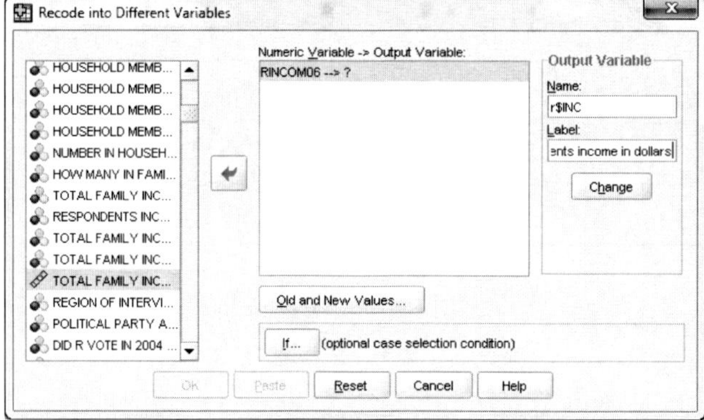

Figure 11.3 SPSS Display for a Recode into Different Variables.

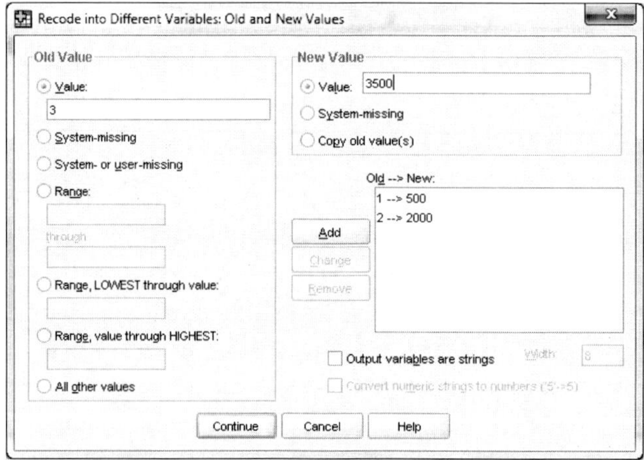

Figure 11.4 SPSS Display for Old and New Values when Recoding into Different Variables.

If you click Continue, the Old and New Values box closes but the Recode into Different Variables box remains open. Click OK, and that box closes and the recodes are executed. It is important to run frequencies on your newly recoded variables, Respondent's Income in Dollars (R$INC) and Total Family Income in Dollars (F$INC). Check to make sure that you did not skip a category. Make certain that you did not make any typos while entering the recodes by checking to see that all the values of the variables match the new value midpoints that you calculated and that frequencies correspond to frequencies prior to the recoding (*e.g.,* the frequencies for 500 was 22 for Respondent's Income in Dollars (R$INC) and 19 for Total Family Income in Dollars (F$INC). If you find that, for example, you typed 350 instead of 3500, you do not need to redo the entire recode. If you go to Transform ⇒ Into the Same Variable and select Respondent's Income in Dollars (R$INC) and Total Family Income in Dollars (F$INC), you can enter 350 as an old value and 3500 as a new value to correct that error.

After all the recoding is done, you can run the regressions. To regress respondents' income on the number of siblings and education, go to Analyze ⇒ Regression ⇒ Linear, and an SPSS Linear Regression box opens. **Figure 11.5** shows that the Respondent's Income in Dollars (R$INC) has been entered in the box marked Dependent and the Number of Brothers and Sisters (SIBS) has been entered in the box labeled Independent(s). If you enter the Highest Year of School Completed (EDUC) in the Independent(s) box, you would get one regression model with the Respondent's Income in Dollars (R$INC) regressed on both Number of Brothers and Sisters (SIBS) and Highest Year of School Completed (EDUC). However, we want two models, one bivariate and one multiple regression. Therefore, click on the Next button; in **Figure 11.5,** an arrow is pointing at the Next button.

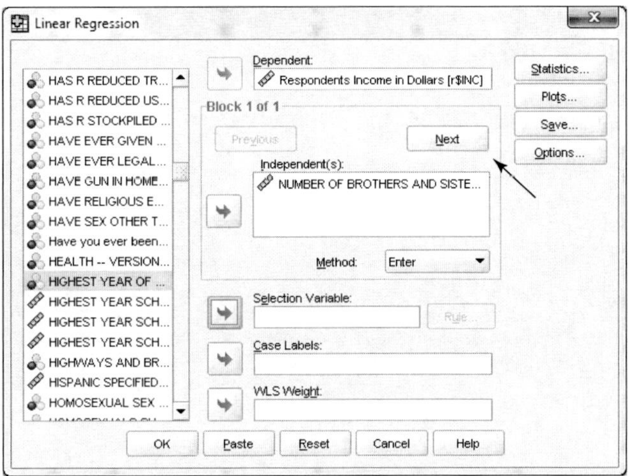

Figure 11.5 SPSS Display for Linear Regression.

After you click Next, the Linear Regression box changes. The Respondent's Income in Dollars (R$INC) remains in the Dependent box, but the Independent(s) box is now empty. However, the Number of Brothers and Sisters (SIBS) is still in the regression. In **Figure 11.6,** the solid arrow is pointing out that the box is showing Block 2 of 2. If you click on Previous, located just below the Block 2 of 2, SPSS will return to Block 1 of 1, which is shown in **Figure 11.5.** While you are in Block 2 of 2, click Highest Years of School Completed (EDUC) in the Independent(s) box.

Next, click on Statistics; in **Figure 11.6,** the arrow with the dotted line is pointing at the Statistics button. SPSS has preselected Estimates and Model Fit, as shown in **Figure 11.7.** Keep this selection, but also click on Descriptives, which provides statistics, such as the means, standard deviations, and correlations. Click Continue, and SPSS will take you back to the SPSS Linear Regression box, where you will click OK.

As you may recall from bivariate regression, SPSS produces a regression printout with numerous statistics. **Table 11.4** shows the R^2 for Models 1 and 2. Model 1 contains only one independent variable: the Number of Brothers and Sisters. Therefore, the R^2 for Model 1 indicates that only a little more than 2 percent (2.3%) of the variance in the respondents' income is explained by the number of siblings they have. In model 2, education

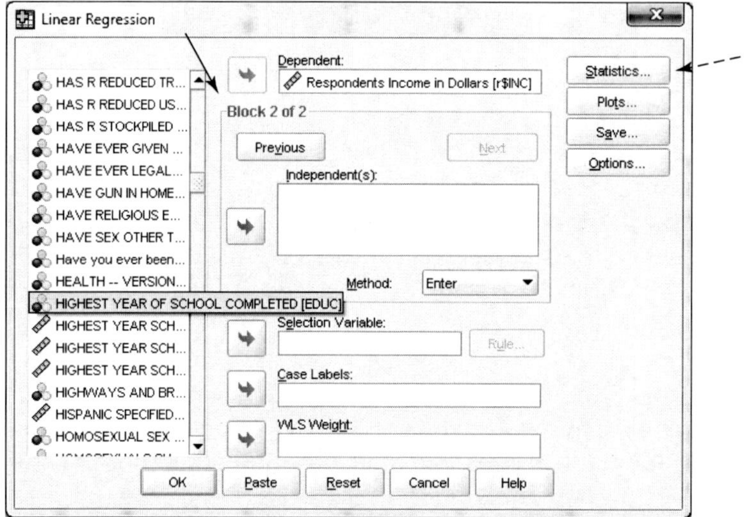

Figure 11.6 SPSS Display of Block 2 of a Linear Regression.

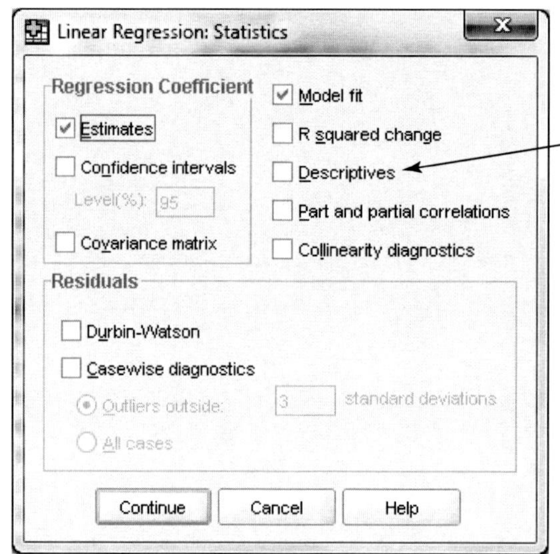

Figure 11.7 SPSS Statistics Display Box
within Linear Regression.

Table 11.4 *SPSS Model Summary of the Regression of Respondents' Income in Dollars (R$INC) on the Number of Brothers and Sisters (SIBS) and Highest Years of School Completed (EDUC)*

Model Summary

Model	R	R Square	Adjusted R Square	Std. Error of the Estimate
1	.153[a]	.023	.022	34864.433
2	.393[b]	.155	.153	32448.180

[a]*Predictors: (Constant), Number of Brothers and Sisters.*
[b]*Predictors: (Constant), Number of Brothers and Sisters, Highest Year of School Completed.*

was added to the analysis. The number of siblings and the highest years of school completed together explain nearly 16 percent (15.5%) of the variance in the respondents' income.

    Table 11.5 provides the statistics that indicate whether the independent variables have an effect on the dependent variable. Model 1 indicates that the Number of Brothers and Sisters has a significant negative effect on the Respondents' Income ($p < .001$). We conclude that the Number of Brothers and Sisters has a significant effect on income because the probability printed in "Sig." is less than .001; see the solid arrow in **Table 11.5.** The effect is negative because the b is negative ($b = -1870.132$; see the dotted arrow in **Table 11.5.** The b means that for each additional sibling the respondent has, his or her income is expected to decrease by $1870.13. The constant (or a) indicates that if the respondent is an only child (i.e., zero brothers and sisters), his or her expected income would be $47,286.22.

Table 11.5   *Coefficients of the Regression of Respondents Income in Dollars (R$INC) on the Number of Brothers and Sisters (SIBS) and Highest Years of School Completed (EDUC)*

Coefficients[a]

Model	Unstandardized Coefficients		Standardized Coefficients		
	B	Std. Error	Beta	t	Sig.
1 (Constant)	47286.220	1583.688		29.858	.000
Number of Brothers and Sisters	−1870.132	351.684	−.153	−5.318	.000
2 (Constant)	−23012.480	5383.117		−4.275	.000
Number of Brothers and Sisters	−326.536	346.492	−.027	−.942	.346
Highest Year of School Completed	4696.752	345.909	.384	13.578	.000

[a]*Dependent Variable: Respondents Income in Dollars.*

Model 2 shows how important multivariate analysis can be. After controlling for the effects of the Highest Year of School Completed, the Number of Brothers and Sisters does not have a significant effect on Respondents' Income ($p = .346$, therefore $p > .05$). The Highest Year of School Completed has a significant positive effect on Respondents' Income, when controlling for the Number of Brothers and Sisters ($p < .001$). For each additional year of school completed, the respondents' income increases by $4696.75, when controlling for number of siblings. We do not interpret the a, as this would suggest that people could have a negative income. In conclusion, this multiple regression example illustrates the importance of including all pertinent variables in our regression model. Not including education in the analysis led us to an erroneous conclusion that the number of siblings affects a person's income.

Dummy Variables and Interaction Effects

Remember that social scientists argue that social outcomes are affected by a wide range of social characteristics, such as education, income, sex , and race. We also argue that these social characteristics interact. For example, would we expect that poor women would have the same social outcomes as wealthy men? We could take this one step further and argue that poor black women have lower outcomes than wealthy white men. This example argues that there is an **interaction** effect between race, sex , and income. An interaction effect is the combined effect of two or more variables on the dependent variable.

    Testing the interaction effect in multiple regressions is relatively simple. All you need to do is create a variable that multiples X_1 by X_2. There are two ways to do this in SPSS. You can open a syntax window and type: Compute [newvarname] = [var1 name] * [var2 name]. Execute. Or, you can go to the "Transform" menu in the SPSS window and select Compute Variable. Once there, you will need to name your new variable and then move the indicator variables to the window and add the mathematical function (*).

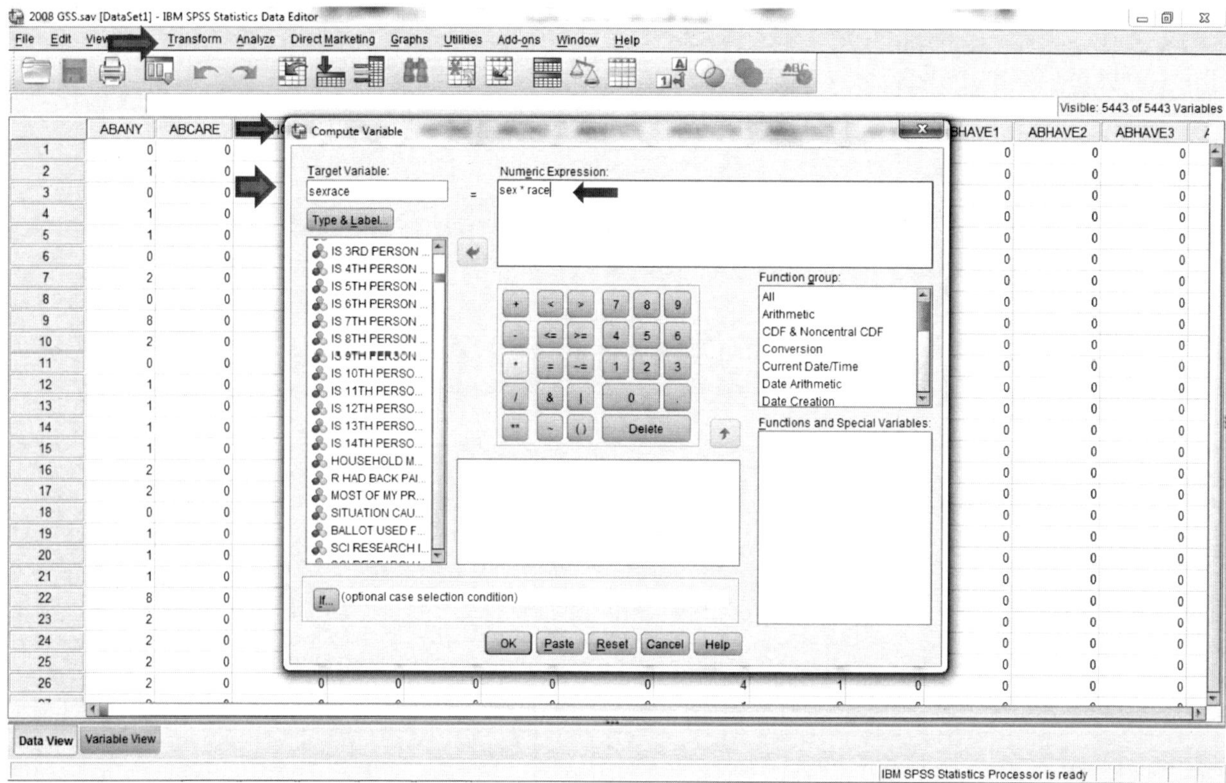

Figure 11.8 SPSS Display for Compute Variable.

This only explains the mechanics of creating the interaction variable and not the theoretical understanding of the data. For example, we have already discussed the use of nominal variables in data. RACE1 and SEX are nominal variables. We assign mathematical values to the data as identifiers in order to use them in our analyses. For them to have mathematical meaning, we have to create variables that convert nominal data into ordinal variables whereby the numbers do more than simply count the data. In order to do this with nominal data, we have to create dummy variables. (See previous chapters for a discussion of recoding variables in SPSS.) Dummy variables allow us to measure the presence or absence of a trait.

Let's use the hypothesis that race and sex interact to affect educational attainment, when all else is equal. Let's also test the theory that social awareness and change have increased educational attainment for African American women, but not for Latina women. In order to test this, we need to be able to examine the interaction of sex and race. This requires us to create an interaction variable that measures the presence of being female AND the presence of African American ethnicity. This means that we need to create two dummy variables to measure these characteristics.

The GSS records SEX as Male = 1 and Female = 2. If we simply use this variable in our analysis, the program mathematically assumes that Female is double the value of Male. When we convert it to a dummy variable [SEXR], where Male = 0 and Female = 1, the analysis is calculated on the basis that a 1 in the variables means presence of a Female.

Likewise, the GSS records RACE as 1 = White, 2 = Black, and 3 = Other. Given this, we are going to use RACECEN1 (What is Respondents Race 1st Mention), which is coded 1 = White, 2 = Black, and 16 = Hispanic. As you have no doubt noticed, a simple dummy variable conversion is not adequate here, as we have 16 nominal categories. The rule for creating dummy variables is that we need to create J-1 variables to represent the data. In the case of RACECEN1, with 16 categories, we would need to create 15variables (16-1) to fully understand the effect of the interaction between Sex and Race. However, for the sake of simplicity in this example, we are only using three categories: white, black and Hispanic. Thus, we need to create only two dummy variables.

When you create the variables, you need to determine what you are interested in measuring. Since we are interested in the effects of being non-white, then we are really looking at the absence of whiteness AND variation within nonwhite groups. Given this, we need to create one dummy variable where Black = 1 and everyone else = 0 [RACEB]. We also need to create a dummy variable where Hispanic = 1 and everyone else = 0 [RACEH]. Whites will have a 0 in both of these categories.

Once we create our dummy variables, we need to create our interaction variables.

$$SEXBL = SEXR * RACEB$$

AND

$$SEXH = SEXR * RACEH$$

Since we are using dummy variables, SEXH will assign a 1 to everyone who is both female and Hispanic. SEXB will assign a 1 to everyone who is both female and African American. Now we can run our multiple regression analysis for our hypothesis that sex and race interact to influence educational attainment. To do this, we need to include both of these terms in the analysis.

We will run two regressions for this model. First, we will run the equation without the interaction terms, and then we will run it with the interaction terms. The first model gives us the following equation, where X_1 = dummy for Females, X_2 = dummy for Blacks, and X_3 = dummy for Hispanics:

$$\hat{Y} = \alpha + \beta_1X_1 + \beta_2X_2 + \beta_3X_3$$

Our first regression provides the results shown in **Table 11.6**.

Table 11.6 *SPSS Printout of Regression of Respondents' Income on Race/Ethnicity and Sex*

Model Summary

Model	R	R Square	Adjusted R Square	Std. Error of the Estimate
1	.295[a]	.087	.085	33804.51423

[a]*Predictors: (Constant), Dummy for Sex-Female, Dummy for Hispanics, Dummy for Blacks.*

Coefficients[a]

Model	Unstandardized Coefficients		Standardized Coefficients	t	Sig.
	B	Std. Error	Beta		
1 (Constant)	51980.175	1450.754		35.830	.000
Dummy for Blacks	−14938.485	2830.300	−.148	−5.278	.000
Dummy for Hispanics	−18758.123	4537.522	−.116	−4.134	.000
Dummy for Sex-Female	−16483.378	1970.666	−.233	−8.364	.000

[a]*Dependent Variable: Respondents' Income in Dollars.*

The results of Model 1 suggest that sex and race/ethnicity have significant effects on respondents' income (all $p < .001$). Women were expected to make $16, 483.38$ less than men. Since Whites are the omitted category for race/ethnicity, they are the group to which blacks and Hispanics are compared. The results indicate that blacks are expected to make \$14,938.49 less than whites, while Hispanics are expected to make \$18,758.12 less than whites. Furthermore, we can calculate an expected income for each sex and race/ethnic group. We only need to include the b for variables, where $X \neq 0$.

$$\hat{Y}_{White\ Males} = a = 51980.175$$

$$\hat{Y}_{Black\ Males} = a + b_{black} = 51980.175 - 14938.485 = 37041.69$$

$$\hat{Y}_{Hispanic\ Males} = a + b_{hispanic} = 51980.175 - 18758.123 = 33222.052$$

$$\hat{Y}_{White\ Females} = a + b_{female} = 51980.175 - 16483.378 = 35496.797$$

$$\hat{Y}_{Black\ Females} = a + b_{female} + b_{black} = 51980.175 - 16483.378 - 14938.485 = 20558.312$$

$$\hat{Y}_{Hispanic\ Females} = a + b_{female} + b_{hispanic} = 51980.175 - 16483.378 - 18758.123 = 16738.674$$

Remember, interaction theories argue that it is not only the presence of race and sex that contributes to the outcome, as Model 1 suggests. Rather, it is the interaction of race and sex. We can now run our second model with our interaction variables and compare the outcomes. When including interaction terms in regression analysis, it is essential to have one model with all the independent variables except the interaction term(s), followed by a model that includes them. Therefore, after the dummy variables are entered, click Next, and add the interaction terms. Also, it is important to select the R squared change option in the Statistics box. The second model gives us the following equation, where $X_1 =$ dummy for Females, $X_2 =$ dummy for Blacks, $X_3 =$ dummy for Hispanics, $X_1X_2 =$ the interaction between the dummy for Female and the dummy for Blacks, and $X_1X_3 =$ the interaction between the dummy for Female and the dummy for Hispanic:

$$\hat{Y} = a + b_1X_1 + b_2X_2 + b_3X_3 + b_4X_1X_2 + b_5X_1X_3$$

Table 11.7 SPSS Printout of Regression of Respondents' Income on Race/Ethnicity and Sex with Interaction Effects for Race/Ethnicity and Sex

Model Summary

Model	R	R Square	Adjusted R Square	Std. Error of the Estimate	R Square Change	F Change	df1	df2	Sig. F Change
					Change Statistics				
1	.295[a]	.087	.085	33804.51423	.087	37.471	3	1177	.000
2	.296[b]	.088	.084	33826.07023	.000	.250	2	1175	.779

[a]Predictors: (Constant), Dummy for Sex-Female, Dummy for Hispanics, Dummy for Blacks.
[b]Predictors: (Constant), Dummy for Sex-Female, Dummy for Hispanics, Dummy for Blacks, Interaction for Hispanic and Sex, Interaction for Blacks and Sex.

Coefficients^a

	Unstandardized Coefficients		Standardized Coefficients		
Model	B	Std. Error	Beta	t	Sig.
1 (Constant)	51980.175	1450.754		35.830	.000
Dummy for Blacks	−14938.485	2830.300	−.148	−5.278	.000
Dummy for Hispanics	−18758.123	4537.522	−.116	−4.134	.000
Dummy for Sex−Female	−16483.378	1970.666	−.233	−8.364	.000
2 (Constant)	52302.231	1523.450		34.331	.000
Dummy for Blacks	−16949.667	4121.913	−.168	−4.112	.000
Dummy for Hispanics	−19852.231	5917.129	−.122	−3.355	.001
Dummy for Sex−Female	−17149.845	2191.552	−.243	−7.825	.000
Interaction for Blacks and Sex	3819.503	5673.230	.029	.673	.501
Interaction for Hispanic and Sex	2564.428	9228.735	.010	.278	.781

[a]Dependent Variable: Respondents Income in Dollars.

Now we can calculate the equations. Remember, we only need to include the variables where $X \neq 0$.

$$\hat{Y}_{White\ Males} = a = 51980.175$$

$$\hat{Y}_{Black\ Males} = a + b_{black} = 51980.175 - 16949.667 = 35030.508$$

$$\hat{Y}_{Hispanic\ Males} = a + b_{hispanic} = 51980.175 - 19852.231 = 32127.944$$

$$\hat{Y}_{White\ Females} = a + b_{female} = 51980.175 - 17149.845 = 34830.330$$

$$\hat{Y}_{Black\ Females} = a + b_{female} + b_{black} + b_{black \times sex} = 51980.175 - 17149.845 - 16949.667 + 3819.503 = 21700.166$$

$$\hat{Y}_{Hispanic\ Females} = a + b_{female} + b_{hispanic} + b_{hispanic \times sex} = 51980.175 - 17149.845 - 19852.231 + 2564.428 = 17542.527$$

For comparison, we can organize these models into one table.

	White		Black		Hispanic	
	Model 1	Model 2	Model 1	Model 2	Model 1	Model 2
Female	35496.797	34830.330	20558.312	21700.166	16738.674	17542.527
Male	51980.175	51980.175	37041.690	35030.508	33222.052	32127.944

Model 1 assumes that sex is experienced regardless of race and that race is experienced regardless of sex. Model 2 assumes that sex and race interact to create unique variations for the sexes and the races. When we look at Model 1, we can see that men in general have higher incomes than women. We can also see that whites, with the exception of black men compared to white women, have better outcomes than blacks and Hispanics. However, when we compare the two models, we see the lowest outcomes for Hispanic females, followed by black women. These data suggest that race and sex do interact to produce lower outcomes for Hispanic women.

There is one final test to determine whether the interaction term explains significant differences in outcomes. We could calculate an F statistic for the two models. The equation for this is:

$$F_{(k_2-k_1)(N-K_2-1)} = \frac{\left(R^2_{m2}-R^2_{m1}\right)/\left(K_2-K_1\right)}{\left(1-R^2_{m2}\right)/\left(N-K_2-1\right)}$$

$m1$ = Model 1

$m2$ = Model 2

K_1 = IVs in Model 1

K_2 = IVs in Model 2

We do not need to calculate F since we have asked SPSS to do it for us. Note that in the Model Summary part of the printout in **Table 11.7,** the second half provides the change statistics that we requested when running the regression. The F Change was only .250 and was not statistically significant ($p > .05$). What the F test indicates is that our results do not allow us to reject the null hypothesis. However, we should not dismiss the idea that sex and race may interact when predicting income. If we look at the distribution of race in our sample, we find that we have 280 blacks and 80 Hispanics. It is more likely that the sample size is contributing to the low F statistic rather than the relationships among the variables.

Curvilinear Relationships

When we began our discussion of regression in Chapter 10, we indicated that the regression model is designed to represent a linear relationship between X and Y. Ordinary Least Squares (OLS) regression specifies a linear, additive relationship (see **Table 11.8**). However, when we discussed scatterplots, we indicated that not all relationship are linear; X could have a curvilinear effect on Y, as depicted in **Figure 11.9.** The coefficients (i.e., the b's) produced from the regression analysis represent how much Y changes for a unit change in X. However, X can affect Y in a non-linear, non-additive way. A unit change in X may also change the effect X has on Y. For example, as X increases in value, the effect that X has on Y may increase or decrease. One of the most common non-linear, non-additive relationships is represented by the quadratic equation in **Table 11.8** in which the nonlinear effect of X is represented in its squared term. The curvilinear relationship represented by the quadratic equation will resemble a U-shape or an inverted U-shape. **Figure 11.9** depicts the inverted U-shaped curve. The inverted U-shape is produced when the quadratic term has a negative effect on Y (i.e., the b for X^2 is negative). The U-shaped curve, on the other hand, is produced when the quadratic term has a positive effect on Y (i.e., the b for X^2 is positive). However, the curves may not always be a complete U-shape or an inverted U-shape. A nonlinear relationship can consist of part of the curve of a U or an inverted U.

Table 11.8 *Comparison of Linear and Curvilinear Effects of X on Y*

Linear Multiple Regression Equation	Nonlinear Quadratic Equation
$\hat{Y} = a + b_1 X_1 + b_2 X_2$	$\hat{Y} = a + b_1 X_1 + b_2 X_2^2$

Unfortunately, one of the possible problems that can occur in multiple regression is increased when a variable and its squared term are included. Multicollinearity occurs when there is a high or nearly perfect correlation between two or more independent variables in a multiple regression. Multicollinearity is problematic in multiple regression because the analysis attempts to separate out the effects that each variable has on the dependent variable. If the variables are highly correlated, there are no unique effects associated with the

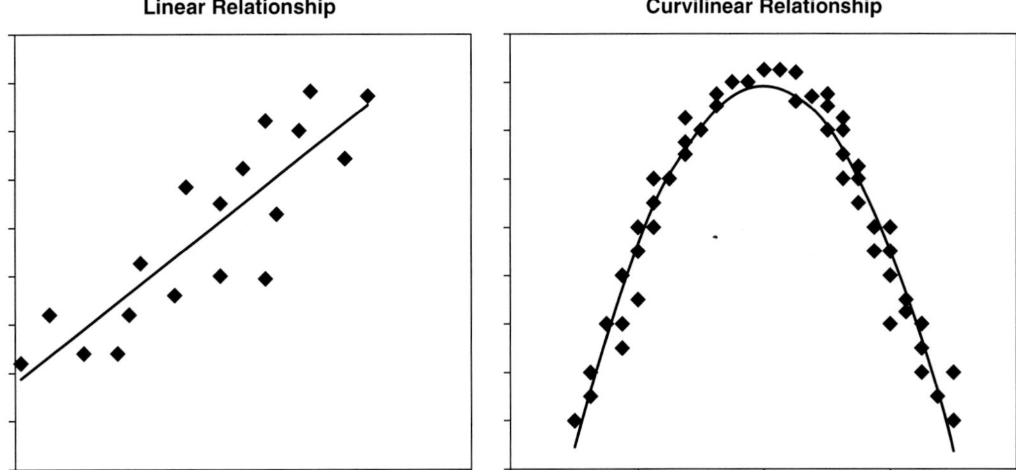

Figure 11.9 Scatterplots of Linear and Curvilinear Relationships.

individual independent variables. If you put two perfectly correlated independent variables into your regression analysis, SPSS will give you an error message. Just because SPSS does not issue an error message, however, does not mean that we don't have to be concerned about multicollinearity. While SPSS will not issue an error message for highly correlated independent variables, their inclusion can cause problems. Because SPSS cannot produce precise estimates of the effects of the independent variables that are highly correlated, their standard error would likely be quite large. Large standard errors make it unlikely that the *b*'s for these variables would be significant. However, if one of these variables were included, it is possible that the effect would be statistically significant. Undetected multicollinearity may cause us to dismiss effects that actually are significant.

There are multiple ways to detect signs of multicollinearity. However, we should not assume that large standard errors are the result of multicollinearity. One way is to examine the correlation matrix for high correlations between independent variables. However, if the multicollinearity is the result of multiple independent variables, we may not uncover the multicollinearity by just looking at bivariate correlations. In SPSS, we can request that statistics that can identify possible multicollinearity be included in the analysis. SPSS produces both the tolerance statistic and the variance inflation factor (VIF). The tolerance statistic ranges between 0 and 1, with low-tolerance values indicating high multicollinearity. It is debatable how low the tolerance statistic has to be. Allison (1999)[1] suggests that one should be concerned about the presence of multicollinearity when the tolerance statistic falls below .40. The VIF is the reciprocal of the tolerance statistic (VIF $= 1 \div$ tolerance). Therefore, we should be concerned if the VIF exceeds 2.50.

There are various options if you detect multicollearity. You may want to delete one of the variables from the model, increase the sample size, create an index that includes the correlated variables, create a latent variable within a structural equation modeling program, or create deviation scores for the correlated variables. Selecting which solution to use depends on the situation. In the case of a variable being correlated with its quadratic term, creating deviation scores is a good solution.

Suppose we are interested in seeing whether education has a nonlinear effect on family income. To create the quadratic term, we go to Transform $\Rightarrow$ Compute Variable, and a Compute Variable box open. As you can see in **Figure 11.10,** we type the name (in this case, educ2) that we want to call the new quadratic term in the Target Variable box. In the Numeric Expression box, we enter the formula for the quadratic term (educ * educ). Note that, to multiply in SPSS, like many other computer programs, you use the symbol *.

---

[1] Allison, Paul D., *Multiple Regression: A Primer* (Thousand Oaks, CA: Pine Forge Press, 1999).

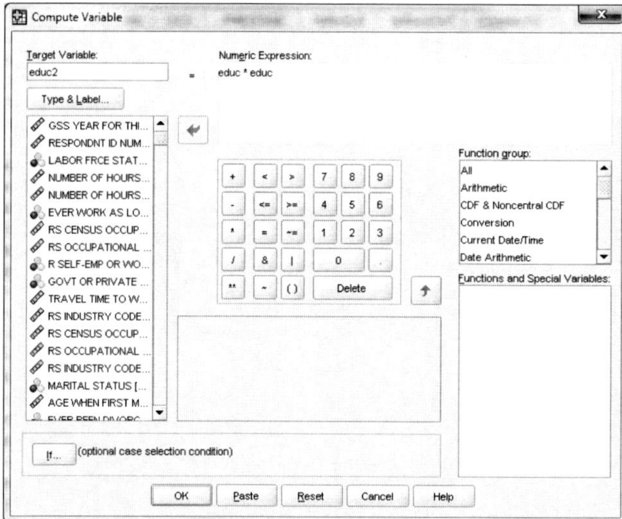

Figure 11.10 SPSS Display for Compute Variables.

Now that education squared has been created, we can run our regression. We go to Analyze ⇒ Regression ⇒ Linear and enter the variables. We can check for signs of multicollinearity by clicking on Statistics. After the statistics box, shown in **Figure 11.11,** opens, we want to click on both Descriptives and Collinearity diagnostics.

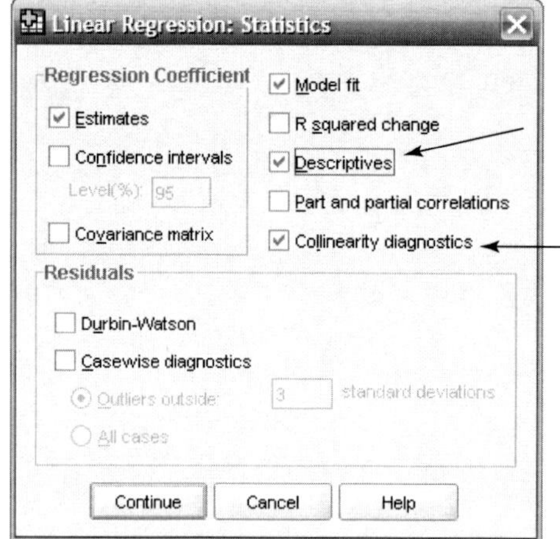

Figure 11.11 SPSS Statistics Display Box within Linear Regression.

From the SPSS output shown in **Table 11.9,** you can see that the tolerance value is below .40 and the VIF is over 2.5. Therefore, we can conclude that multicollinearity exists between the Highest Year of School Completed and its quadratic term.

To resolve the problem of multicollinearity, we can create a new variable with the deviation score from its mean for the variable, Highest Year of School Completed. From the Descriptive Statistics printed before the regression results, we can see that the mean for Highest Year of School Completed was 13.49. Therefore, we can use 13 for the mean years of school completed. To create the deviation score, we go to Transform ⇒ Compute Variable. In the Target Variable box, we type the name of our new variable (in this case, educdev). Since we want to calculate deviation scores from the mean, in the Numeric Expression box, we click over the Highest Year of School Completed and then also type − 13 (the value of the mean). Now we can create a quadratic term for the deviation, similar to the creation of the quadratic term in **Figure 11.10.** However,

Table 11.9   SPSS Output of the Regression of Family Income Regressed on Highest Year of School Completed and Its Quadratic Term

Coefficients[a]

Model	Unstandardized Coefficients		Standardized Coefficients	t	Sig.	Collinearity Statistics	
	B	Std. Error	Beta			Tolerance	VIF
1 (Constant)	5538.675	10432.239		.531	.596		
Highest Year of School Completed	1136.553	1587.464	.077	.716	.474	.040	24.983
$educ^2$	194.319	60.229	.347	3.226	.001	.040	24.983

[a]*Dependent Variable: Family Income in Dollars.*

in this case, we might call the quadratic term educdev2, which should be typed in the Target Variable box. In the Numeric Expression box, we type educdev * educdev to create the quadratic term. Once the deviation score and quadratic term have been created, we can now re-run the regression analysis with them (i.e., educdev and educdev2) entered as the independent variable. That regression produces the results shown in **Table 11.10**. The results show that the tolerance value of .999 is above .40 and the VIF of 1.001 is below 2.5. Therefore, using the deviation score for Highest Year of School Completed and its quadratic term solved our problem of multicollinearity.

The results also indicate that Highest Year of School Completed has a significant curvilinear effect on Family Income. Since the b for the quadratic term is positive, we can also conclude that the curve will be U-shaped, rather than being an inverted U-shaped curve. However, when interpreting the effect that the Highest Year of School Completed has on Family Income, we must be mindful that we used deviation scores. We must remember that the regression is based on how people deviated from the mean years of education, not the actual number of years of school completed.

Table 11.10   Output of the Regression of Family Income Regressed on the Deviation Score for Highest Year of School Completed and Its Quadratic Term

Coefficients[a]

Model	Unstandardized Coefficients		Standardized Coefficients	t	Sig.	Collinearity Statistics	
	B	Std. Error	Beta			Tolerance	VIF
1 (Constant)	53153.690	1137.303		46.737	.000		
educdev	6188.834	317.777	.420	19.475	.000	.999	1.001
$educdev^2$	194.319	60.229	.070	3.226	.001	.999	1.001

[a]*Dependent Variable: Family Income in Dollars.*

Therefore, the predictive equation is:

$$\hat{Y} = a + b1(X - \bar{X}) + b2(X - \bar{X})^2$$

Suppose we want to compare the effects that 8 years versus 12 years and 16 years of education has on family income. Using the values for *a* and *b* in **Table 11.10** for 8, 12, and 16 years of education, the predictive equations are:

$$\hat{Y}_8 = 53153.690 + 6188.834(8 - 13) - b2(8 - 13)^2$$

$$\hat{Y}_8 = 27{,}067.50$$

$$\hat{Y}_{12} = 53153.690 + 6188.834(12 - 13) + b2(12 - 13)^2$$

$$\hat{Y}_8 = 47{,}159.18$$

$$\hat{Y}_{16} = 53153.690 + 6188.834(16 - 13) + b2(16 - 13)^2$$

$$\hat{Y}_8 = 73{,}469.06$$

The additional 4 years of education from 8 to 12 years increased Family Income by \$20,091.68. However, because of the curvilinear effect, another 4 years of education does not increase family income by \$20,091.68, but by \$26,309.89. Thus, the effects of Years of School completed are nonlinear and nonadditive in nature.

Graphing the Results

You can use the Microsoft Excel program to graph the results. Once you open the Excel spreadsheet, create the following headings: EdDev for column A, Education for column B, and Family Income for column C. You should only graph within the range of the data. Since the frequency of Years of School Completed shows that respondents range from zero years of formal education to 20 years, we will use those as the low and high values for Years of School Completed in our graph. Enter the Years of School Completed in column B, starting with 0 in increments of 2 and ending with 20. Then, enter the deviation scores for each year of education (education − 13) in column A. Excel can calculate the deviation scores for each cell. Click on cell A2 and then in the command line (where the solid arrow is pointing in **Figure 11.12**), type =, click on cell B2, and then type − 13. If you click on the checkmark that the dotted arrow is pointing to, the typing in cell A2 changes from the equation that you typed to the solution to that equation. The program will calculate the remaining deviation score if you copy cell A2, which was calculated. Then, place the cursor in cell A3 and while left-clicking and holding, scroll down to cell A12. Click Paste. The complete column of deviation scores should appear.

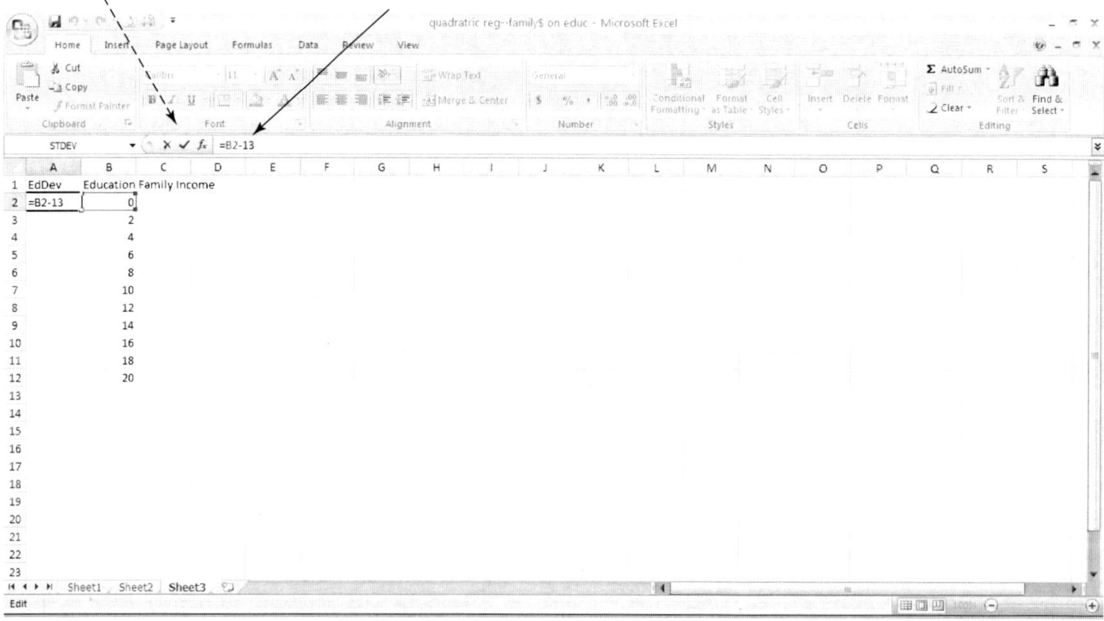

Figure 11.12 Excel Spreadsheet Showing the Calculation of Initial Deviation Score.

To calculate the predictive Family Income which we will use in the graph, place the cursor in row 2 of column C. Type in the command line above the columns an equal sign (=), followed by the regression equation using the *a* and the *b*'s from the regression and cell A2 for *X* and A2 * A2 for X^2. In this case, while your cursor is in cell C2, type = 53153.69 + (6188.834 * A2) + (194.319 * A2 * A2) in the command line. After you have entered the equation, click on the checkmark to the left of the equation. The predicted value of *Y* for those who have zero years of formal education should now appear in the cell C2. To calculate the remaining predictive values, place the cursor in cell C2 and click Copy. Then place the cursor in cell C3, and while still left-clicking and holding, scroll down to cell C8. Click Paste. The complete column of predictive values of *Y* should appear.

Once the data are entered, you can create a graph by highlighting the data in columns B and C, as shown **in Figure 11.13** by the solid arrow. Click on the Insert tab, which the dotted arrow is pointing to. Click on Scatter, shown by the dashed arrow, and choose your desired design. Excel produces a graph that can be modified. If you click on Layout 1 under Chart Layouts, which should appear just above the command line, the Axis Title and Chart Title should appear in the graph. For the axis titles, replace with the names of your variables and provide the graph with an appropriate title. If the legend is uninformative, as it is here, delete it. The completed graph will look like the one shown in **Figure 11.14.**

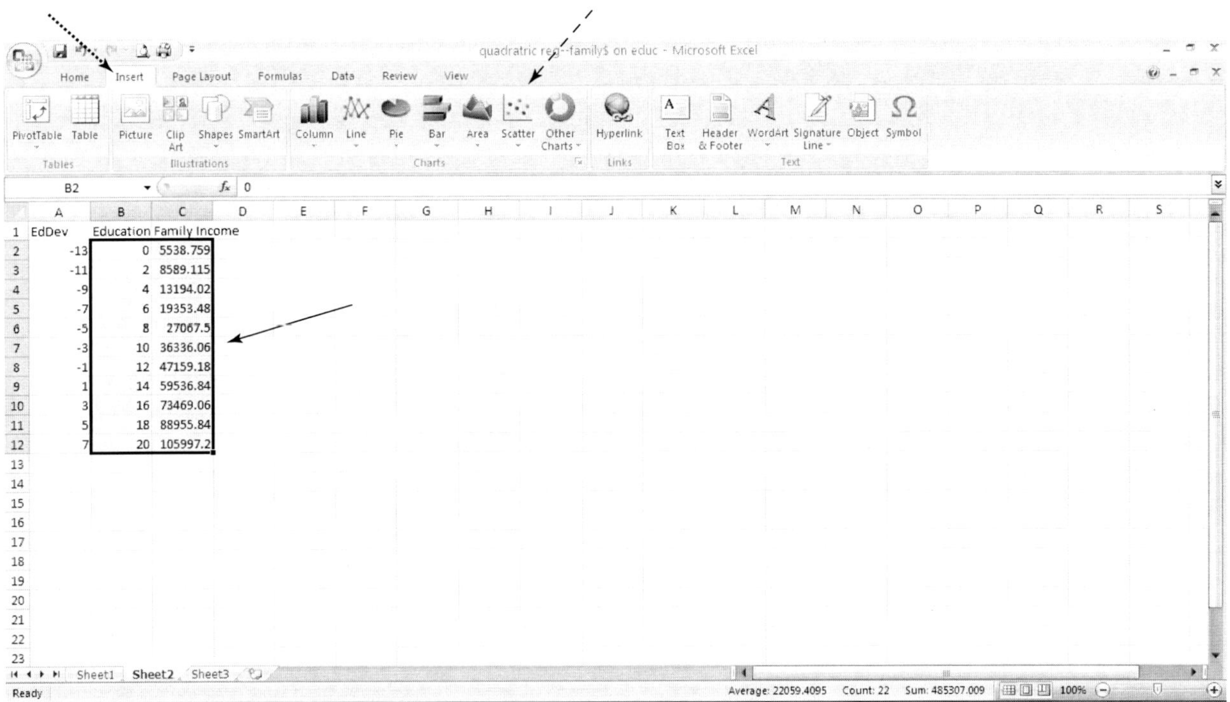

Figure 11.13 Excel Spreadsheet Showing How to Create a Scatterplot.

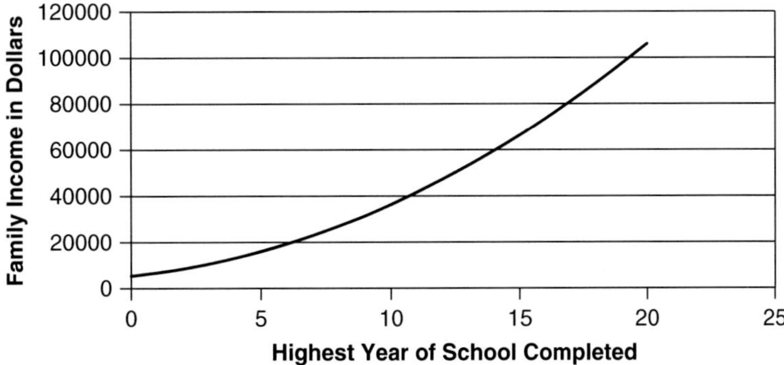

Figure 11.14 Regression of Family Income on the Highest Year of School Completed.

Figure 11.14 visually displays the curvilinear effect that Highest Year of School Completed has on Family Income. When we looked at the b for the quadratic term, we concluded that the shape of the curve would be more U-shaped than shaped like an inverted U. The curve in **Figure 11.19** shows us that the curve does not have to be a perfect U or an entire U. Thus, knowing whether the b is positive or negative gives you an idea of what the shape of the curve will look like. However, by graphing the relationship using the lowest to the highest value of the independent variable, you obtain a clear indication if the relationship between the variables resembles an entire U or only part of a U.

Exercises and Notes

Multiple regression is a statistical technique that allows researchers to determine the relationship between ________________ and ______________.

Three-Variable Multiple Regression

Bivariate and Three-Variable Regression Equations

Type of Equation	Bivariate Equations	Three Variable Equations
Sample regression equation		
Predictive equation		
Population predictive equation		
Standardized predictive equation		

Formulas for the Unbiased Estimates of Population Parameters

$b_1 =$

____

$b_2 =$

____

$a =$

____

Calculating the Unbiased Estimates

Table of Correlation, Means, and Standard Deviations of Workers' Hours Worked, Education, and Weekly Pay (N = 1189)

	Hours	Education	Weekly Pay	Means	Standard Deviations
Hours	1.000			36.92	19.498
Education	.023	1.000		13.83	2.881
Weekly Pay	.332***	.392***	1.000	784.70	678.123

$$b_1 =$$

____

Interpret b_1 _______________________________________

____

____

$$b_2 =$$

Interpret b_2 ____

____

____

$$a =$$

Interpret a ____

____

Write the predictive equation using the calculated coefficients.

What would we expect the weekly pay to be of someone who works 35 hours per week and has a college degree (i.e., 16 years of education)?

____

____

Standardizing the b's

$$\beta_1^W =$$

____

____

$$\beta_2^W =$$

____

____

Multiple Regression in SPSS

In SPSS, go to Analyze $\Rightarrow$ Regression $\Rightarrow$ Linear, and the Linear Regression box will open.

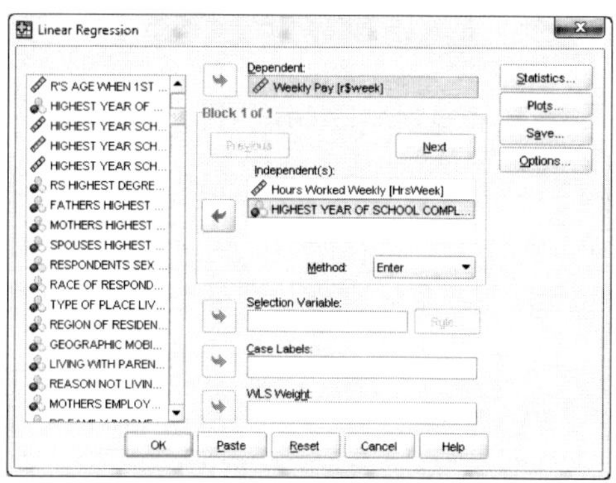

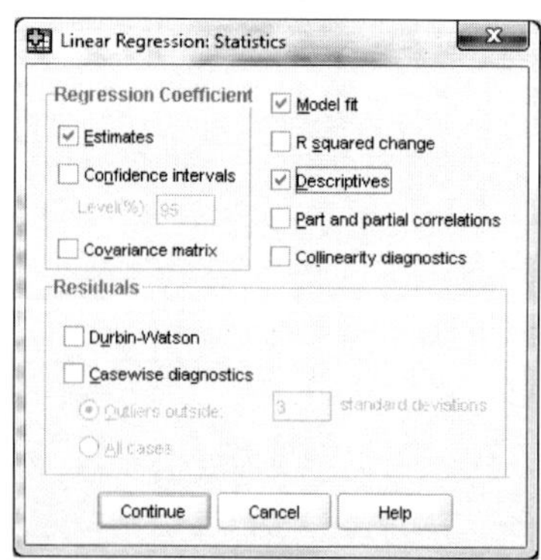

SPSS Printout

Descriptive Statistics

	Mean	Std. Deviation	N
Weekly Pay	784.7011	678.12343	1189
Hours Worked Weekly	36.9243	19.49832	1189
Highest Year of School Completed	13.83	2.881	1189

Correlations

		Weekly Pay	Hours Worked Weekly	Highest Year of School Completed
Pearson Correlation	Weekly Pay	1.000	.332	.392
	Hours Worked Weekly	.332	1.000	.023
	Highest Year of School Completed	.392	.023	1.000
Sig. (1-tailed)	Weekly Pay	.	.000	.000
	Hours Worked Weekly	.000	.	.213
	Highest Year of School Completed	.000	.213	.
N	Weekly Pay	1189	1189	1189
	Hours Worked Weekly	1189	1189	1189
	Highest Year of School Completed	1189	1189	1189

Model Summary

Model	R	R Square	Adjusted R Square	Std. Error of the Estimate
1	.508[a]	.258	.257	584.53706

[a]*Predictors: (Constant), Highest Year of School Completed, Hours Worked Weekly.*

How much variance in weekly pay is explained by hours and education?

____

____

____

<table>
<tr><td colspan="6" align="center">Coefficients[a]</td></tr>
<tr><td rowspan="2" align="center">Model</td><td colspan="2" align="center">Unstandardized Coefficients</td><td align="center">Standardized Coefficients</td><td rowspan="2" align="center">t</td><td rowspan="2" align="center">Sig.</td></tr>
<tr><td align="center">B</td><td align="center">Std. Error</td><td align="center">Beta</td></tr>
<tr><td>1 (Constant)</td><td align="center">−883.054</td><td align="center">88.475</td><td></td><td align="center">−9.981</td><td align="center">.000</td></tr>
<tr><td>Hours Worked Weekly</td><td align="center">11.228</td><td align="center">.870</td><td align="center">.323</td><td align="center">12.906</td><td align="center">.000</td></tr>
<tr><td>Highest Year of School Completed</td><td align="center">90.621</td><td align="center">5.888</td><td align="center">.385</td><td align="center">15.391</td><td align="center">.000</td></tr>
</table>

[a]*Dependent Variable: Weekly Pay.*

What effect do hours worked have on weekly pay, controlling for education?

____

____

What effect does education worked have on weekly pay, controlling for hours worked?

____

____

Do hours worked have a significant effect on weekly pay, controlling for education?

____

____

How do you know?

____

____

Does education have a significant effect on weekly pay, controlling for hours worked?

____

____

What is the standardized effect of hours worked on weekly pay, controlling for education?

____

____

What is standardized effect education has on weekly pay, controlling for hours worked?

____

____

Dummy Variables

- Regression requires that variables are ____ level.

- We can transform _________________ data into dummy variable for their inclusion in a regression analysis.

- **Dummy variables** are variables that are coded as ___ and ___.

  - ___ indicates the _______________ of the attribute.

  - ___ indicates the _______________ of the attribute.

For example, Religion is measured in the GSS as:

1 Protestant

2 Catholic

3 Jewish

4 None

5 Other

- In order to use Religion, a ___________-level variable, in a regression equation as an independent variable, we construct dummy variables that are coded ___ and ____.

- We create ________ variables.

- We included ________ variables in the equation because the inclusion of ___ dummy variables would be _________________.

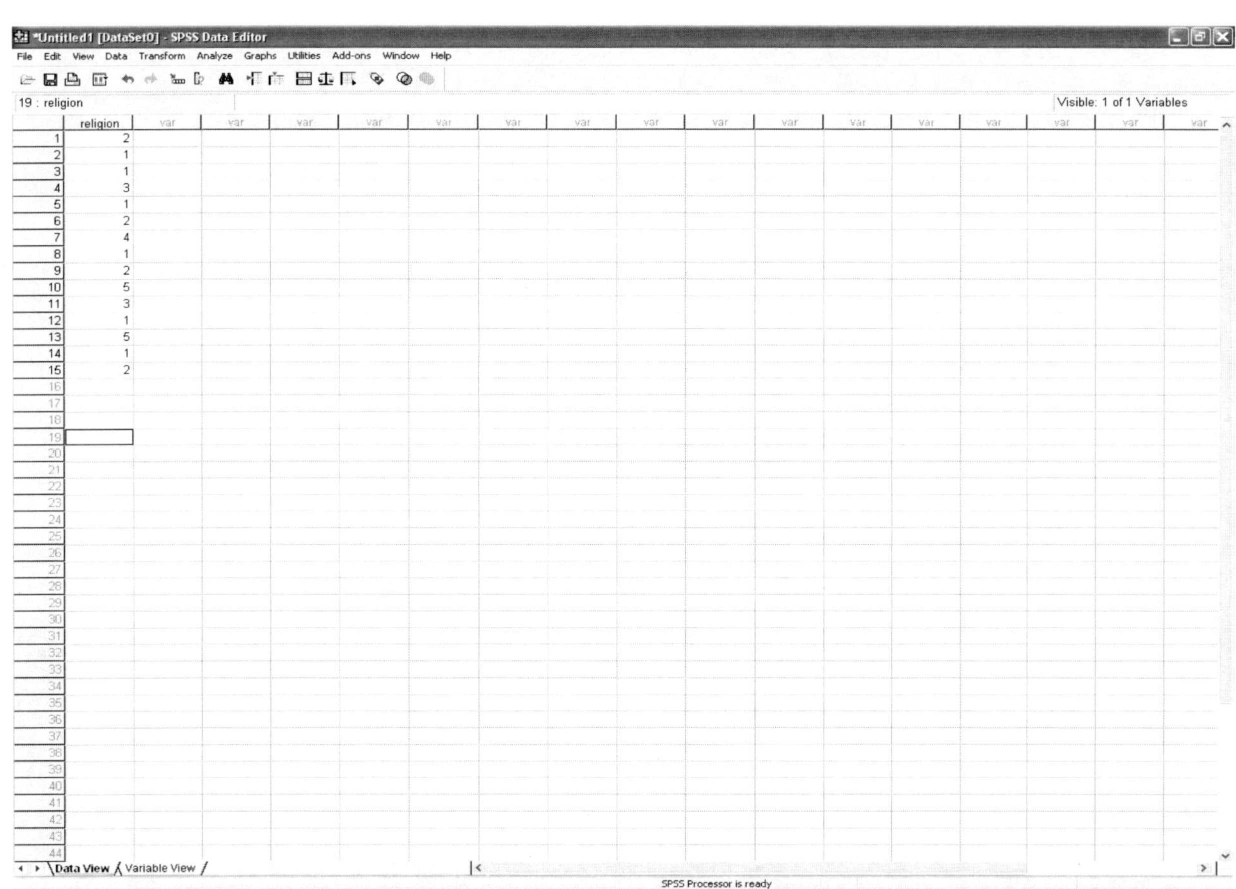

- In this example, we did not include the dummy variable for ______________ in the analysis.

- However, we could have chosen any of the other categories as the one eliminated from the analysis.

- If we regress a dependent variable on a set of dummy variables, the results are identical to those obtained from __________________.

Dummy Variables and Interaction Effects

- Interaction effects are intended to capture the ____________________ effects of ______________ variables on the ____________________ variable.

- If there is an interaction, the effect of one ______________ variable depends on ________________________ ____________________________ variable.

- To capture the interaction effect, an interaction variable is constructed by _________________________________ ____.

- The regression equations must include the ____ and ______________ ____.

A criminology student thinks that whites serve less prison time than blacks and Hispanics do (data on Whites, Blacks, and Latinos), regardless of their sentence lengths. He also believes that the effect of race/ethnicity on time served is less when offenders have shorter sentences compared to longer sentences.

- Dependent Variable

  -

- Main Effects

  -

  -

- Interaction Effect

  -

- How many dummy variables will you include in the regression equation?

- Which categories of race do you choose to make into dummy variables?

- To test for the interaction effects, you must have one equation (or model) with just the ____________________ and another regression equation with the ____________________ and the ________________________________.

- Write the first equation with just the ________________________________.

- Write the second equation with ____.

- Although we include ________ dummy variables in the second equation, we actually have ______ regression equations.

- Write the regression equations for each category of the variable, ________.

  When the offender is ____________, the equation is:

  When the offender is ____________, the equation is:

  When the offender is ____________, the equation is:

Testing the Significance

- To determine the significance of the ____
  _____________, we use ___________ to test the ______.

- To determine the significance of the **interaction effect** (in this case, _____________________________________),
  we cannot use _______________. For example, would we conclude that there is a significant interaction,
  if __________ significantly _differed_ from __________, but ____________ _did not differ_ significantly from
  _______________?

To test for an interaction effect:

- We compare the ______ of the equation _with_ the interaction term(s), in this case:

- … with the ______ of the equation _without_ the interaction terms, in this case:

To compare the _______, we use the following _________:

K_2 = The number of IV in the equation used to estimate ______.

K_1 = The number of IV in the equation used to estimate ______.

- The underlying null hypothesis is H_0: ___________________.

- If we can reject H_0, we conclude that ___________________________.

- If we cannot reject H_0, we conclude that _______________________________.

Interaction Effects in SPSS

How to Make Dummy Variables for Race/Ethnicity

Go to Transform $\Rightarrow$ Recode into Different Variables. The first dummy variable is Black.

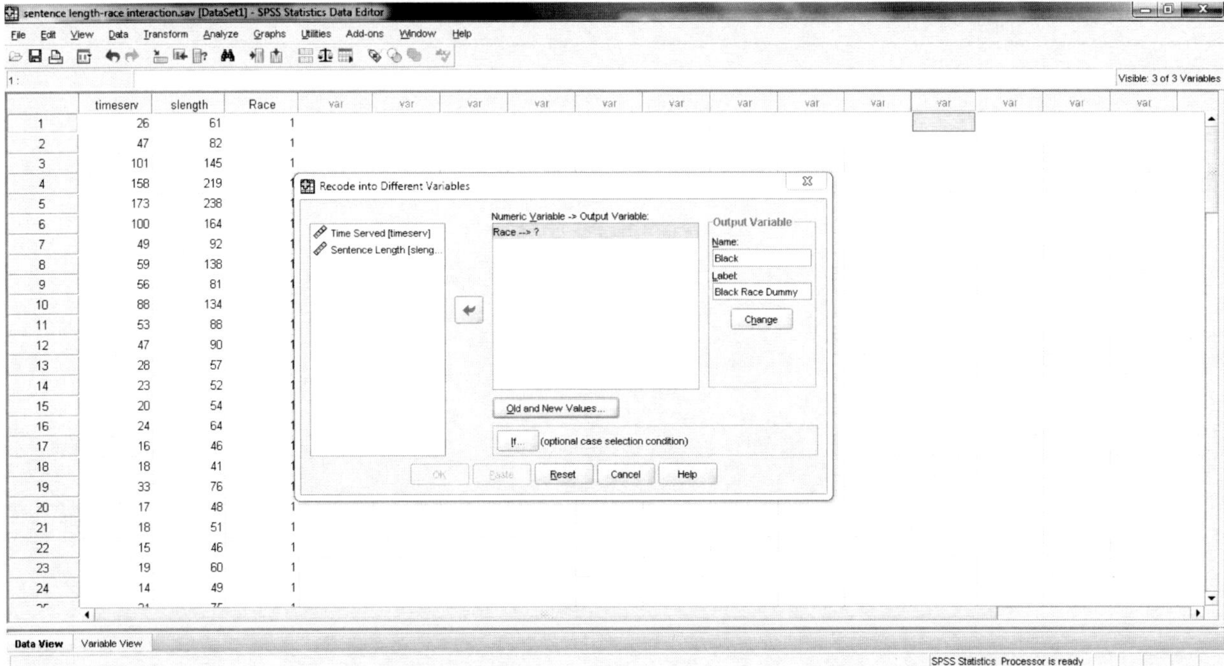

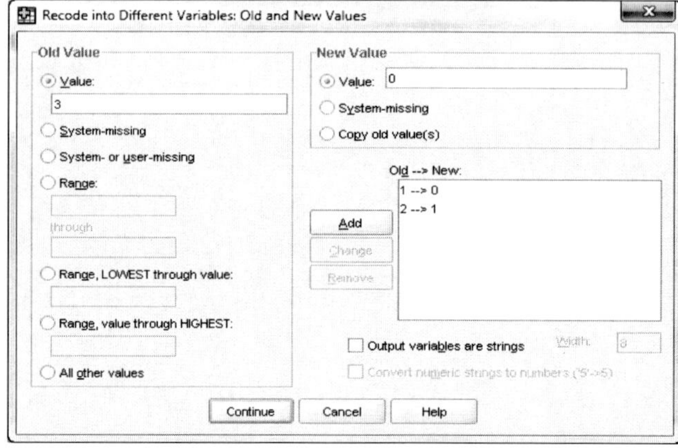

The process is repeated to create the second dummy variable, Latino.

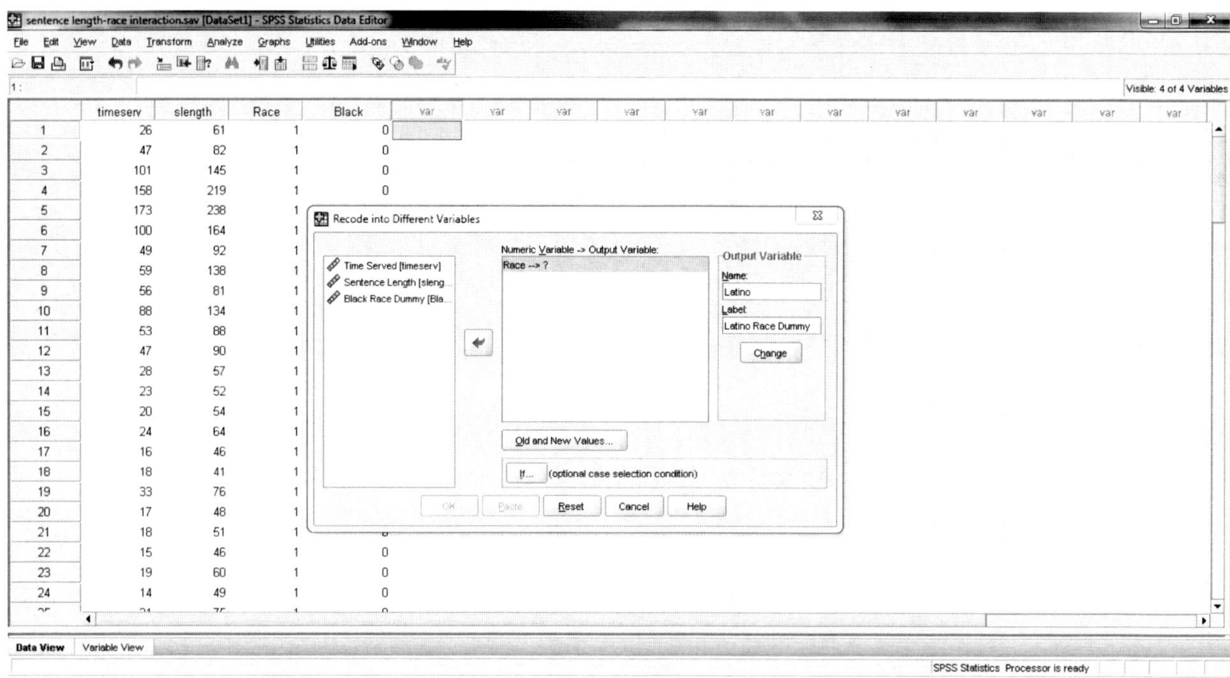

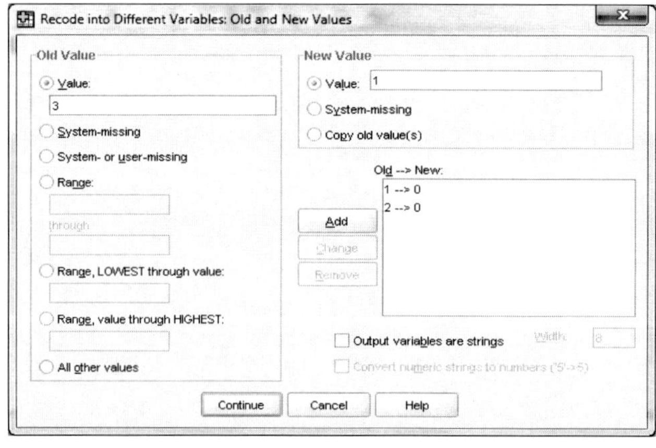

How to Create the Interaction Term in SPSS

Go to Transform ⇒ Compute Variable. First, create an interaction between Sentence Length and Black.

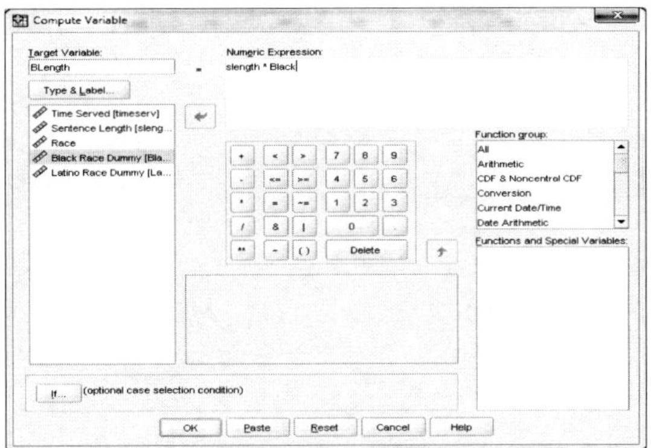

Next, create an interaction between Sentence Length and Latino.

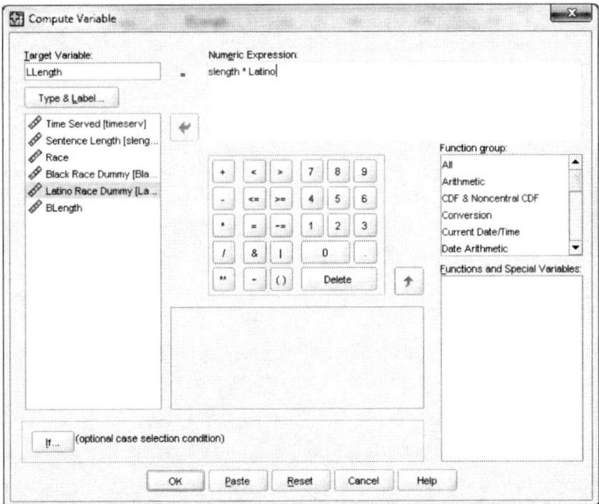

How to Run a Regression with Interaction Terms

Go to Analyze ⇒ Regression ⇒ Linear. First, enter the dependent variable and the main effects. Then click Next. Enter the interaction effects.

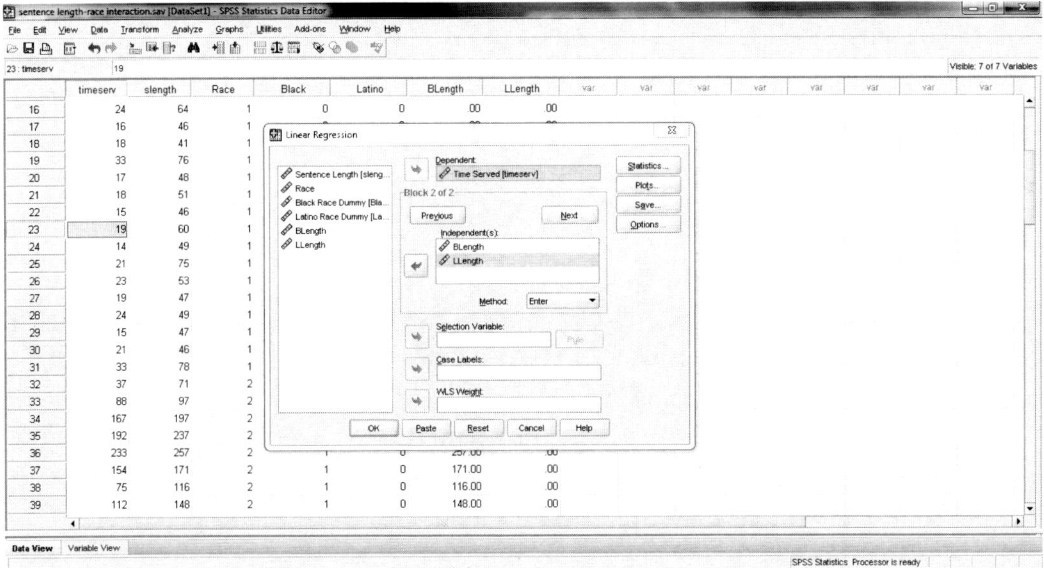

Click Statistics. After the Statistics Dialogue box opens, click *R* squared change, as shown below.

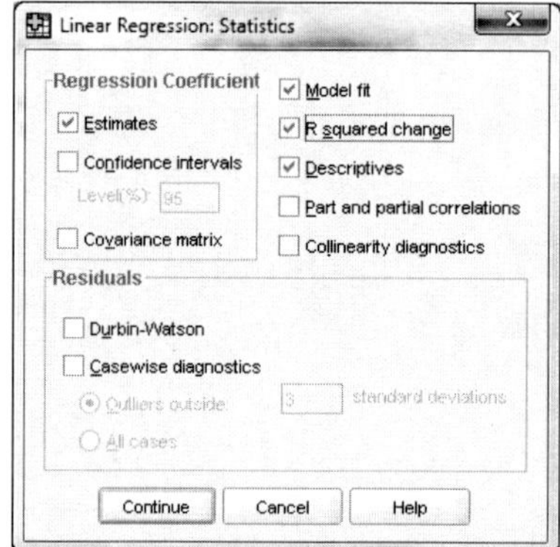

When you run the regression, the following is part of the output that will be produced.

Coefficients[a]

Model		Unstandardized Coefficients		Standardized Coefficients		
		B	Std. Error	Beta	t	Sig.
1	(Constant)	-26.516	1.636		-16.212	.000
	Sentence Length	.848	.014	.954	61.177	.000
	Black Race Dummy	16.921	1.651	.175	10.248	.000
	Latino Race Dummy	2.187	2.016	.018	1.085	.280
2	(Constant)	-21.410	2.088		-10.252	.000
	Sentence Length	.786	.022	.885	36.410	.000
	Black Race Dummy	4.591	2.961	.047	1.550	.123
	Latino Race Dummy	1.999	3.535	.017	.565	.573
	BLength	.139	.029	.169	4.798	.000
	LLength	.006	.035	.005	.159	.874

a. Dependent Variable: Time Served

Model Summary

Model	R	R Square	Adjusted R Square	Std. Error of the Estimate	R Square Change	F Change	df1	df2	Sig. F Change
1	.982[a]	.964	.963	9.162	.964	1330.254	3	151	.000
2	.985[b]	.969	.968	8.455	.006	14.156	2	149	.000

a. Predictors: (Constant), Latino Race Dummy, Sentence Length, Black Race Dummy

b. Predictors: (Constant), Latino Race Dummy, Sentence Length, Black Race Dummy, LLength, BLength

Write the equation from the model with the interaction terms.

____

____

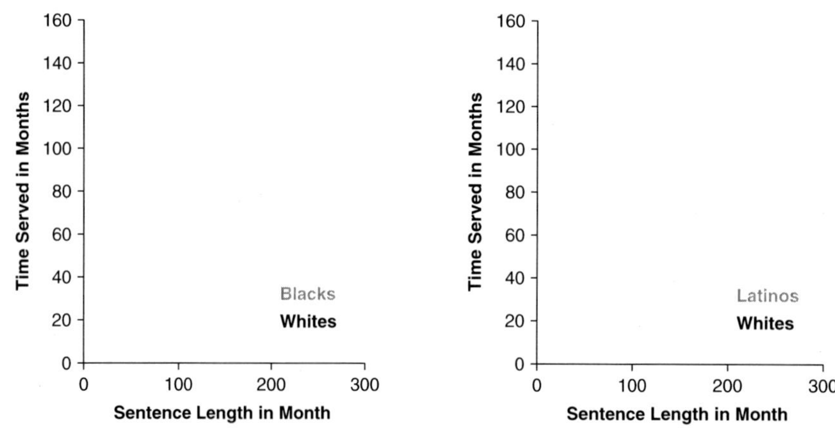

Nonlinear Relationships

- In ordinary multiple regression, a coefficient represents how much ___________________ changes ___________________, ___________________ other variables ___________________.

- Ordinary Least Squares (OLS) regression specifies ___________________ relationship:

- However, X can affect Y in a ___________________ way.

- A unit change in ___________________ may also change the effect ___________________ has on ___________________. One of the most common nonlinear, nonadditive relationships is represented by this quadratic equation:

- This ___________________ relationship may resemble a ___________________ or ___________________.

  - ____
    ____.

  - ____
    ____.

- Having both X and X^2 in the equation can lead to ___________________.

- You can check for ___________________ by

  - examining the ___________________ and the ___________________.

  - using a statistical package.

    - ____.

    - ____.

- If there is ___________________ between ___________________ and ___________________, it can be reduced by calculating ___________________ and ___________________ variables using ___________________.

Suppose a city council wants to determine how the city's residents feel about urban sprawl. One councilperson suggests that middle-income residents would be most concerned. The council asks you to see if income has a curvilinear effect on people's attitudes towards urban sprawl.

- ___________________, the ___________________ variable would be the person's ___________________
  ____.

- The ___________________ would be ___________________.

- A ___________________ and its ___________________ are not as highly correlated as the original variables. Why?

- If ___________________ are used, remember that X is no longer ___________________ but ___________________. Be sure to use the new variable, ___________________, when doing such things as predicting values of Y. Thus, predicting Y when ___________________ equals ___________________ would be:

- How would you predict Y for someone who ___________________?

How to Create a Quadratic Term in SPSS

Go to Transform $\Rightarrow$ Compute Variable. Enter a name for a new quadratic variable (in this case, income2) in the Target Variable box, and in the Numeric Expression box, enter the original variable multiplied by itself (in this case, income\$ * income\$).

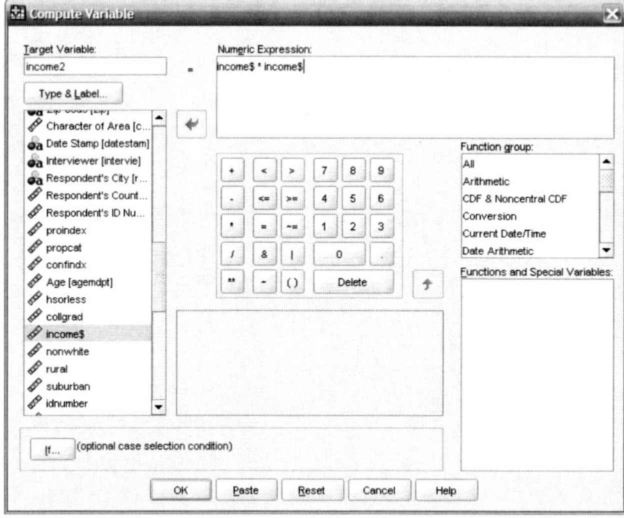

With the quadratic term created, you can now regress attitudes about urban sprawl by income and its squared term. When running the regression, you need to test for multicollinearity between income and its squared term.

Click on the Statistics button in the Regression dialogue box. Click Descriptives and Collineary Diagnostics.

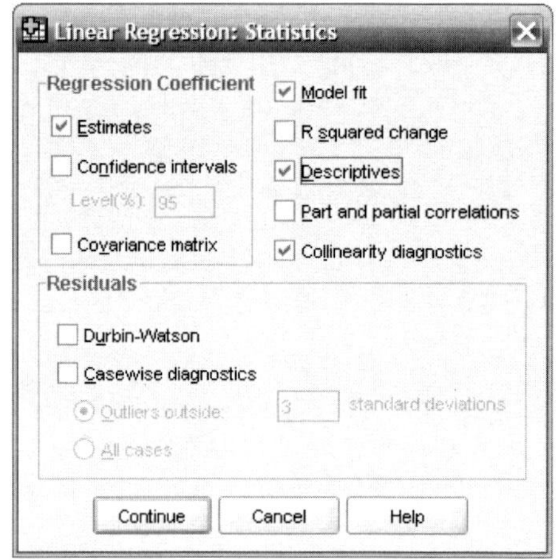

The tolerance levels of .036 are below the .40 criterion and the VIF of 27.478 are above the 2.5 criterion. Therefore, you can conclude that multicollineary _____________________ between the variables of income and income^2.

Coefficients[a]

Model	Unstandardized Coefficients		Standardized Coefficients			Collinearity Statistics	
	B	Std. Error	Beta	t	Sig.	Tolerance	VIF
1 (Constant)	71.015	3.804		18.669	.000		
income$	.537	.203	.766	2.641	.009	.036	27.478
income2	−.007	.002	−.907	−3.128	.002	.036	27.478

[a]Dependent Variable: Sprawl.

Creating Deviation Scores and Its Quadratic Terms

- Go to Transform ⇒ Compute Variable. Enter a name for a new deviation score variable (in this case, incomdev) in the Target Variable box, and in the Numeric Expression box, enter the original variable minus approximately its mean (in this case, income$ − 40).

- Next, Go to Transform ⇒ Compute Variable. Enter a name for a new quadratic variable (in this case, incomdv2) in the Target Variable box, and in the Numeric Expression box, enter the deviation score for income multiplied by itself (in this case, incomdev * incomdev).

- Now, you can re-run the regression. Replace income and income2 with incomdev and incomdv2. The regression will produce the following printout. As you can see, the tolerance level of _____ is greater than _____ and the VIF of _____ is less than _____.

Coefficients[a]

Model	Unstandardized Coefficients		Standardized Coefficients			Collinearity Statistics	
	B	Std. Error	Beta	t	Sig.	Tolerance	VIF
1 (Constant)	81.604	1.605		50.848	.000		
incomdev	−.045	.041	−.064	−1.091	.276	.891	1.122
incomdv2	−.007	.002	−.183	−3.128	.002	.891	1.122

a. Dependent Variable: Sprawl.

How to Create a Graph of the Curvilinear Relationship in Excel

- In an Excel spreadsheet, create the following headings: $Dev for column A, Income for column B, and Sprawl for column C.

- Enter the Incomes in column B starting with 15,000 and increments of 5,000, ending with 75,000, while entering the deviation scores for each income (income − 40) in column A. Note: Income in the data was measured in thousands of dollars, so the first deviation score is $15 − 40 = −25$.

- Place the cursor in row 2 of column C. Type in the command line above the columns an equal sign (=), followed by the regression equation using the a and the b's from the regression and cell A2 for X and A2 * A2 for X^2.

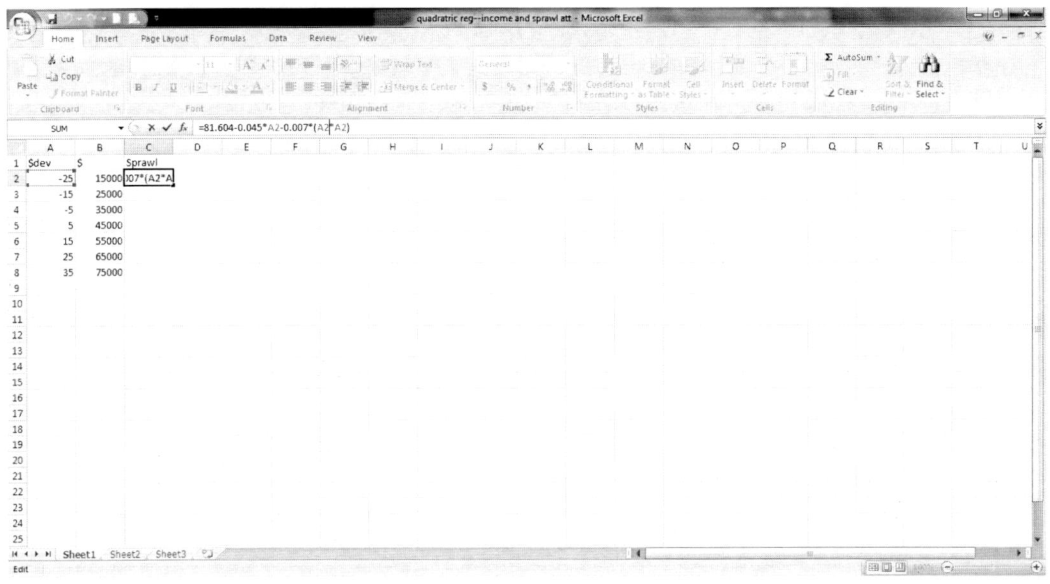

- After you have entered the equation, click on the checkmark to the left of the equation. The predicted value of Y for those who make $15,000 should now appear in cell C2.

- To calculate the remaining predictive values, place the cursor in cell C2 and click Copy. Now, place the cursor in cell C3 and while still left-clicking, scroll down to cell C8. Click Paste. The complete column of predictive values of Y should appear.

Once the data are entered, you can create a graph.

- Highlight the data in columns B and C as shown below.

- Click on the Insert tab.

- Click on Scatter and choose your desired design.

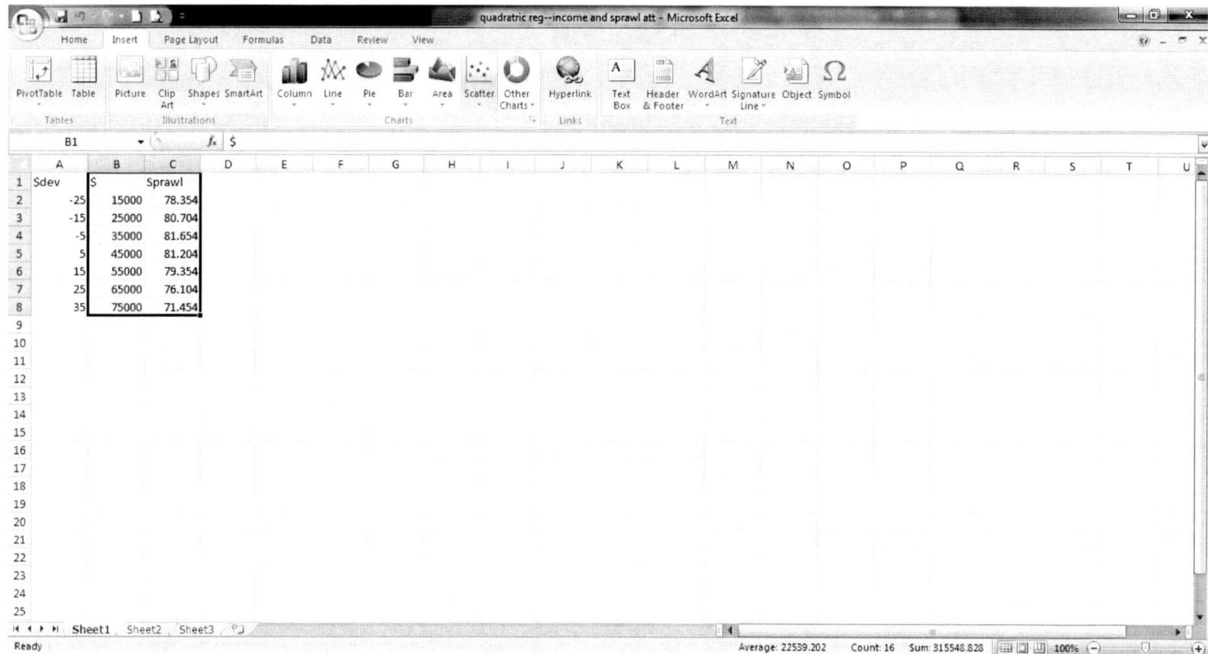

- An initial graph is produced that can be modified. Click on Layout 1 of the Chart Layouts. Axis Title and Chart Title should now appear in the graph. For the axis titles, replace with the names of your variables and provide the graphic with an appropriate title. If the legend is uninformative, as it is here, delete it.

- The numerous categories of Income make the graph look cluttered, and you can reduce the number of categories by right-clicking on any of the income amounts. From the menu that appears, click on Format Axis. Change the Major unit from Auto to Fixed. Change 10,000 to 20,000. The completed graph will look like the following graph.

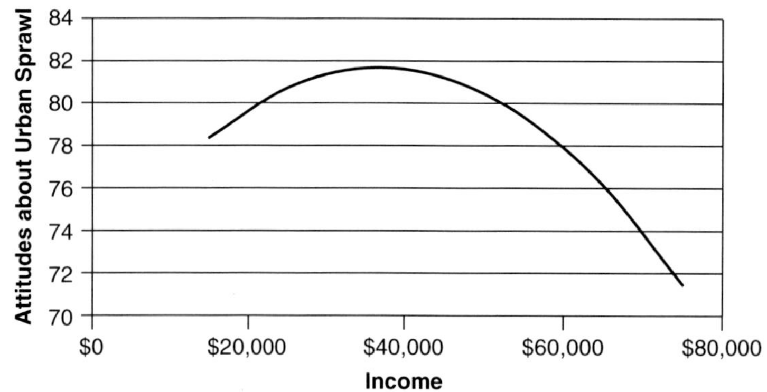

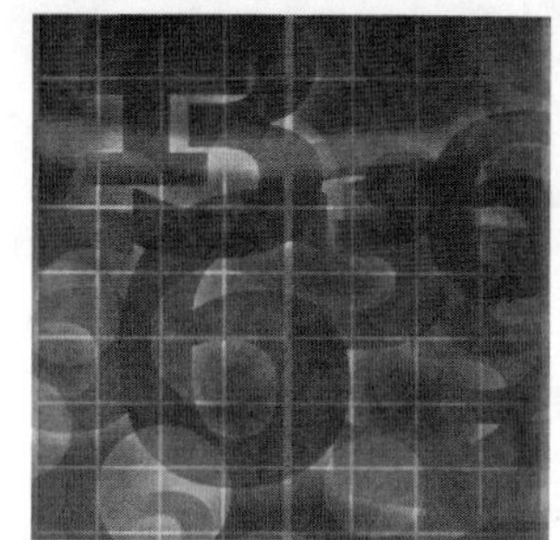

Learning Check

1. With the data provided below, calculate the a, b_1, and b_2. Y = Monthly Grocery Bill, X_1 = Household Size, and X_2 = Number of Times They Eat Out

	Grocery Bill	Household Size	Times Eat Out	$\bar{X}$	S
Grocery Bill	1.000			445.00	189.077
Household Size	.538	1.000		3.93	2.243
Times Eat Out	−.623	−.338	1.000	5.00	4.594

2. With the data from problem 1, calculate the standardized coefficients.

3. An urban sociologist analyzing the property values in a depressed section of the city regressed Property Values on the Age of Building, Type of Building, and Number of Times Sold ,with Interactions between Type of Building and Number of Times Sold. The types of buildings were: multiunit housing, nonresidential, vacant lot, and single-family home.

   (a) Set up *two* equations necessary to test whether Type of Building and Number of Times Sold interact in the prediction of Property Values. Make Single-Family home the reference category for the dummy variables.

   (b) Write the separate equations for the different types of buildings using the following results of the regression:

   $$\$ = 28{,}150.76 - 130.28\text{Age} + 4{,}454.82\text{Multi} - 27{,}563.26\text{Non-R} - 26{,}644.43\text{ VLot} \\ - 542.51\text{Sold} - 1005.28\text{MultiSold} + 1067.51\text{Non-RSold} + 544.47\text{VLotSold}$$

4. In the section of the chapter on Multiple Regression in SPSS, the recode of Respondent's Income (RINCOM06) and Total Family Income (INCOME06) was discussed. Complete the abbreviated worksheet for recoding these variables (see **Table 11.3**) and recode the variables within SPSS. Be sure to save the data after you recode the variables. With the income variables recoded, regress the recoded respondents' income measured in dollars on Highest Year of School Completed (EDUC) and the Number of Hours Worked Last Week (HRS1) by entering Highest Year of School Completed (EDUC), clicking Next, and then entering the Number of Hours Worked Last Week (HRS1). Be sure to define any missing values prior to running the regression.

5. Using the printout from problem 4, answer the following:

   (a) How much variance in the respondents' income does Highest Year of School Completed (EDUC) explain?

   (b) How much variance in the respondents' income do Highest Year of School Completed (EDUC) and the Number of Hours Worked Last Week (HRS1) explain?

   (c) Interpret the b in Model 1.

   (d) What happens to the b for Highest Year of School Completed (EDUC) when the Number of Hours Worked Last Week (HRS1) is added to the analysis?

   (e) Interpret the b for the Number of Hours Worked Last Week (HRS1) from Model 2.

6. Because of changing societal mores, one might suspect that having never married and age could interact when predicting the number of children a person has. Regress the Number of Children (CHILDS) on a dummy variable for Never Married, the Age of Respondent (AGE), and an interaction term for Never Married and Age. Check to be sure that missing values are defined. Use Marital Status (MARITAL) to create a dummy variable for Never Married and an interaction term for the dummy variable Never Married and Age. In SPSS, when doing a regression with an interaction term, be sure to include the appropriate number of models and request the needed statistics.

7. We expect age to have a nonlinear effect on respondents' income; it should rise but then decline. However, is the effect the same for men and women? To answer this question, regress income on age, its quadratic term, and sex. Use the recoded respondent income in dollars variable and create a dummy variable for Sex. Check for possible multicollinearity. If you run into multicollinearity problems, you may have to compute a deviation score.

Chapter Twelve

Nonlinear Regression

Earlier chapters have focused on helping you learn the mathematics involved in statistical analysis, as well as providing information on how to interpret statistical relationships when using social science data. This chapter will shift the focus away from a mathematical approach to statistics and towards an applied approach that is more useful for sociology students. We will discuss the mathematical procedures involved in these more advanced techniques but will not provide in-depth examples or problems that need to be solved. This chapter will focus more on the use of SPSS and AMOS to create advanced analytical models and on the proper uses for and interpretation of these models.

Chapters 10 and 11 introduced you to the concept of bivariate and multivariate regression. These tests allow us to predict the value of Y when X is held at 0. We are then able to calculate the change in Y for every increase of 1 in the value of X. Linear regression equations calculate the intercept of Y and the slope of the regression line. Consider the data in Figure 12.1, for example.

It is easy to draw a regression line between our values for X and Y when we use two related continuous variables, such as education and income.

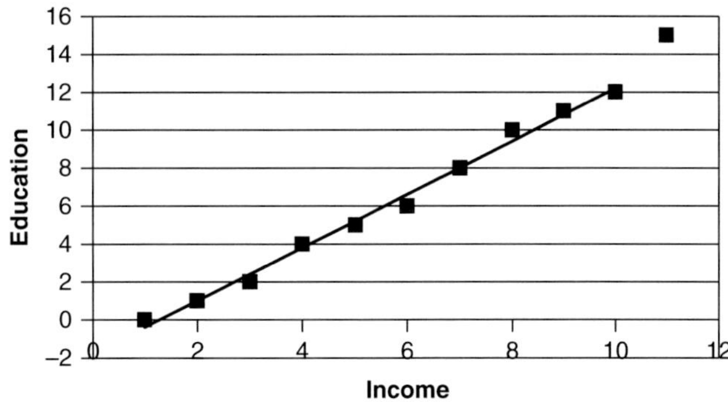

Figure 12.1

These tests become less useful, however, when our dependent variable is noncontinuous, for example, if it is a dummy variable (0, 1). Many of the variables that we assess as social scientists measure the presence or absence of a quality. For example, criminologists may be interested in studying arrests (1 = yes; 0 = no) or recidivism (1 = yes, 0 = no).

As you can see from the second scatterplot, a dummy variable graphs along 0 and 1, so there is no clear linear relationship between the two variables, even though they are likely related to one another.

Another issue with using a linear regression is that it is used to predict the value of Y for every increase in X. A linear regression assumes that the line has an infinite extension both above and below zero. For example, if we plot the relationship between income and wealth, we see that our regression line can extend into the negative and positive grids, and neither is capped. There is no mathematical limit to the amount of income a

Scatter Plot for Prior Arrests and Recidivism

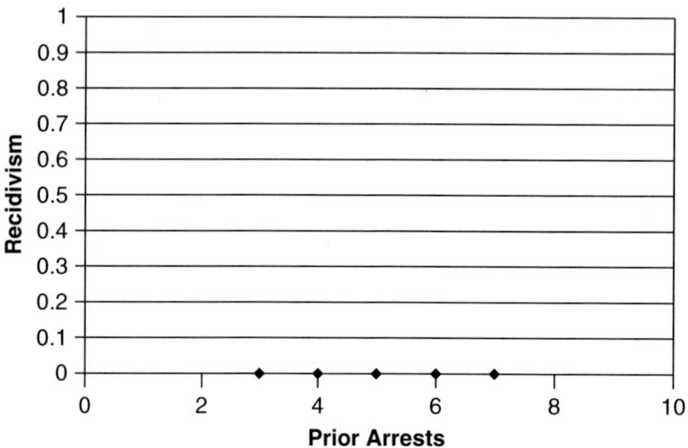

Figure 12.2

person can earn or the amount of debt a person can incur. (See **Figure 12.3**) When our dependent variable is a dummy variable, we can only have values of 0 and 1. These value limits make plotting a linear regression line problematic.

Finally, when we have a dummy variable as our dependent variable, we can't assume that an increase in X will have the same effect on Y as we near the higher and lower values of X. For example, think about the relationship between education and income. We are likely to see little difference in income between someone with an 8th-grade education and someone with a 9th-grade education. We might, however, see more income for someone with a high school diploma when compared to someone with an 11th-grade education or for someone with a college degree compared to someone with a high school diploma. On the higher end of the scale, we will have some variation in income for advanced degrees when we control for discipline, but we are likely to see little variation in income for those with 20 to 21 years of education.

Income and Wealth

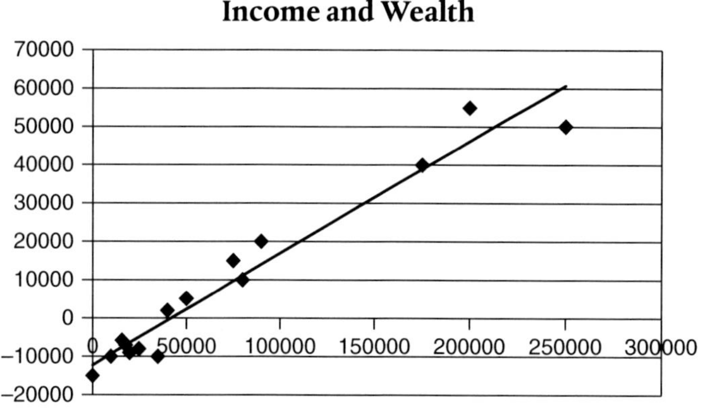

Figure 12.3

When we plot this relationship, we will get an S-shaped curve, which shows less variation at the high and low levels of X and more variation in the middle ranges.

Given the distribution of possible responses for a dummy variable, we do not attempt to predict how much change in Y is caused by every increase in X. Rather, we attempt to predict the odds that Y will occur with a change in X. Let's think back to the example used in **Figure 12.2**. We might hypothesize that the number of times someone has been arrested will have a predictive value for future recidivism. Since recidivism is a dummy variable (0 = no; 1 = yes), we want to predict the odds that someone with prior arrests will be arrested.

As stated previously, this chapter focuses on a practical application of logistic regression using SPSS and not on the mathematics used to calculate the results. However, it is important that you understand the theory behind the mathematics to fully recognize the differences between OLS regression and logistic regression. OLS assumes that data are linearly distributed, while logistic regression assumes a curvilinear distribution. As

OLS assumes the linear distribution, it calculates the constant (α = amount of Y that is present when $X = 0$) and the slope (β = amount that Y increases for every increase in X.) Logistic regression calculates the odds that an event will happen, or that Y will equal 1. In order to do this, we first calculate the odds ratio that Y will occur by dividing the probability by 1 minus the probability. In other words, for the example in the above paragraph, we want to divide the probability that a person will be rearrested by the probability that he or she will not be rearrested. This is expressed mathematically in the following equation.

$$O_j = \frac{P_j}{1-P_j}$$

(We have covered probability in a previous chapter.) Once we have found the odds ratio of Y occurring, we transform it to its logit by taking the natural log of those odds:

$$Logit_p = log\left(\frac{P_j}{1-P_j}\right) = log(P) - log(1-p)$$

The benefit of using the logit is that it can be greater than or less than 1, meaning that the distribution, as with nondummy variables, can be assumed to have a linear relationship that can extend into infinity in both directions. For example, if the odds ratio for recidivism for a person with three prior convictions is 2.5, this means that he or she is 2.5 times more likely to be rearrested than someone with 0 prior convictions. Suppose that the odds are -2.5 in this example, however. This would mean that the person is 2.5 times less likely to return to prison after three prior convictions.

Calculating logits allows us to produce an α for the variables in our model. It also transforms a nonlinear relationship into a linear one, which enables us to predict the effect that changing values of X will have on Y. When we apply the logistic function through logistic regression, however, we take the inverse of the logit, or the exponentiated form of the logit.

Think back to the logic of using the sum of squares when we computed a standard deviation. Without taking the sum of squares (SS), our differences ($x - \bar{x}$) would always equal zero. Because we used SS, we need to take the square root in our final step to transform our variance into the standard deviation that will adequately reflect the relationship of the mean to the distribution of the data. We can apply this same logical understanding to the logistic function in logistic regression. We want to predict the probability that Y will occur when we use logistic regression. Since probabilities are linear, we convert probabilities to logits. In order to transform our logits back to probabilities, we need to use the anti-log or the exponentiated form of our logit, as illustrated in the equation:

$$logit^{-1}(\alpha) = \frac{1}{1+exp(-\alpha)} = \frac{exp(\alpha)}{1+exp(\alpha)}$$

Our logistic regression equation is as follows:

$$logit = \alpha + \beta_1 x_1 + \beta_i x_i$$

When we covert our logits back to probabilities, however, we have the following equation, where $\hat{Y}$ is the probability that an event will occur:

$$\hat{Y} = exp\beta_0 * exp\beta_1 x_1 * exp\beta_i x_i$$

Interpretation and SPSS

The good news for most sociology students is that SPSS not only calculates our logits but also converts them back to the exponentiated form in logistic regression. As long as you are able to read the output and appropriately apply it to the regression equation, interpreting a logistic regression is no more difficult than working with OLS.

This next example uses data from a prison system rather than the GSS 2008 sample provided above in order to follow through with an example of using SPSS to test recidivism. Running a logistic regression equation in SPSS is similar to running an OLS regression. Open the data file and click on the Analyze menu. Once there, select Regression, and from that menu, select Binary Logistic Regression.

This will open the Analysis window so that you can select variables. As you can see below, we have moved recidivism into the Dependent Variable window and the number of prior prison terms served into the Independent Variable window. For now, we'll use the default settings and click on OK.

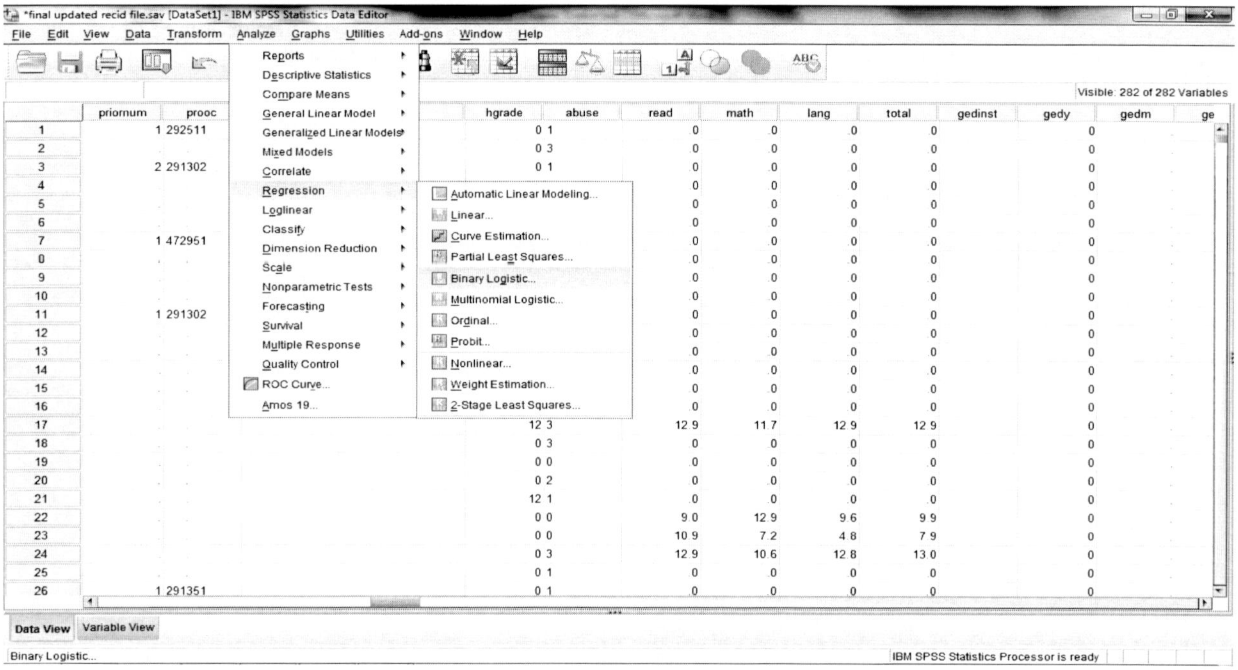

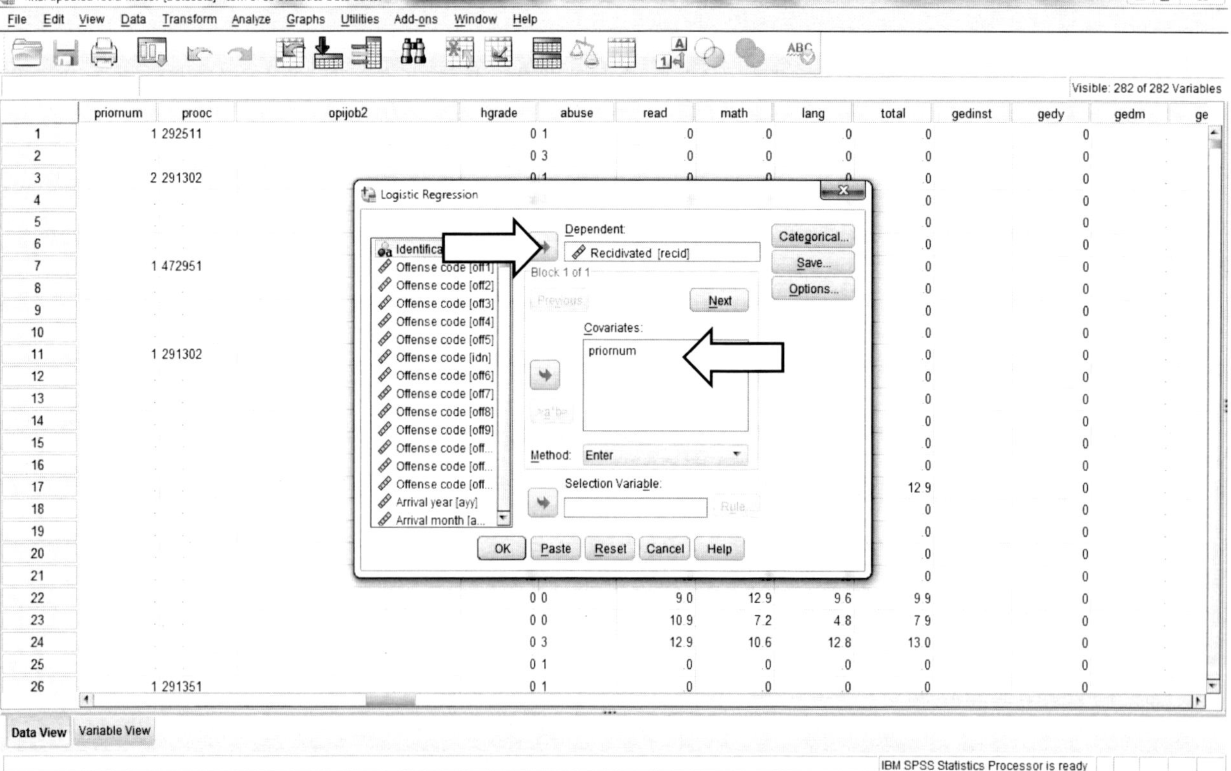

By default, SPSS first tests the null hypothesis. That analysis is called the block analysis. It then uses the variables in a step equation, which is similar to the stepwise equation in OLS. This enters one variable at a time into the equation to let you test various models. Finally, it runs the model with all of the entered variables. SPSS gives us the results shown in **Table 12.1** and described below.

The first thing to examine is the χ^2 for the null and the final model. We want to pay attention to the last column (Sig.). This is our significance level for the models. As you can see, the null and the research model are both significant ($\alpha < .001$). This is because we are dealing with recidivism data and the high probability of being rearrested after serving time in prison. **Table 12.2** gives the exponentiation of the β; this is our odds ratio. As you can see, it is 1.753. That means that the probability of spending a second term in prison is roughly 75% for everyone in the sample.

Table 12.1 *Omnibus Tests of Model Coefficients*

		Chi Square	*df*	Sig.
Step 1	Step	76.077	1	.000
	Block	76.077	1	.000
	Model	76.077	1	.000

Table 12.2 *Variables in the Null Equation*

	B	SE	Wald	*df*	Sig.	Exp.(β)
Constant Step 0	.561	.028	415.796	1	.000	1.753

Logistic regression doesn't calculate an R^2 like an OLS regression. However, SPSS computes several equations that attempt to function in the same way. (See **Table 12.3**.) As you can see, the Cox & Snell results are much lower than the Nagelkerke results. If we choose to interpret these, the best language to use would be conservative estimation. We could say that the prior number of arrests likely accounts for between 1.3% and 1.8% of recidivism.

Table 12.3 *Model Summary*

Step	-2 Log Likelihood	Cox & Snell R Square	Nagelkerke R Square
1	7405.972	.013	.018

Now we can turn to the results of our model summary and begin the interpretation process. Our model is summarized in **Table 12.4** below. Again, we want to pay attention to the significance level and exponentiation of our βs. We can see that the prior number of arrests is a significant predictor for recidivism ($\alpha < .001$), as is the first arrest ($\alpha < .05$). To determine the odds or recidivism for someone with four prior arrests, we simply need to plug our results into our regression equation.

Remember that the rules of probability apply here too. Since our outcomes are related and we are using the exponentiated form of our logged odds ratio, then we must apply the multiplicative rule here.

$$\hat{Y} = exp\beta_0 * exp\beta_1 x_1 * exp\beta_i x_i$$

$$\hat{Y} = 1.146 * (4*1.309) = 6.0$$

Table 12.4 *Variables in the Equation*

		β	SE	Wald	*df*	Sig.	Exp(β)	95% CI for Exp.(β)	
								Lower	Upper
Step 1[a]	Priornum	.270	.032	69.761	1	.000	1.309	1.229	1.395
	Constant	.136	.057	5.728	1	.017	1.146		

[a]*Variable(s) entered in step 1: priornum.*

This means that a person with four prior prison terms is 6 times more likely to return to prison than someone with no prior prison terms.

Regression with an Ordinal Dependent Variable

In previous chapters, we discussed the difficulty in using parametric tests when we have an ordinal dependent variable. A form of logistic regression, called ordinal regression, works well when we have an ordinal dependent variable (DV). There are several approaches that we can use to examine the odds ratio for an ordinal dependent variable. The first is to convert the ordinal variable into a dummy variable and run a simple logistic regression.

Suppose we are interested in looking at quality of life and age. Our hypothesis is that middle-aged adults will report higher qualities of life than younger and older adults. This means that we are hypothesizing an S-shaped curve between X and Y. Our dependent variable is:

How satisfied are you with your quality of life?

1. Very dissatisfied
2. Somewhat dissatisfied
3. Somewhat satisfied
4. Very satisfied

We can use a recode command in SPSS to create a new satisfaction variable whereby all responses 1 and 2 are recoded as zero (dissatisfied), and 3 and 4 are recoded as 1 (satisfied). Then we can use this new dummy variable to test the odds ratio that a person will say that he or she is satisfied versus dissatisfied. While this is a legitimate analysis plan, we also have the option of entering the variable as it is into an ordinal regression. The advantage of using an ordinal regression is that it allows us to calculate cumulative odds. In other words, how much more likely is someone to choose 1 in response to the question above than they are to give any other response? This is an advantage over just determining whether a person places himself or herself in a satisfied versus dissatisfied category.

Ordinal regression assumes that your categories of dependent variable are ordered so that categories can be compared using $>$ and $<$ symbols. It is not appropriate for a nominal dependent variable. It also assumes that $-\infty < Y < \infty$. This means that it assumes that Y can be plotted in a linear fashion, even though the ordinal variable has the same limitation of the dummy variable in its nonlinear distribution. Finally, ordinal regression calculates the cumulative odds ratio of the dependent variable. In the ordinal regression ,Y is unobserved and must be calculated through a series of complex equations. In other words, we can't look at our variable and "see" the cumulative odds ratio.

As previously mentioned, the equations used to calculate the cumulative odds ratio are complex. Given that, we will discuss the mathematics required to obtain these odds ratios but will focus on the use of SPSS and the interpretation of the results.

Ordinal regression calculates the following:

$$\theta = \frac{Probaility\ of\ j}{Probaility\ of > j}$$

This means that it calculates a probability for every value of Y. It calculates the probability that someone will select 1 over the probability that they will select >1, the probability that someone will select 2 over the probability that they will select >2, and so on. There is no calculation for the highest level of Y, in this case 6, because there is no probability of selecting a value higher than 6 in this dependent variable. In every ordinal regression, you will have $j - 1$ calculations of θ. This basically means that ordinal regression calculates an α for every value of Y.

Once we have our coefficients, we can plug them into our regression equation:

$$ln(\theta_j) = \alpha_j - \beta X$$

You will notice that we are using a negative relationship here. This is because ordinal regression predicts the probability of selecting lower values of Y. A positive coefficient means that as X increases, values of Y are also likely to increase. This means that lower scores of Y are less likely to occur when values of X are high. We have to subtract βX to reflect this relationship.

For this example, we are going to test the relationship between sex (SEX) and highest degree earned (DEGREE). We have added category 7 (IAP) to the missing values list in DEGREE. This gives us the following response categories:

Degree	SEX
0–"Less than high school"	1–Male
1–"High School"	2–Female
2–"Junior College"	
3–"Bachelor"	
4–"Graduate"	

Before moving on, there are a few things to remember.

1. Since we are using the logit in this formula, we will need to convert our probabilities back to the exponentiated form before interpreting them.
2. Ordinal regression calculates cumulative odds. This means that our thresholds, or α's, predict the probability of obtaining our score of interest or a lower score. In other words, our α for a score of 2 in our variable tells us the probability of an individual answering 1 OR 2 on our question.
3. We need to subtract our β in our regression equation because we are predicting the probability of selecting lower probabilities. A positive coefficient means that as X increases, Y also increases. Since this process predicts the probability of lower answers, we would say that as our coefficient increases, the probability of selecting a lower score decreases. This is why we subtract rather than add in ordinal regression.

Ordinal Regression in SPSS

For this example we will create our output in SPSS and then discuss each part and how to interpret it. Open SPSS and GSS 2008. Click on the Analyze type and select Ordinal from the Regression menu. This will open the Analysis window. Move Degree into the Dependent slot and Sex into the Factor slot.

Before we click OK to run the analysis, we want to change two default settings. On this same window, click on Output and check Estimated Response Probabilities under the Saved Variables menu and Test of Parallel Lines under Display.

Now click Continue and then OK to run the analysis. We have one final analysis to run before leaving SPSS. Click on the Descriptive Statistics tab and run Descriptives on your estimated response probabilities. (See Chapter 2.) SPSS saves these at the end of the data file. Since you have five categories in DEGREE, you will have four estimated responses. Probability is not calculated for the last response, which will be explained below.

SPSS puts the goodness of fit tests first in the output, but we'll come back to those. First we want to examine the thresholds and β for our model.

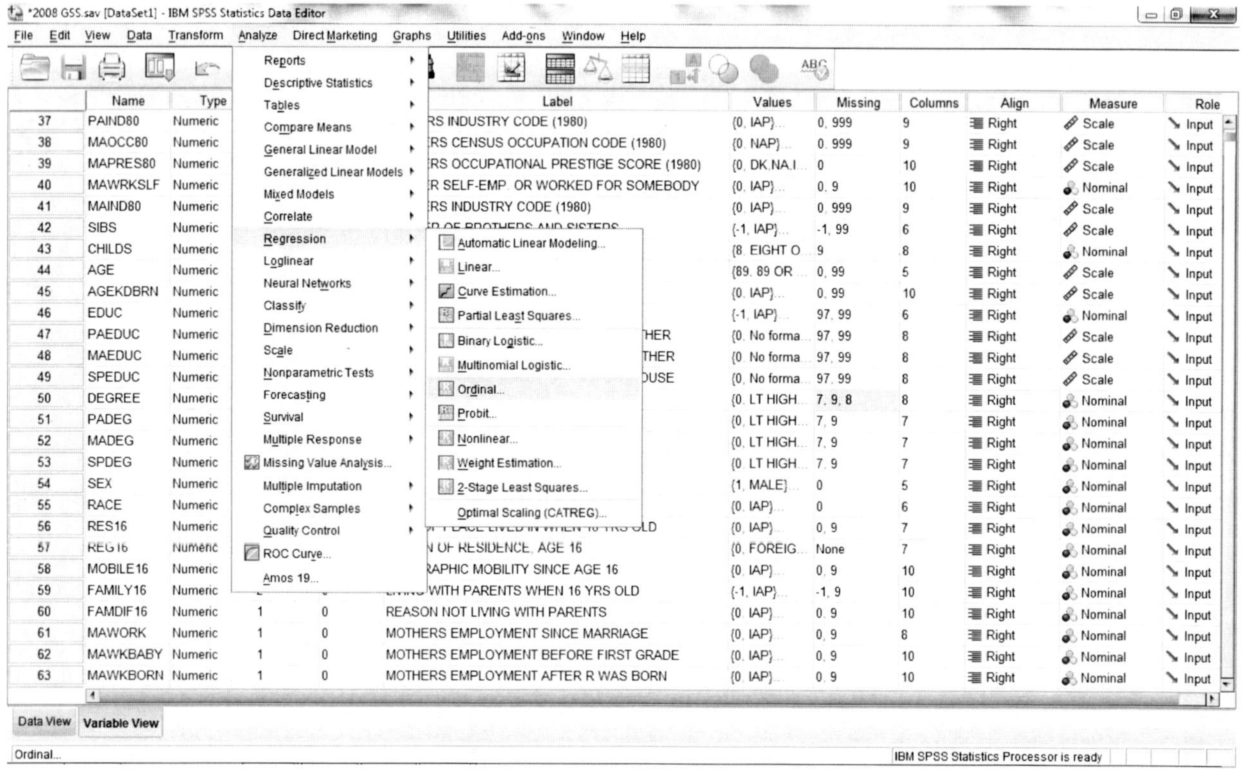
*2008 GSS.sav [DataSet1] - IBM SPSS Statistics Data Editor
File Edit View Data Transform Analyze Direct Marketing Graphs Utilities Add-ons Window Help
Reports
Descriptive Statistics
Tables
Compare Means
General Linear Model
Generalized Linear Models
Mixed Models
Correlate
Regression
Loglinear
Neural Networks
Classify
Dimension Reduction
Scale
Nonparametric Tests
Forecasting
Survival
Multiple Response
Missing Value Analysis...
Multiple Imputation
Complex Samples
Quality Control
ROC Curve...
Amos 19...
Automatic Linear Modeling...
Linear...
Curve Estimation...
Partial Least Squares...
Binary Logistic...
Multinomial Logistic...
Ordinal...
Probit...
Nonlinear...
Weight Estimation...
2-Stage Least Squares...
Optimal Scaling (CATREG)...
Data View   Variable View
Ordinal...
IBM SPSS Statistics Processor is ready

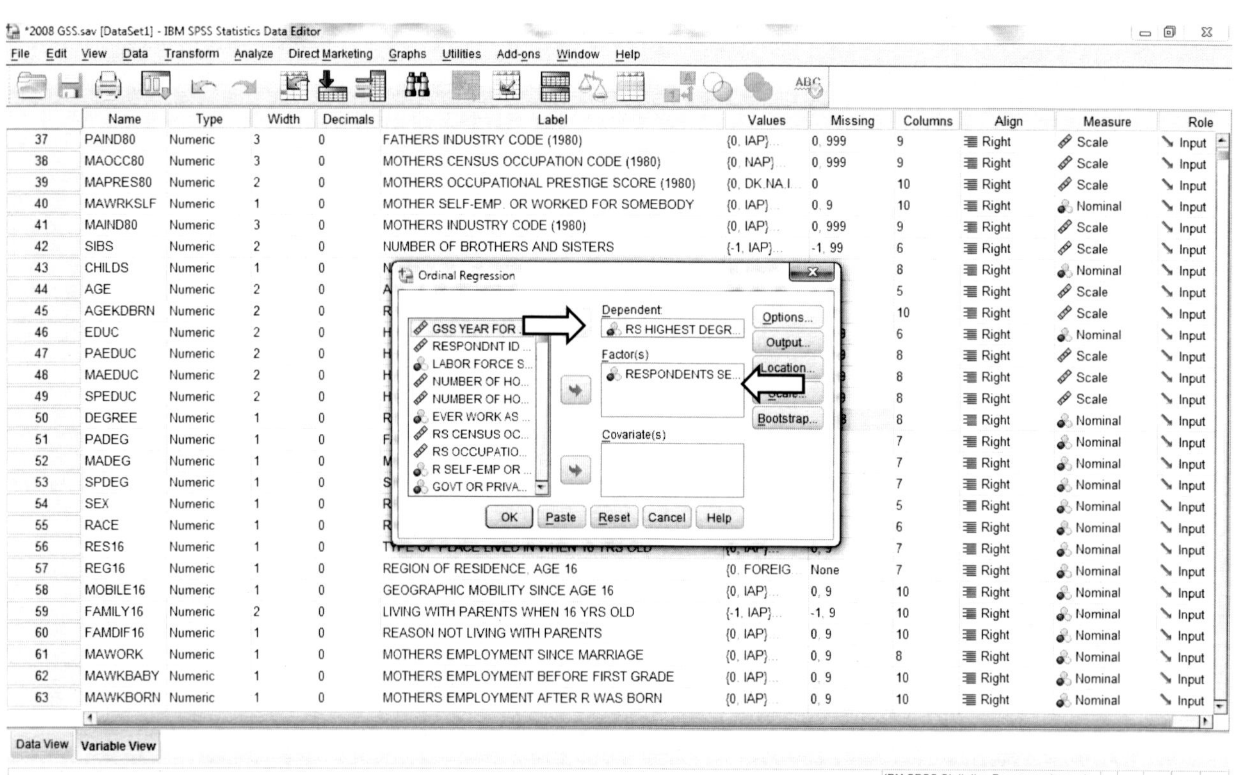
*2008 GSS.sav [DataSet1] - IBM SPSS Statistics Data Editor
File Edit View Data Transform Analyze Direct Marketing Graphs Utilities Add-ons Window Help
Ordinal Regression
GSS YEAR FOR
RESPONDNT ID
LABOR FORCE S...
NUMBER OF HO...
NUMBER OF HO...
EVER WORK AS
RS CENSUS OC...
RS OCCUPATIO...
R SELF-EMP OR
GOVT OR PRIVA...
Dependent:
RS HIGHEST DEGR...
Factor(s):
RESPONDENTS SE...
Covariate(s):
Options...
Output...
Location...
Scale...
Bootstrap...
OK   Paste   Reset   Cancel   Help
Data View   Variable View
IBM SPSS Statistics Processor is ready

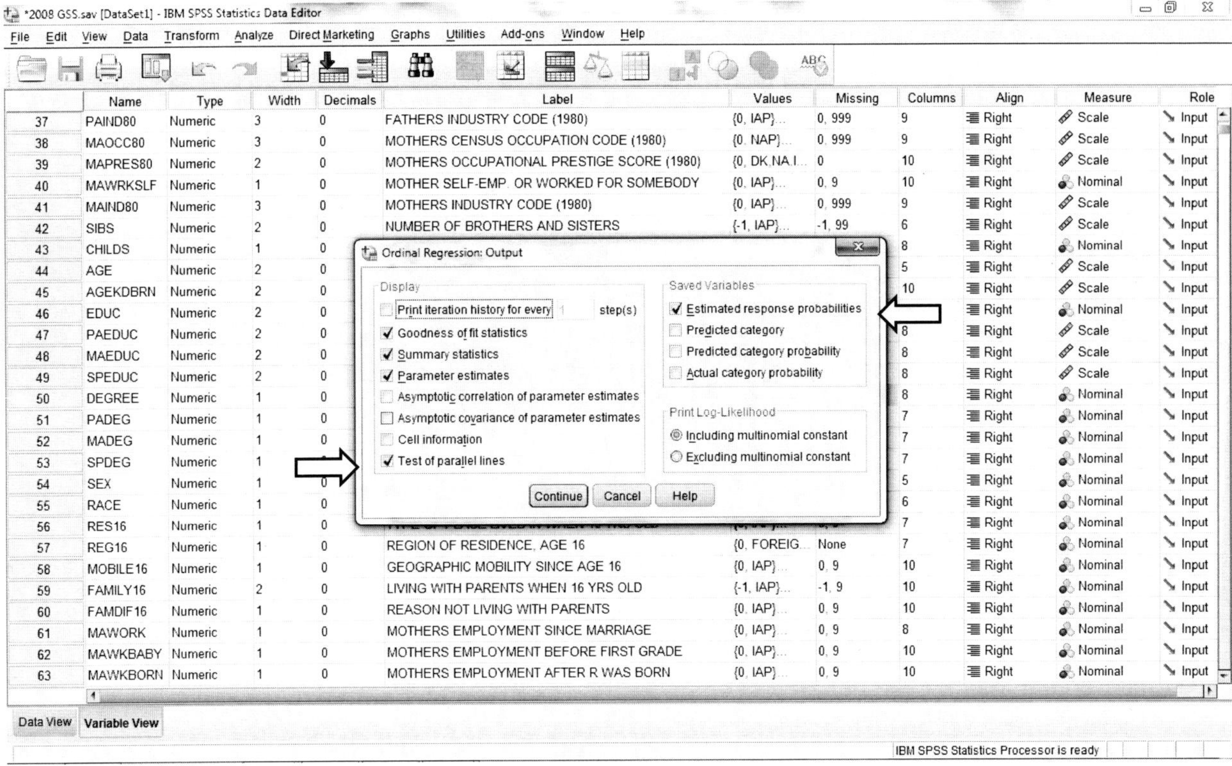

Table 12.5 *Parameter Estimates*

		Estimate	Std. Error	Wald	*df*	Sig.	95% Confidence Interval	
							Lower Bound	Upper Bound
Threshold	[DEGREE = 0]	−1.734	.073	561.092	1	.000	−1.877	−1.590
	[DEGREE = 1]	.614	.060	103.742	1	.000	.496	.732
	[DEGREE = 2]	1.013	.063	257.250	1	.000	.889	1.137
	[DEGREE = 3]	2.269	.085	714.175	1	.000	2.103	2.436
Location	[SEX = 1]	.056	.083	.452	1	.501	−.107	.218
	[SEX = 2]	0[a]	.	.	0	.	.	.

Link function: Logit. This parameter is set to zero because it is redundant.

Remember that the thresholds for the variable are similar to the α's that we get when we run an OLS regression. For the time being, we're going to ignore these and discuss the calculations under Location. These are our βs. We have a zero for Sex = 2 because this is our reference category and a coefficient isn't calculated. We need to interpret the direction, strength, and significance of the β. As we can see, there is not a strong or significant relationship between SEX and DEGREE ($\beta = .056$, $p = .501$). When we interpret this we would say that women tend to have higher levels of degrees; however, the difference is small and nonsignificant. We know that women tend to be more educated because our coefficient is positive and our category 2 in SEX = female. As the X increases, Y increases. We can take the exponentiated form of our $\beta(e^{-\beta})$ to determine the odds ratio for males indicating lower categories of degrees versus females ($e^{-\beta} = 1.945$). As you can see, males are less than 1% less likely to report higher levels of education than females. We can say from this analysis that sex is not an important predictor of the level of degree that an individual attains.

While this is not a significant relationship, it is an adequate example to illustrate how to interpret an ordinal regression. Our next step is to calculate our expected values for our cumulative predicted probabilities. The formula for this is:

$$Probablity_i = \frac{1}{1 + exp^{-(\alpha)}}$$

In this formula, "i" refers to every category in our dependent variable. Remember, each category has its own threshold, or α, so we have to calculate a probability for each category. We don't calculate a probability for category 4 because it is our highest level. The final category always has a probability = 1 because the probability of selecting that category or one lower is 100%.

$$Prob_0 = \frac{1}{1 + exp^{1.734}} = .15$$

$$Prob_1 = \frac{1}{1 + exp^{-.614}} = .649$$

$$Prob_2 = \frac{1}{1 + exp^{-1.013}} = .734$$

$$Prob_3 = \frac{1}{1 + exp^{-2.269}} = .906$$

These calculations give us the cumulative probabilities. That means that $Prob_0$ is the probability of indicating "Less than High School" as the highest degree achieved. $Prob_1$ is the probability of indicating "Less than High School" **OR** "High School" as the highest degree achieved. In order to determine the probability for each category, we simply subtract. Our first category doesn't need to be altered.

$$Prob_{Less\ than\ High\ School} = \mathbf{.15}$$

$$Prob_{High\ School} = .649 - .15 = .5$$

$$Prob_{Junior\ College} = .734 - .649 = .085$$

$$Prob_{Bachelors} = .906 - .734 = .17$$

$$Prob_{Graduate} = 1 - .906 = .094$$

This means that there is a 15% probability that, when asked, a person in the United States would report that he or she had less than a high school diploma in 2008 and so on. Notice too. that the probabilities are lower at the lowest and highest ends of our scale (15% for less than high school and slightly less than 1% for graduate degrees). The majority of responses fall within the mid ranges, which would produce an S-shaped curve. In practice, you will not need to calculate these.

As you can see, our predicted responses are listed in the Mean for the variables. You have a standard deviation of zero because the program simply assigns the value to every case in the sample.[1]

---

[1] Note that newer versions of SPSS may assign variations of the equation by case. For example, SPSS 19 assigned .49 to 50% of the cases for $prob_2$ and .5 to the other 50%, which produced a range and a standard deviation that was less than .01. This is not a problem for your analysis if you get these results, but rather a function of the program.

Table 12.6 *Descriptive Statistics*

	N	Minimum	Maximum	Mean	Std. Deviation
Estimated Cell Probability for Response Category: 0	2023	.15	.15	.1500	.00000
Estimated Cell Probability for Response Category: 1	2023	.50	.50	.5000	.00000
Estimated Cell Probability for Response Category: 2	2023	.09	.09	.0850	.00000
Estimated Cell Probability for Response Category: 3	2023	.17	.17	.1700	.00000
Estimated Cell Probability for Response Category: 4	2023	.09	.09	.0940	.00000
Valid N (listwise)	2023				

The next step is to determine our probabilities when we take into account our β. The only difference in what you've already done here is that you add the β to the calculations.

$$Prob_0 = \frac{1}{1 + exp^{1.734 - .056}} = .157$$

$$Prob_1 = \frac{1}{1 + exp^{-.614 - .056}} = .662$$

$$Prob_2 = \frac{1}{1 + exp^{-1.013 - .056}} = .744$$

$$Prob_3 = \frac{1}{1 + exp^{-2.269 - .056}} = .911$$

Since sex is not a significant factor for education in our sample, the probabilities when we include β are similar to the probabilities without β.

Goodness of Fit Tests

There are several goodness of fit tests for ordinal regression. Several of these tests may feel counterintuitive to what you've learned about goodness of fit up to this point in our analyses. The first test that we concern ourselves with is the χ^2. This test tells us whether the model with the predictors is better than the model without the predictors. In this example, our IV did not produce a significant β, so we have a nonsignificant χ^2. This indicates that this model does not have a good fit.

Table 12.7 *Model Fitting Information*

Model	−2 Log Likelihood	Chi-Square	*df*	Sig.
Intercept Only	53.070			
Final	52.618	.452	1	.501

Link function: Logit.

The next goodness of fit tests are the Pearson's and Deviance χ^2. Unlike the model χ^2 above, these tests tell us how well the expected and observed frequencies are correlated. Remember, the χ^2 examines the relationship between expected and observed frequencies. For these χ^2s, we want to have **nonsignificant** differences. As you can see in **Table 12.8**, both results are nonsignificant, so we can argue that the ordinal regression is an appropriate model for our dependent variable. Remember that χ^2 is a function of sample size, so this test works best with variables that do not have a lot of empty cells or missing data.

Table 12.8 *Goodness-of-Fit*

	Chi-Square	*df*	Sig.
Pearson	.9033	3	.825
Deviance	.9033	3	.825

Link function: Logit.

The next test we should discuss is the pseudo-R^2. Remember that logistic regression does not have a calculation that is similar to the R^2, so a variety of substitutes have been developed. As you can see below, all of our pseudo-R^2s are zero here because there is no significant relationship between the DV and IV in our model. If you interpret these at all, you should interpret conservatively and with caution.

Table 12.9 *Pseudo R-Square*

Cox & Snell	.000
Nagelkerke	.000
McFadden	.000

Link function: Logit.

Finally, we test the parallel lines for our models. Ordinal regression assumes that the IV has the same effect on every category of our DV. That is why you only have one β. This means that when we plot our data by category, we expect to have a series of parallel lines. The test of parallel lines determines whether that is true. In this test, our null is that our data are parallel.

Table 12.10 *Pseudo R-Square*

Cox & Snell	.000
Nagelkerke	.000
McFadden	.000

Link function: Logit.

Test of Parallel Lines[a]

Model	-2 Log Likelihood	Chi-Square	*df*	Sig.
Null Hypothesis	52.618			
General	51.716	.903	3	.825

The null hypothesis states that the location parameters (slope coefficients) are the same across response categories.
[a]Link function: Logit.

Given this, we are interested in proving the null, which means we want to have a nonsignificant χ^2. If our χ^2 is significant, it means that the IV has a different effect on some or all of the categories of the DV and ordinal regression is not appropriate for the data. As you can see, we have a nonsignificant χ^2, which tells us that our lines are in fact parallel.

Interaction Effects in Logistic Regression

The last thing that we will discuss in this chapter is the use of interaction effects. Since we use social data, we recognize that outcomes are often influenced by several variables, as well as the interaction of those variables. Consider, for example, the roles that religion and gender play in shaping our values and beliefs. We know, for example, that females tend to be more religious than males and that those who are very religious tend to be politically conservative. Finally, we know that those who report themselves to be strongly religious are more likely to answer "no" to the question: Should a woman be allowed to have an abortion for any reason?(ABANY) The example that we will use to explore interaction effects is the relationship between sex and religiosity in determining whether or not an individual will respond "yes" to this question. Our null hypothesis for this example is: Sex and level of religiosity will not interact to influence responses to ABANY.

In order to test this relationship, we must first create dummy variables for SEX (SEX_D ,where Female $= 1$) and for levels of religiosity. For this example, we are going to use the variable PRAY in GSS 2008. PRAY asks: How often do you pray? and measures a private behavior that may be a better reflection of spiritual beliefs than: How often do you attend religious services? or How religious are you? We need to calculate three dummy variables for our calculations. $VERY_D$ will code those who respond "more than once a day" and "once a day" as 1 and everyone else as 0. $SOME_D$ will code those who say "several times a week,", "once a week," and "less than once a week" as 1 "Never" is our comparison level and so we don't need to create a variable for that level. Finally, we need to recode ABANY to create a dummy variable ($ABANY_D$) where "yes" $= 1$ and "no" $= 0$.

Calculating an interaction is simply a function of multiplying $VAR_{D1} * VAR_{D2}$. While you can use the SPSS windows to compute this new variable, the simplest way to do so is to open a new syntax window and type the following command:

```
Compute SEXVERY=VERY * SEXD.
Compute SEXSOME = SOME * SEXD.
```

This creates our new interaction terms SEXVERY and SEXSOME as variables at the end of the data set. We are now ready to run our logistic regression analysis. $ABANY_D$ is our dependent variable and all of our dummy variables and interaction terms are our independent variables. (See the first section of this chapter for instructions on running a logistic regression is SPSS.)

Interpreting the Results

Table 12.11 provides the results of our regression analysis. We can tell at first glance that there is no significant relationship for our interaction term (SEXVERY $e^\beta = .793, p = .595$; SEXSOME $e^\beta = .504; p = .793$). However, we do want to determine if there is a difference in the way religious males and females answer the question: Do you believe a woman should be allowed to have an abortion for any reason?

The calculations explained below will allow us to determine if there are differences by gender for those who indicate that they pray often, occasionally, or never. Since "Never" is our reference group, we can start with that.

First, compute the odds ratios of males and females that say "Never," This is a simple issue for males; since they are our reference group, their odds ratio equals the constant. Remember that the constant represents individuals who have a zero for every value. In this example, this means males who never pray. Thus, the odds ratio for males who never pray is 2.273. To calculate the odds ratio for females, we simply add the β to our regression equation.

$$OR_{female} = 2.273 * 1.48 = 3.364$$

Table 12.11 *Variables in the Equation*

		B	SE	Wald	df	Sig.	Exp(B)	95% C.I. for EXP(B)	
								Lower	Upper
Step 1[a]	SEXD	.392	.402	.951	1	.329	1.480	.673	3.254
	very	−1.722	.248	48.074	1	.000	.179	.110	.291
	some	−.752	.252	8.924	1	.003	.471	.288	.772
	sexvery	−.231	.435	.283	1	.595	.793	.338	1.861
	sexsome	−.302	.452	.446	1	.504	.740	.305	1.793
	Constant	.821	.209	15.446	1	.000	2.273		

[a]*Variable(s) entered in step 1: SEXD, very, some, sexvery, sexsome.*

Now we can calculate the difference between males and females who never pray. If there is no difference between the genders, our difference would be 1, taking into account error.

$$\frac{OR_f}{OR_m} = \frac{3.364}{2.273} = 1.48$$

Since this equation does not take into account the effect of religion, you will see that the difference between males and females who never pray is equal to the exponentiated form of our β for SEX_D. Our significance levels and confidence intervals let us know that there is not a significant gender difference in the ways nonreligious individuals answer our ABANY question. In other words, we can't say that the difference between 1.48 and 1 can be attributed to anything other than sampling error.

Now we need to calculate the differences for males and females who are very religious and those who are somewhat religious.

Very Religious:

$$\frac{OR_{vrf}}{OR_{vrm}} = \frac{2.273*1.48*.179}{2.273*.179} = \frac{.6022}{.4069} = 1.48$$

Somewhat Religious:

$$\frac{OR_{sf}}{OR_{sm}} = \frac{2.273*1.48*.471}{2.273*.471} = \frac{1.5844}{1.0705} = 1.48$$

As we can see, the odds ratio for males and females by religion are equal at all levels. This indicates that sex is not an important predictor for attitudes on ABANY regardless of religious practices. In other words, religion is important with the odds ratio of the very religious being lower than the somewhat religious (.471:.179), but the sex of the respondent is not an important factor.

Name: ____   Date: ________________

Exercises and Notes

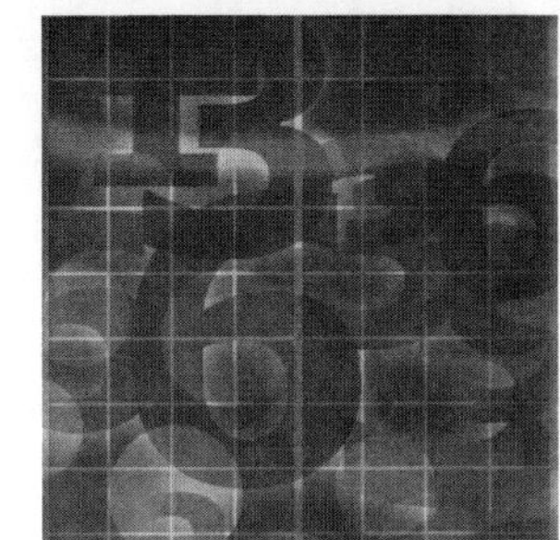

1. A researcher is interested in examining the relationship between political participation and gender. Political participation is coded:

   0–Never vote

   1–Always vote

   2–Campaign for office

   3–Run for office

   4–Elected official

   Which regression analysis would the researcher need to run? _________________________

2. A researcher is interested in examining the relationship between pet ownership and age. Pet ownership is coded:

   0–Do not own a pet

   1–Own a pet

   Which regression analysis would the researcher need to run? _________________________

3. What three things to do we need to remember when interpreting ordinal regression?

   a. _______________________________

   b. _______________________________

   c. _______________________________

4. We hypothesize that females are more likely to have jobs that require the use of a computer. Consider the following SPSS output:

Variables in the Equation

		B	SE	Wald	*df*	Sig.	Exp(B)	95% CI for EXP(B) Lower	Upper
Step 1[a]	SEXD	.409	.160	6.555	1	.010	1.506	1.101	2.060
	Constant	−.389	.109	12.702	1	.000	.678		

[a]*Variable(s) entered in step 1: SEXD.*

Dependent variable

Could you do your job without a computer?

Independent variable

Sex

0 Yes   0 Male
1 No   1 Female

Parameter Estimates

		Estimate	Std. Error	Wald	df	Sig.	95% Confidence Interval	
							Lower Bound	Upper Bound
Threshold	[XMARSEX1 = 1]	1.655	.100	275.248	1	.000	1.459	1.851
	[XMARSEX1 = 2]	2.822	.129	480.146	1	.000	2.570	3.074
	[XMARSEX1 = 3]	4.119	.203	411.739	1	.000	3.721	4.516
Location	[SEXD = .00]	.564	.136	17.314	1	.000	.299	.830
	[SEXD = 1.00]	0[a]	.	.	0	.	.	.

Link function: Logit.
[a]*This parameter is set to zero because it is redundant.*

Calculate the following probabiliites:

Males who need computers to do their job _______________________

Females who need computers to do their job _______________________

Is there a significant difference between males and females? _______________________

Which sex is more likely to need a computer to do their job? _______________________

5. We hypothesize that females will be more likely to report less acceptance of extramarital sex. Consider the following SPSS output:

Dependent Variable   **Independent variable**

Is extramarital sex wrong?   Sex

1. Always Wrong   0 Male

2. Almost Always Wrong   1 Female

3. Wrong Only Sometimes

4. Not Wrong at All

Calculate the expected probabilities for each category of response.

1. _______________

2. _______________

3. _______________

4. _______________

Calculate the expected probabilities for females in each response category.

1. _______________

2. _______________

3. _______________

4. _______________

Variables in the Equation

		B	SE	Wald	df	Sig.	Exp.(B)	95% CI for EXP(B)	
								Lower	Upper
Step 1[a]	SEXD	.487	.371	1.720	1	.190	1.627	.786	3.368
	very	1.602	.260	37.968	1	.000	4.965	2.982	8.265
	some	.626	.243	6.618	1	.010	1.870	1.161	3.013
	sexvery	−.455	.429	1.124	1	.289	.634	.274	1.471
	sexsome	.182	.443	.168	1	.682	1.199	.503	2.857
	Constant	.206	.194	1.127	1	.288	1.229		

[a]*Variable(s) entered in step 1: SEXD, very, some, sexvery, sexsome.*

6. We are interested in examining the interaction effect between gender and religiosity in responses to the question: Is extramarital sex wrong? We hypothesize that religious females will be more likely to report that it is always wrong. We will use the same religiosity variable that we created in the text example. Consider the following SPSS output:

Dependent variable

Is extramarital sex always wrong?

0 No

1 Yes

Independent variables

Sex

0 Male

1 Female

Very

    1 prays every day or more than once a day

Some

    1 prays, but less than once a day

Is the interaction between sex and religiosity important?

Calculate the odds ratios for gender for each level of religiosity:

Very ______________________

Somewhat ______________________

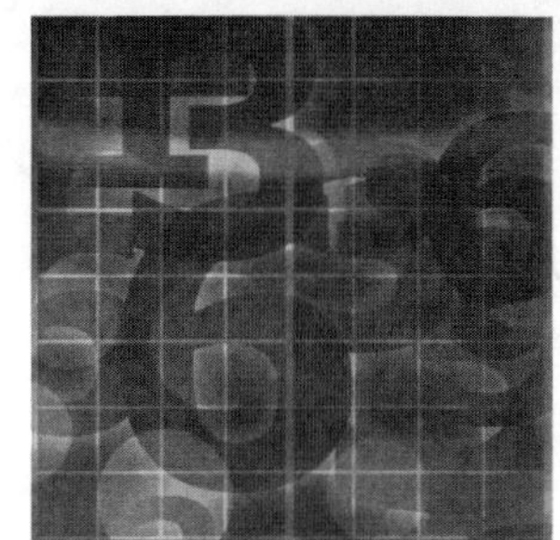

Learning Check

1. Discuss the limitations of OLS regression with nonparametric data.

2. Which types of variables require logistic regression?

3. Ordinal regression is a form of logistic regression. Discuss when you use ordinal regression instead of a binary logistic regression.

4. What is the function of the threshold in ordinal regression?

5. Why do we use the exponentiated β in logistic and ordinal regressions?

6. Why do we need to subtract β in ordinal regression?

7. What is the purpose of using an interaction term in logistic regression?

8. We hypothesize that women will be more likely to be pro-choice if a pregnancy is the result of rape. Consider the following SPSS output:

Variables in the Equation

		B	SE	Wald	*df*	Sig.	Exp.(B)	95% CI for EXP(B)	
								Lower	Upper
Step 1[a]	SEXD	−.348	.135	6.673	1	.010	.706	.542	.919
	Constant	1.391	.104	179.607	1	.000	4.017		

[a]*Variable(s) entered in step 1: SEXD.*

Dependent Variable

Should a woman be allowed to have abortion if pregnancy is the result of rape?

0 No

1 Yes

Independent Variable

Sex

0 Male

1 Female

Are there significant differences in the opinions of males and females?

Who is more likely to say yes?

What is the odds ratio for saying yes to this variable when comparing the sexes?

9. We would like to know if gender mediates religiosity when making a pro-choice argument if pregnancy is a result of rape. Our codes for ABRAPE are 0 = No, 1 = Yes. Consider the following SPSS output:

Variables in the Equation

		B	SE	Wald	df	Sig.	Exp.(B)	95% C.I. for EXP(B)	
								Lower	Upper
Step 1[a]	SEXD	.434	.822	.280	1	.597	1.544	.309	7.726
	very	−1.940	.412	22.132	1	.000	.144	.064	.322
	some	−.572	.451	1.604	1	.205	.565	.233	1.367
	sexvery	−.433	.838	.267	1	.605	.649	.126	3.351
	sexsome	−.816	.878	.864	1	.353	.442	.079	2.471
	Constant	2.679	.391	47.015	1	.000	14.571		

[a]*Variable(s) entered in step 1: SEXD, very, some, sexvery, sexsome.*

Does sex mediate religiosity?

Calculate the odds ratios for gender for each level of religiosity:

Very _______________________

Somewhat _______________________

10. We are interested in determining differences in response levels on the question: Is it wrong for a woman to get an abortion if she has a low income? We also are interested in determining if males and females respond differently to this question. Consider the following SPSS output:

Parameter Estimates

		Estimate	Std. Error	Wald	df	Sig.	95% Confidence Interval	
							Lower Bound	Upper Bound
Threshold	[ABPOORW = 1]	.028	.072	.152	1	.697	−.114	.170
	[ABPOORW = 2]	.439	.073	35.771	1	.000	.295	.583
	[ABPOORW = 3]	.915	.077	141.924	1	.000	.765	1.066
Location	[SEXD = .00]	.150	.103	2.101	1	.147	−.053	.352
	[SEXD = 1.00]	0[a]	.	.	0	.	.	.

Link function: Logit.
[a]*This parameter is set to zero because it is redundant.*

Dependent variable	**Independent variable**
Is it wrong for a woman to get an abortion if she has a low income?	Sex

1. Always Wrong	0 Male
2. Almost Always Wrong	1 Female
3. Wrong Only Sometimes	
4. Not Wrong at All	

Calculate the expected probabilities for each category of response.

1. _______________

2. _______________

3. _______________

4. _______________

Calculate the expected probabilities for females in each response category.

1. _______________

2. _______________

3. _______________

4. _______________

Chapter Thirteen

Data Reduction

As social researchers, we rely on the reports of individuals to assess their social attitudes and beliefs. However, as social researchers, we also recognize that beliefs are rarely written in black and white. Rather, they are a complex system of interrelated beliefs. This means that we often use constructs or identify latent variables to determine patterns of attitudes and beliefs in our data. Creating a construct is a process known as **data reduction.** Data reduction simply means that we are reducing the number of variables in our analysis by combining related variables. There are several reasons why we would want to do this.

One of the reasons we may want to reduce our data is to avoid overidentification in our model. For example, if our sample size is only 100, we are limited to using a maximum of ten variables in our analysis. As our sample increases, our variable limit increases, but adding more than ten variables often creates a model that is difficult to interpret. Also, highly correlated independent variables can increase our R^2 to the point of multi-collinearity or, alternatively, it can make an insignificant relationship appear to be significant.

The next reason that we use data reduction is to identify a **latent variable.** Latent variables are not observed but are inferred from the relationships that we believe exist in our data. Recall from Chapter 1 that a variable measures one attitude, such as: Do you feel safe at work? We may be more interested, however, in the variability of how safe people feel. How do we determine the attitude of a person who answers "yes" to this question but then pauses and says, "Well, mostly"? Given this type of scenario, we may decide to use a construct. A construct combines related variables to create an overall scale. A construct is a form of a latent variable. For example, if we are interested in measuring how safe people who work in the service industry, such as gas station attendants and restaurant employees, feel during night shifts, we might ask a series of questions. Suppose we ask the following questions and assign a value of 1 to "yes," and a value of 0 to "no":

S-1   Is the parking lot well lit and secure?
S-2   Do you feel safe being alone for much of your shift?
S-3   Do you work in a safe neighborhood?
S-4   Do you feel safe with your coworkers?
S-5   Do you feel safe with your customers?

Consider the following response patterns to the questions in **Table 13.1.**

By adding the number of the answers "yes" and "no" for all of our variables, we can create a construct "safety" that gives a cumulative score for how safe a person feels. A person such as John answers "yes" to every question and has a Safety score of 5. Bob, who does not feel safe at all, answers "no" to every question and has a safety score of 0. The other respondents in our sample vary in their responses and have levels that fall between Safe and Unsafe.

Look back at the safety questions above. It is logical to assume that someone who feels that he or she works in an unsafe neighborhood would not feel safe working alone for much of his or her shift. In other words, we would expect these variables to have a strong correlation. By combining the individual scores, we

Table 13.1 *Response Patterns*

Question	John	Jane	Mary	Tina	Bob
S-1	1	0	1	1	0
S-2	1	1	0	0	0
S-3	1	1	0	0	0
S-4	1	1	1	1	0
S-5	1	1	1	0	0
Safety Score	5	4	3	2	0

maximize the covariance by creating one underlying scale. We have also created a latent variable, safety, which allows us to determine how safe these workers feel overall. This can be much more useful in our analysis than a simple discussion of how many people feel that their parking lot is safe.

Another benefit of using the latent variable of safety is that we now have an interval variable that ranges from 0 to 5, which means that it can easily be used as a dependent variable in an OLS regression analysis if we are interested in how other variables influence feelings of safety, such as gender or race.

Creating a Construct in SPSS

For this example, we are going to use the questions in GSS 2008 that measure pro-choice attitudes. To simplify our analysis, we will use only ABDEFECT, ABHLTH, ABNOMORE, ABPOOR, and ABRAPE. Remember, these variables are coded 1 = yes and 2 = no. Since we have five variables, this means our new variable ABBEL (the score for pro-choice beliefs) will have a range of 5 to 10. Before we collapse our variables, it is useful to run a correlation matrix to determine if our variables are, in fact, correlated. (See Chapter 9 for help with the correlation if needed.) Since our data are nominal, we need to use Spearman's correlation. As you can see from **Table 13.2** below, all of our variables significantly correlate with one another. Furthermore, all of our correlations are positive, which means that they move in the same direction.

Table 13.2 *Correlation Matrix X*

	ABDEFECT	ABHLTH	ABNOMORE	ABPOOR	ABRAPE
ABDEFECT	1.000				
ABHLTH	.521**	1.000			
ABNOMORE	.510**	.299**	1.000		
ABPOOR	.474**	.305**	.803**	1.000	
ABRAPE	.605**	.498**	.455**	.453**	1.000

***Correlation is significant at the 0.01 level (two-tailed).*

Now we are ready to create our new variable, ABBEL. Select Compute for the Transform menu on the analysis bar.

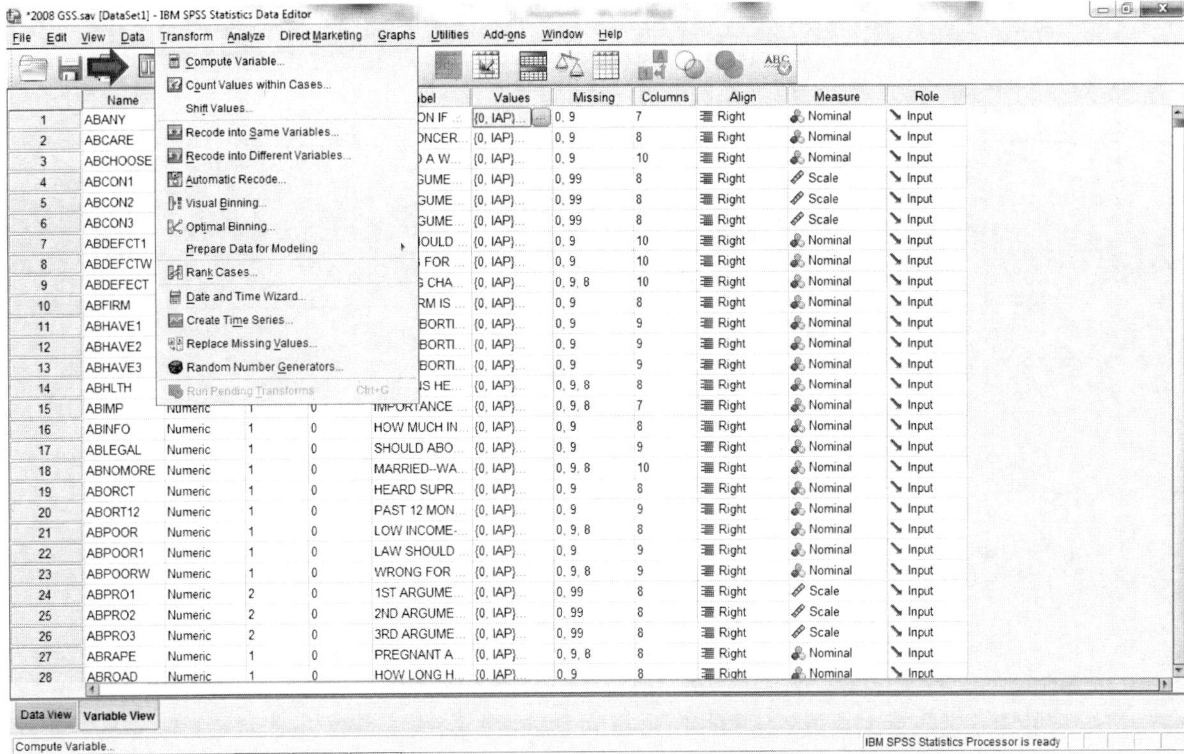

Figure 13.1

When the Compute window opens, we will create our new variable by adding our five target variables and clicking OK. This creates a new variable, ABBEL, in our data set.

We can now run a frequency on our new variable, ABBEL.

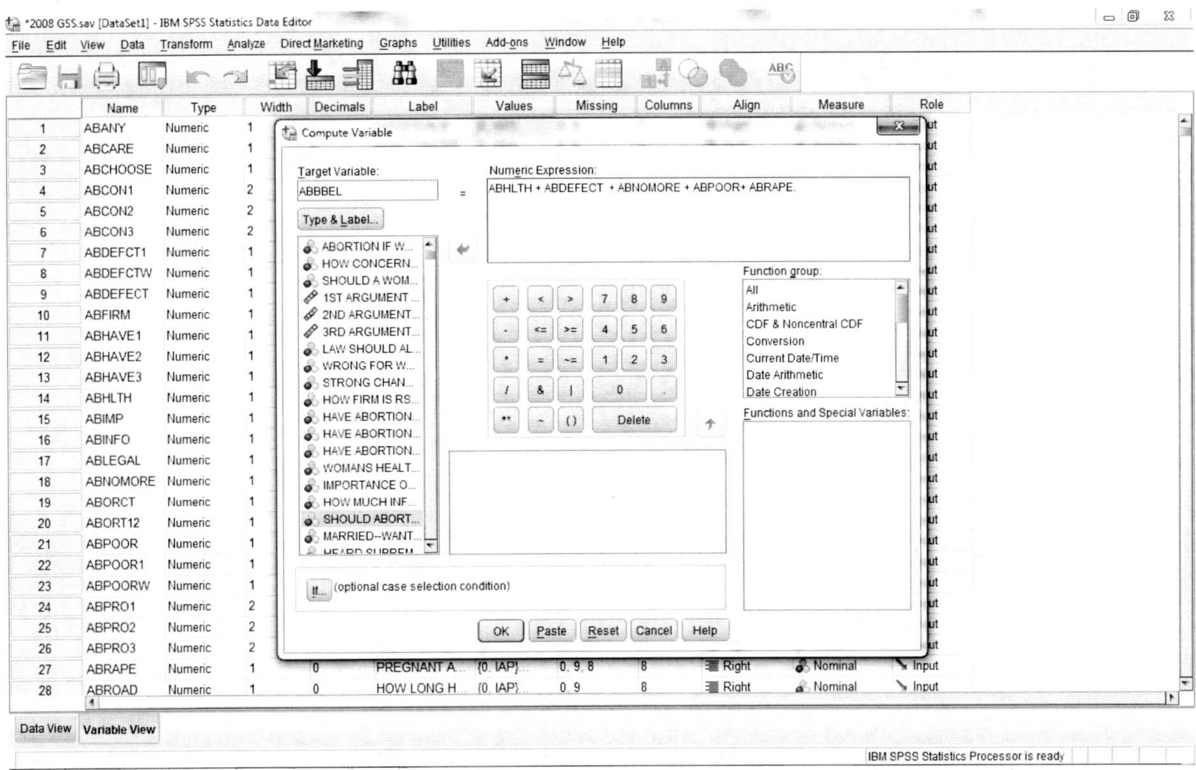

Figure 13.2

Table 13.3 *Overall Abortion Belief Scale*

		Frequency	Percent	Valid Percent	Cumulative Percent
Valid	5.00	490	24.2	41.7	41.7
	6.00	93	4.6	7.9	49.7
	7.00	246	12.2	21.0	70.6
	8.00	141	7.0	12.0	82.6
	9.00	104	5.1	8.9	91.5
	10.00	100	4.9	8.5	100.0
	Total	1174	58.0	100.0	
Missing	System	849	42.0		
Total		2023	100.0		

As we can see from our new scaled variable, 490 respondents said "yes" to every question and 100 said "no" to every question. Extreme ends of our scale (5 and 10) account for 49% of our sample, with the remaining 51% having scores that vary between the high and low ends of the scale. We can now include this variable in a more detailed analysis, such as OLS regression, if we are interested in how much education, sex, race and/or income influence the variability in our scores.

There are a few problems with creating a construct in this way. First, when we use this method, we are assuming that a relationship exists between our variables. If we run a correlation matrix, as we did above, we may find that this is true. However, suppose there had been a high correlation between ABHLTH, ABRAPE, and ABDEFECT but these two variables were weakly correlated with ABNOMORE and ABPOOR? What if some of our variables were negatively correlated and some were positively correlated? Finally, this variable assumes that each of our observed variables has the same weight in our unobserved variable. If we collapse these variables into one scale, we could be combining variables that are not measuring our latent variable correctly. To test the accuracy of our latent variable identification, we would use **factor analysis.**

There are two basic applications of factor analysis. The most commonly used is **exploratory factor analysis (EFA).** EFA is used when we wonder whether our measures involve a latent variable but we are not certain. We input our variables into our analysis and EFA generates a description of how well our variables "fit" our supposed model. EFA is typically undertaken without having a theoretical framework or hypothetical test in mind. It is exploratory in nature because it allows us to discover if there are latent variables at play in our data. **Confirmatory factor analysis (CFA)** uses an identical mathematical process. The difference

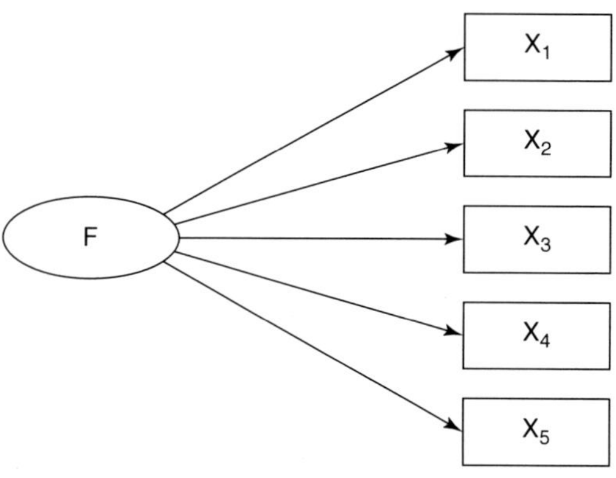

Figure 13.3

between the two applications is that we already have a theoretical framework and hypothetical model that presupposes that we have latent variables at play in our data. Suppose, for example, that you have used a well-known instrument, such as a depression scale, as part of your data analysis. Other researchers have already used this instrument to identify latent variables. You would not be exploring the possibility of latent variables (EFA) but instead confirming that they exist in your data (CFA).

The Logic of Factor Analysis

Factor analysis assumes that there is a latent variable that is a covariate of observed data. Think about the pro-choice attitude variables in the example above. When we look at our new scale, we can see that few people always say "yes" and even fewer always say "no." More than half of the respondents give a combination of yes and no responses on the abortion attitude variables. Given this, we can assume that there is some underlying variable that influences our respondens' answers to these variables, which we will call ABBEL. In other words, each measure that we observe is likely measuring some element of an underlying belief system or overall opinion of self as pro-choice or pro-life.

Factor analysis allows us to test the assumption that our variables are related to an unmeasured or latent variable. Also, it also allows us to see the weight, or overall effect, that each variable has on our overall score or on an individual's pro-choice beliefs. We can visualize this as a path diagram, where F represents our latent factor and X_i represents our independent observed variables, as in **Figure 13.3.**

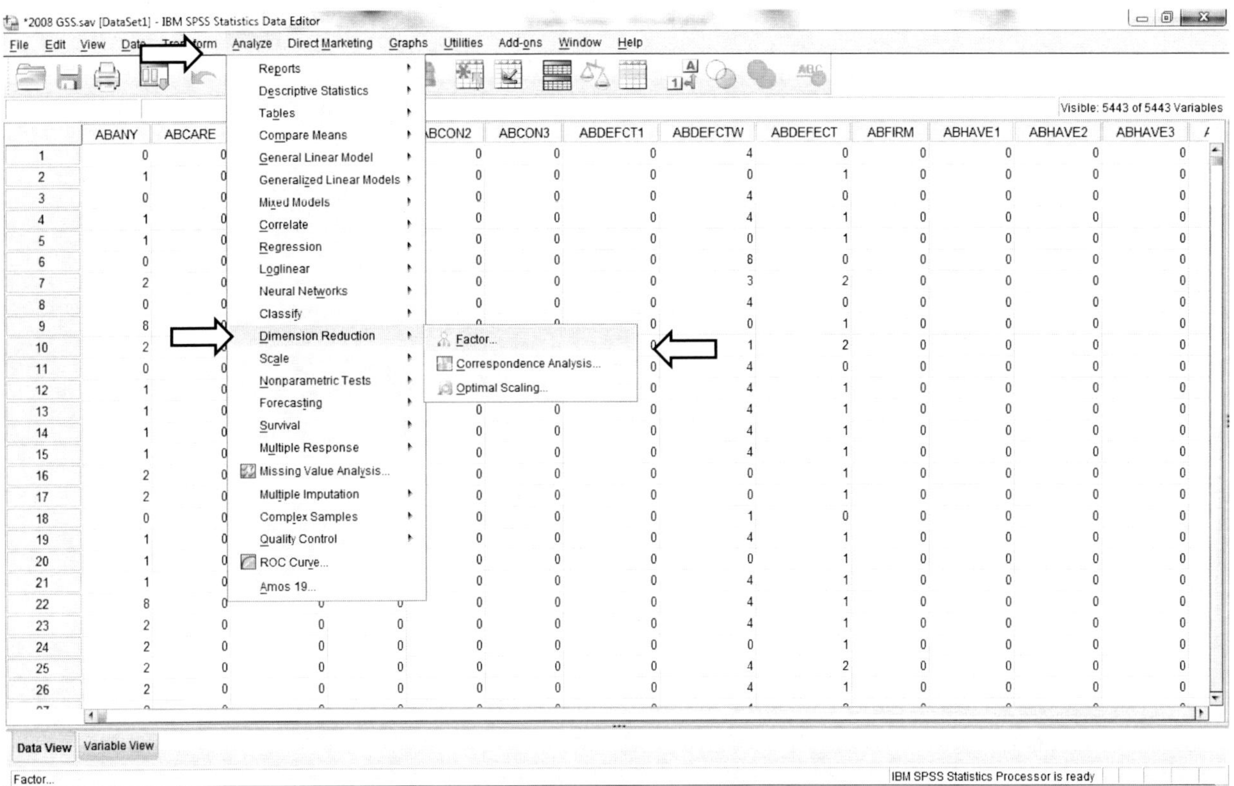

Figure 13.4

Figure 13.4 illustrates that our measures are reflecting some underlying measure that we cannot observe. Think back to the pro-choice questions again. When we interpret responses on these questions, we recognize that there are many factors that influence attitudes and beliefs about a woman's choices in terminating an unwanted pregnancy. An ultraconservative Christian, for example, may score high on this index because that person is staunchly pro-life and opposed to all abortions. A person who is to the far left politically may believe that any infringement on personal choice is unconstitutional, which would contribute to a lower score on this scale. However, most people fall in the middle ranges, which means there are a variety of personal values and beliefs that prompt responses on this question. For example, some may believe that abortion should be a choice if a pregnancy places a woman's health risk but not if she can't afford another child. Factor analysis also allows us to determine which individual scores account for the highest level of contribution to our latent variable. We will discuss this in detail in the next section when we turn to SPSS.

This chapter will not discuss the mathematical equations necessary to compute a factor analysis. The process is complex and requires a series of calculations. Rather, we will focus on computing a factor analysis in SPSS and interpreting the results. However, it may be useful to discuss the logic of factor analysis. When we think of our observed variables in a factor analysis, a spatial analogy makes sense. Imagine that we can create a scatterplot of our variables within a see-through plastic box, in which each data point is suspended in the space of the box. We want to draw a regression line and get it as close to as many data points as possible within the box. We hold the box up and rotate it in various directions and at various degrees until we find the best place to draw our line. Basically, this is what factor analysis does. It finds the location of the most shared variation and creates a slope. This process is called **principle component analysis.**

Using SPSS in Factor Analysis

For the purpose of this discussion, we are going to use SPSS's default settings for our factor analysis. We will discuss the various options for running a factor analysis later in this chapter.

Refer to **Figure 13.5.** Click on the Analysis menu and then on Dimension Reduction and then on Factor. If you are using an older version of SPSS, Dimension Reduction may be called Data Reduction.

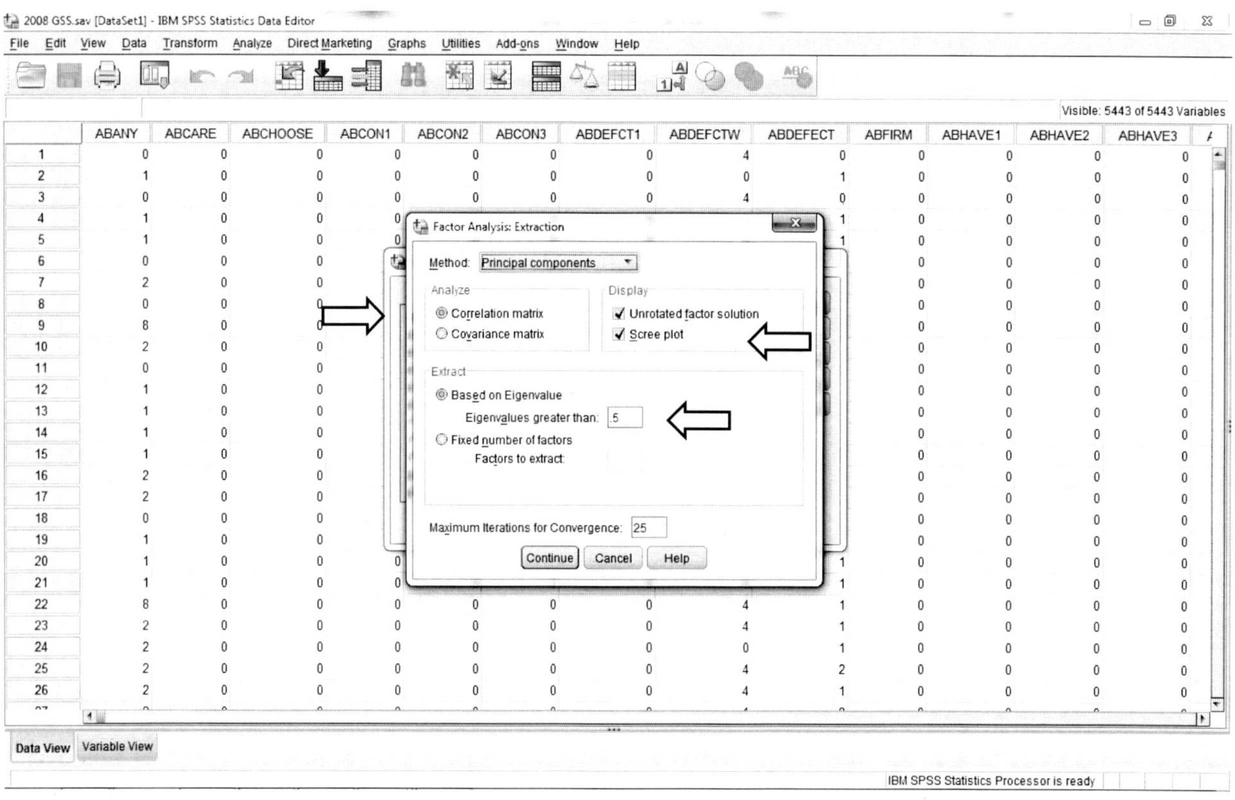

Figure 13.5

This will open the Factor Analysis menu. Move ABDEFECT, ABHEALTH, ABNOMORE, ABPOOR, and ABRAPE into the Variable list menu and click OK.

The first table in the output is labeled Communalities. (See **Table 13.4.**) We will first discuss the definition of the term *commonalities* and then the interpretation. **Tables 13.5** through **13.7** provide the results of our factor analysis for ABBEL.

Table 13.4 *Communalities*

	Initial	Extraction
Woman's Health Seriously Endangered	1.000	.444
Married-Wants no More Children	1.000	.660
Pregnant as a Result Of Rape	1.000	.605
Low Income-can't Afford More Children	1.000	.639
Strong Chance of Serious Defect	1.000	.659

Extraction Method: Principal Component Analysis.

The **communality** tells us how much a factor accounts for the variability of responses in our data. The initial communality in our analysis should be 1. A higher value than 1 indicates a problem in our data. Communalities are interpreted in much the same way that we interpret correlations. For example, our results show us that the communality for ABHLTH is .444. We would say that this is a moderate relationship, meaning that our latent variable (ABBEL) explains less than half of the variation in responses for this variable across our sample.

Next we turn to the factor loadings, which require some explanation before we proceed. Since we used the SPSS default settings, we have asked the program to remove any latent variables that had an initial Eigenvalue of < 1. When we look at the final relevant table, we will discuss the value in doing this versus other options. SPSS discounted Eigenvalues of < 1 but first had to calculate them. Our output provides the results of those calculations in the Total Variance Explained section of the output shown in **Table 13.5**.

As you read **Table 13.5,** Total Variance Explained, remember that its presentation is similar to that of a frequency table. The first column, labeled Initial, gives us the Eigenvalues. An Eigenvalue measures the total amount of variation across our sample that can be attributed to our component. In other words, it gives us the amount of common variability in our observed variables that can be accounted for by our latent variable. The second column of the table, % of Variance, tells us the percentage of the common variability in all of our observed variables that is accounted for by the latent variable. The cumulative percntage of variance is calculated in the same way as the cumulative frequency. Thus, in the example in **Table 13.5**, we know that the first component identified accounts for 60.1% of the common variability in our observed variables. Notice that only the first of these components is listed on the far right side of this table. That is because we have our Eigenvalue cutoff set to 1. Since component 2 has an Eigenvalue of .932, SPSS does not continue to extract the sums of squared loadings for this component. The Eigenvalue is a ratio of common variability to independent variability. If it is less than 1, it means that the component accounts for less variation than does each variable independently. Using a cutoff of 1 for Eigenvalues is standard practice in most social science disciplines and is known as the **Kaiser criterion.**

Table 13.5 *Total Variance Explained*

Component	Initial Eigenvalues			Extraction Sums of Squared Loadings		
	Total	% of Variance	Cumulative %	Total	% of Variance	Cumulative %
1	3.007	60.131	60.131	3.007	60.131	60.131
2	.932	18.641	78.772			
3	.476	9.522	88.295			
4	.390	7.806	96.101			
5	.195	3.899	100.000			

Extraction Method: Principal Component Analysis.

As researchers, we need to decide if we will rerun the analysis and instruct SPSS to use Eigenvalues of $>.5$ or if we will stay with the default. There are legitimate reasons to support either choice. First, the standard practice is to use a cutoff of 1 as our criterion. However, note that the second component accounts for an additional 19% of the common variability in our observed variables. Ultimately, the final decision rests with the researcher. Don't let the multiple components confuse you. Remember, we are able to rotate the box in any direction. After we have identified where to place the slope, SPSS rotates the data to look for the next slope of common variability. There are as many components as variables in the factor analysis. Since we included five variables, we have five components. SPSS lists the variable that accounts for the most-shared variance first and the one that accounts for the least-shared last. Notice that by the time SPSS gets to the final component, it has already accounted for 96% of the shared variation. This analysis continues until it has accounted for 100% of the shared variability between our variables.

For now we will use the Kaiser criterion and proceed with one component.

Table 13.6 provides **factor loadings**. Factor loadings can be interpreted like correlations. For example, the factor loading for ABHLTH is .666. We use the following equation to calculate the amount of variation in ABHLTH that is shared with our latent variable:

$$\% \text{ shared variation} = FL^2 * 100 = .666^2 * 100 = .44$$

Thus, we know that ABHLTH shares 44% of its variability with our latent variable. Also note here that .44 is our communality for ABHLTH.

Table 13.6 *Extraction Method: Principal Component Analysis*

Component Matrix[a]	
	Component
	1
Woman's Health Seriously Endangered	.666
Married-Wants no More Children	.812
Pregnant as a Result of Rape	.778
Low Income-can't Afford More Children	.799
Strong Chance of Serious Defect	.812

If we continue calculating percentages with this formula, we get the results presented in **Table 13.7**.

Table 13.7

Variable	% of Common Variability
ABHLTH	44.4
ABNOMORE	65.9
ABRAPE	60.5
ABPOOR	63.8
ABDEFECT	65.9

When we examine these values, we see that our latent variable shares greater than 60% of common variation with every variable other than ABHLTH. This means that we can call our variable ABBEL and label it Pro-Choice Beliefs. Remember, all of our correlations were positive and all of all our factor loadings are positive. This means that if we include this variable in a regression, a negative ß would reflect someone who is more pro-choice and not less pro-choice.

Explaining Rotation

We assume that X and Y have an **orthogonal** relationship. This means that X and Y are perpendicular to each other. You are used to seeing this relationship because it is nothing more or less than the X and Y axes on a scatterplot or regression line. The default equation in SPSS calculates a factor analysis on this assumption. Factor analysis can draw multiple orthogonal plots even if our variables are not correlated with each other. Think of this in terms of the following equations, where r is our correlation, X is any variable, and F is our latent variable.

$$r_{X_1 X_2} = 0 \text{ but } r_{X_1 f_1} > 0 \text{ and } r_{X_2 f_1} > 0$$

This means that while X_1 and X_2 are not related to each other, they are related to a third, unseen factor.

A simple practical example of this logic is to consider that we ask our sample subjects the following questions:

Can you play a musical instrument?
Are you a good singer?

There would not necessarily be a correlation with singing ability and ability to play an instrument. It is possible to play an instrument without having a good singing voice and vice versa. We could assume an orthogonal relationship between these variables or $r_{X_1 X_2} = 0$. However, we can assume that both variables are measuring some form of musical talent or aptitude, or $r_{X_1 f_1} > 0$ and $r_{X_2 f_1} > 0$. While we can make this assumption, our variables don't really measure talent as the questions are asked. Factor analysis allows us to identify the underlying concept of talent.

Remember that it was previously suggested that you visualize the data points spatially. When we think of rotation, we are more or less looking at our scattered data from different angles. The most commonly used form of rotation in factor analysis is the **Varimax rotation**. The Varimax rotation, put simply, allows us to maximize the variance in the data. If we visualize this, it is the mathematical equivalent to rotating our box of data and looking at it from different angles until we identify a slope that intersects the most data points with little variation around that slope. This first slope is Component 1. Since Component 1 won't intersect with all the variability in our data points, we rotate the box again until we can identify the next slope that does the same thing with the leftover variance. We keep doing this until all the variability or the variance is captured by one of our slopes.

Other rotation options that we will not discuss in this chapter are the Direct Oblimin, Quartimax, Equamax, and Promax. It is, however, important to note that regardless of the type of rotation you use, factor analysis assumes an orthogonal relationship.

Let's go back to SPSS and make a few changes to our previous run. Reopen the Dimension Reduction screen. If you have not closed your program, the variables will still be in the window. If they are not, move ABHEALTH, ABPOOR, ABRAPE, ABNOMORE, and ABDEFECT into the Variable list window. Click on the Extraction menu and change the cutoff point to .5 from 1. Also, check the option for Scree Plot on this menu. (See **Figure 13.5.**)

Next, click on Rotation, then select Varimax, and click Continue. (See **Figure 13.6.**)

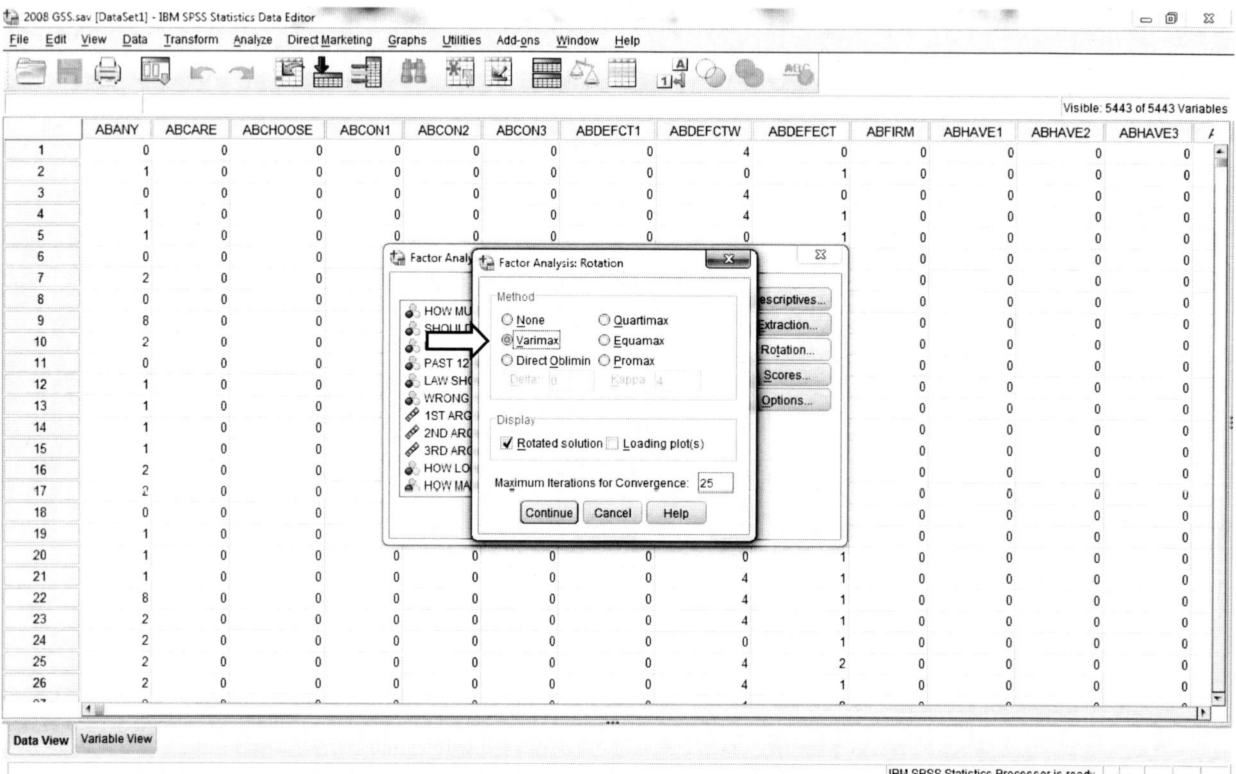

Figure 13.6

Finally, click on Scores and check Save as Variables. (See **Figure 13.7.**)

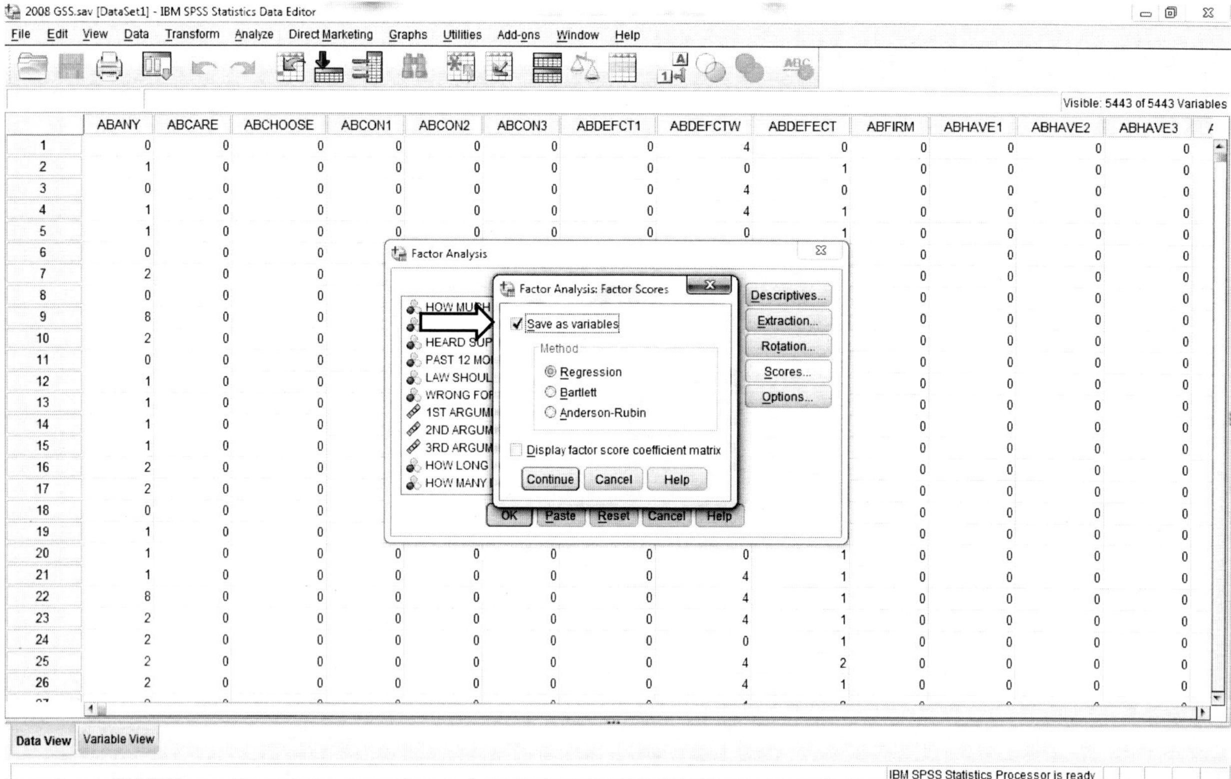

Figure 13.7

The only significant change to our output is that we now have factor loadings for component 2. We need to determine if we are going to use it. Of course, we still have the option of applying the Kaiser criterion and rejecting it because it has an initial Eigenvalue < 1. However, if we examine our factor scores, we see that two of our variables have negative loadings on component 2. (See **Table 13.8.**) The shared variation for component 2 means that the individual is likely to give a "yes" response. These negative responses are much weaker than the positive loadings on factor 1, however. We can calculate the percentage of shared variance to compare them. (See **Table 13.9.**)

Table 13.8. *Two-Component Model*

Component Matrix[a]		
	Component	
	1	2
Strong Chance of Serious Defect	.812	.241
Woman's Health Seriously Endangered	.666	.561
Married-Wants no More Children	.812	−.483
Low Income—can't Afford More Children	.799	−.497
Pregnant as a Result of Rape	.778	.281

[a]*Two components extracted.*

Table 13.9 *Percent of Common Variability*

Variable	Component 1	Component 2
ABHLTH	44.4	5.8
ABNOMORE	65.9	31.5
ABRAPE	60.5	23.3
ABPOOR	63.8	24.7
ABDEFECT	65.9	7.9

As you can see, Component 2 does not include as much variation, but that is to be expected. When we decide to include components, we need to consider the amount of variation that is being measured; however we also need to consider what the subsequent variables mean. All five variables load high on Component 1, so we know that it is a good measure for an underlying latent "belief" variable. However, two variables have negative loadings with Component 1, ABNOMORE and ABPOOR. Note that ABHLTH has a modest positive loading with this variable as well. We could argue that there is a second underlying value that causes a person to slightly favor abortion as an option when the mother's health is at risk and definitely favor abortion as a choice when the woman is married and wants no more children or when she is too poor to afford a child. This still doesn't provide an indication of whether or not we should include this second component.

Our final point of consideration is our scree plot. (See **Figure 13.8**.)

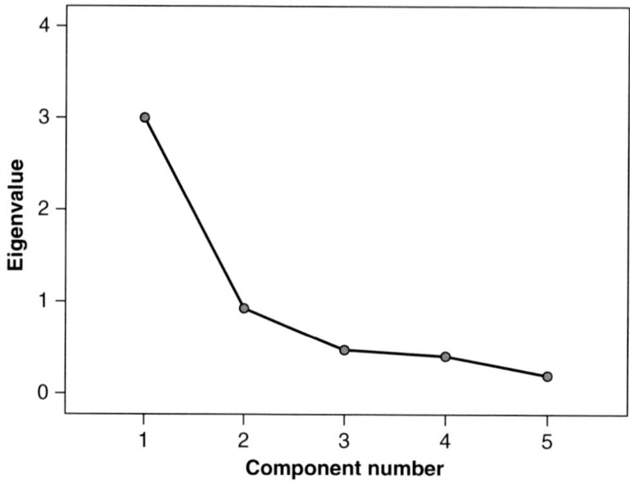

Figure 13.8 Scree Plot.

The scree plot provides a simple plot of our Eigenvalues. We interpret this as a downhill slope of our variation. In other words, we look at where the change in variation stops its downhill slope. As we can see from this plot, the slope diminishes at Component 3. If we use this method to decide on our number of factors to maintain, we would keep Factor 2.

Ultimately, the researcher makes the final decision on which rotation to use and on how many components he or she will include in the final analysis. In the examples above, the Kaiser criterion results in the most conservative interpretation of the components. This conforms more to standard practices in the social sciences. Also, Varimax rotation is the most often used formula in this analysis. Our second analysis, which is less conservative and accepts lower Eigenvalues, reveals that there may in fact be a second latent variable that accounts for roughly 20% of the shared variability in our responses. As you determine the components that you will use, you must decide whether the error introduced by using Eigenvalues < 1 is offset by the potential explanation of shared variance that additional factors account for.

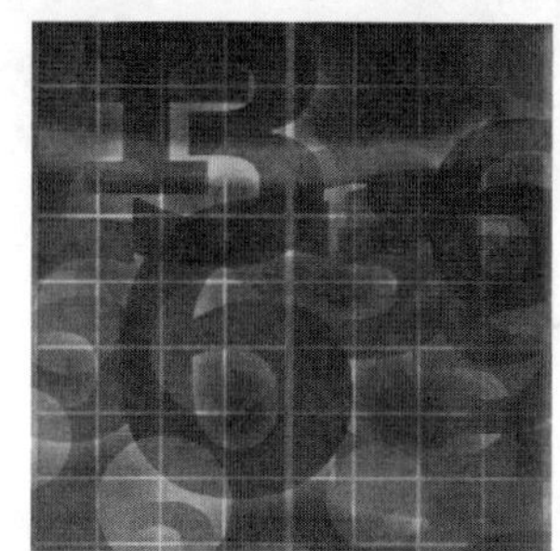

Exercises and Notes

1. List the primary reasons why we use data reduction methods:

   a. ______________________________________

   b. ______________________________________

   c. ______________________________________

2. How does a construct differ from a latent variable?

3. When do we use exploratory factor analysis (EFA)?

4. When do we use confirmatory factor analysis (CFA)?

5. X has a factor loading of .876 on a component. What is the percentage of the shared variability in the component that is accounted for by X?

6. What is the Kaiser criterion?

7. Factor analysis assumes orthogonal relationships between identified variables. What does this mean?

8. What is the logic behind using a Varimax rotation?

Communalities

	Initial
Allow Anti-Religionist to Teach	1.000
Should Communist Teacher be Fired	1.000
Allow Homosexual to Teach	1.000
Allow Militarist to Teach	1.000
Allow Anti-American Muslim Clergymen Teaching in College	1.000
Allow Racist to Teach	1.000

Extraction Method: Principal Component Analysis.

Total Variance Explained

Component	Initial Eigenvalues			Rotation Sums of Squared Loadings		
	Total	% of Variance	Cumulative %	Total	% of Variance	Cumulative
1	2.458	40.960	40.960	1.466	24.437	24.437
2	.996	16.596	57.556	1.050	17.507	41.943
3	.809	13.480	71.036	1.003	16.716	58.659
4	.632	10.540	81.576	1.001	16.690	75.350
5	.617	10.278	91.854	.990	16.504	91.854
6	.489	8.146	100.000			

Extraction Method: Principal Component Analysis.

9. The above data uses Varimax rotation and a cutoff of .5 for Eigenvalues. Based on this information only, how many of these components would you include in your analysis? Why?

10. The following table provides factor loadings for these components. Examine them carefully and decide how many components you would use. Does knowing the loadings change the number of components that you decided to keep?

Rotated Component Matrix[a]

	Component				
	1	2	3	4	5
Allow Anti-Religionist to Teach	.883	.032	.131	.032	.185
Should Communist Teacher be Fired	.059	.011	.010	.997	.033
Allow Homosexual to Teach	.159	.156	.971	.011	.080
Allow Militarist to Teach	.744	.368	.097	.064	.143
Allow Anti-American Muslim Clergymen Teaching in College	.208	.930	.164	.007	.150
Allow Racist to Teach	.246	.156	.084	.038	.952

Extraction Method: Principal Component Analysis. Rotation Method: Varimax with Kaiser normalization.
[a]Rotation converged in 5 iterations.

11. How much of the common variability does Allow Homosexual to Teach explain in Component 1? How much does it explain in Component 3? What does this tell us about these components?

12. Allow Anti-American Muslim Clergymen Teaching in College and Allow Homosexual to Teach are both coded in a way that positive responses mean less tolerance for these individuals. Examine the factor loadings for Components 2 and 3. What latent variable is being measured by these components?

13. Examine the following scree plot for this analysis. Based on this plot, how many components should you keep from the factor analysis? Does this agree with your previous decisions?

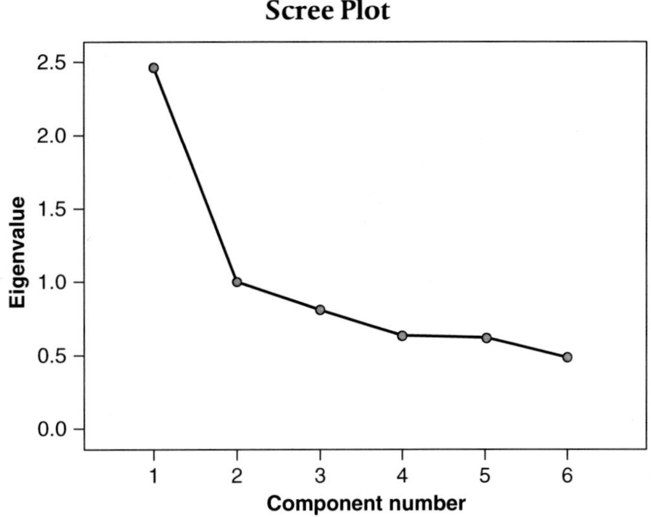

Scree Plot

14. Consider all of the data above. How many components will you keep? What are the benefits and drawbacks of your decision?

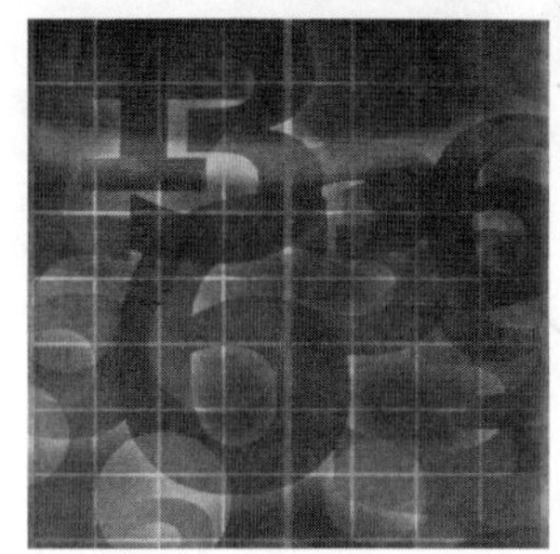

Learning Check

1. Discuss the primary reasons why we use data reduction methods.

2. Discuss the differences between constructs and latent variables. What are their strengths and weaknesses?

3. What distinguishes exploratory factor analysis (EFA) from other methods of dimension reduction?

4. What distinguishes confirmatory factor analysis (CFA) from other methods of dimension reduction?

5. X has a factor loading of .219 on a component. What is the percentage of the shared variability in the component that is accounted for by X?

6. Discuss the validity of the Kaiser criterion as a standard for component identification.

7. Factor analysis assumes orthogonal relationships between identified variables. What does this mean?

8. What is the logic behind using a rotated solution rather than a nonrotated solution in factor analysis?

Communalities

	Initial
Allow Anti-Religionist to Teach	1.000
Should Communist Teacher be Fired	1.000
Allow Homosexual to Teach	1.000
Allow Militarist to Teach	1.000
Allow Anti-American Muslim Clergymen Teaching in College	1.000
Allow Racist to Teach	1.000

Extraction Method: Principal Component Analysis.

Total Variance Explained

Component	Initial Eigenvalues			Rotation Sums of Squared Loadings		
	Total	% of Variance	Cumulative %	Total	% of Variance	Cumulative %
1	2.458	40.960	40.960	1.466	24.437	24.437
2	.996	16.596	57.556	1.050	17.507	41.943
3	.809	13.480	71.036	1.003	16.716	58.659
4	.632	10.540	81.576	1.001	16.690	75.350
5	.617	10.278	91.854	.990	16.504	91.854
6	.489	8.146	100.000			

Extraction Method: Principal Component Analysis.

9. In the above table, Component 2 has an Eigenvalue of .996. Discuss the disadvantages and advantages of including this component in an analysis.

10. The following table provides factor loadings for these components. Examine them carefully and decide how many components you would use. Does knowing the loadings change the number of the components that you decided to keep?

Rotated Component Matrix[a]

	Rotation Sums of Squared Loadings				
	1	**2**	**3**	**4**	**5**
Allow Anti-Religionist to Teach	.883	.032	.131	.032	.185
Should Communist Teacher be Fired	.059	.011	.010	.997	.033
Allow Homosexual to Teach	.159	.156	.971	.011	.080
Allow Militarist to Teach	.744	.368	.097	.064	.143
Allow Anti-American Muslim Clergymen Teaching in College	.208	.930	.164	.007	.150
Allow Racist to Teach	.246	.156	.084	.038	.952

Extraction Method: Principal Component Analysis. Rotation Method: Varimax with Kaiser Normalization.
[a]Rotation converged in 5 iterations.

11. How much of the common variability does Allow Militarist to Teach explain in Component 1? How much does it explain in Component 2? What does this tell us about these components?

12. Allow Anti-American Muslim Clergymen Teaching in College and Allow Homosexual to Teach are both coded in a way that positive responses mean less tolerance for these individuals. Examine the factor loadings for Components 2 and 3. What name would you give to Components 1–3 for future analysis?

## Chapter Fourteen

Structural Equation Modeling

Structural equation modeling (SEM) is a powerful tool for social scientists. It combines the data reduction strengths of factor analysis with the hypothesis testing of regression analysis and allows us to test a variety of interrelated behaviors. We will not focus on the math that is required to perform SEM in this chapter. While there are a variety of programs that will allow you to build your analysis through structural equation modeling, we will focus on the use of Analysis of Moment Structures (AMOS), an add-in for SPSS that has been designed for this reason. Finally, this chapter will provide only a basic introduction to the concepts of SEM using a simple model. Most graduate students who are interested in learning SEM take a specialized class at the doctoral level that provides a more in-depth understanding.

Understanding Human Behavior

We often discuss how one variable affects another. For example, we might argue that sex influences drug use. It has been well-documented that men use drugs with more frequency than do women. We could test this relationship with a logistic regression model wherein drug use is recorded as a present/absent dummy variable and sex is also recorded as a dummy variable. But what if we are interested in the motivations for drug use? For example, we know that individuals with mental health issues are more likely to self-medicate with various types of drugs (such as nicotine, street drugs, and alcohol). We also know that women are more likely to seek mental health treatment and report more episodes of depression than men are. How would we test the relationship between sex and drug use when we examine these variables? Does sex have a direct relationship with drug use, such as that men are more likely to use drugs for the sole reason that they are men? Or is sex indirectly related to drug use, that is, are men more likely to use drugs because they are less likely to seek mental health treatment? We will discuss direct and indirect relationships more thoroughly below.

SEM allows us to test the relationships that exist between several variables. Since it builds on factor analysis, it allows us to determine the role of latent variables versus observed variables. It also allows us to draw a path model between our variables so that we can visualize the interconnectedness between the variables in our equation.

Important Vocabulary

Before we can discuss how we build our model using SEM, there are several terms that should be defined. We already discussed some of these definitions in Chapter 13, but we will repeat them here. SEM uses **observed variables** to create **latent constructs.** Observed variables are the data that we have actually collected. We used the observed pro-choice questions from the GSS in Chapter 13. These were the actual questions that respondents were asked to answer. Referring back to Chapter 13, recall that we used factor analysis to determine how much of the variation in each of our observed variables can be attributed to some unmeasured characteristic. This unmeasured characteristic is our latent construct, or the measure that is inferred from the relationships that we believe exists in our data. SEM uses this same process to identify latent variables in our model.

SEM also uses **endogenous** and **exogenous** variables. Endogenous variables are those that are predicted by other variables. Drug use would be an example of an endogenous variable because we can use other variables to predict the probability that it will occur. Exogenous variables are those that we use to predict other variables but are not predicted in the model. Sex and age are two examples of exogenous variables. While both can be used to predict the likelihood of Y, Y cannot be used to predict a person's age or sex.

SEM tests both **causal** and **noncausal relationships.** Causal relationships are those relationships that have an effect on Y. In the above example, we may determine that sex accounts for 40% of the variation in drug use. This means that sex has a causal relationship with drug use, or it affects the choice to use drugs. Noncausal relationships are those that have a spurious effect or a suppressor effect on Y. For example, we could determine that when we add mental health to our model, the relationship between sex and drug use is spurious, or nonexistent. We might also determine that mental health suppresses the effect of sex. Sex is an important predictor, but it becomes less significant when we control for mental health.

SEM also allows us to test **mediating variables.** Mediating variables, as the name suggests, are those that mediate the relationship between X and Y. In the above example, we might find that sex (X_1) does influence drug use (Y) but that the strongest predictor is mental health (X_2), which is also associated with drug use. SEM would set up our equation as $X_1 \rightarrow X_2 \rightarrow Y$. Mediating variables are those that we find in the middle of the path.

SEM also tests **direct** and **indirect relationships.** A direct effect refers to the amount of variation in Y that can be attributed to X. In our example, it would refer to the effect of mental health (X_2) on drug use. The indirect effect refers to the amount of variation that a variable causes through another variable. In the same example, sex would have an indirect effect because it flows through the mediating variable, mental health.

SEM uses both observed variables and latent constructs to discover the **total association** and the **total effect** of the model. The total association refers to the noncausal effect plus the total effect. Total effect refers to the direct effect plus the indirect effect.

Finally, SEM uses both **recursive** and **nonrecursive** models. In a recursive model, all of the causal relationships are unidirectional, meaning that our Xs are not correlated, and neither are their error terms. When a model is nonrecursive, it means that two of the variables are related or that their error terms are related. AMOS will tell you if a model is nonrecursive. Ideally, you want a recursive model and would modify your path to achieve this, if feasible.

Reading the Path

Just as we have definitions for SEM, we have symbols within the model that need to be understood. Consider the following model.

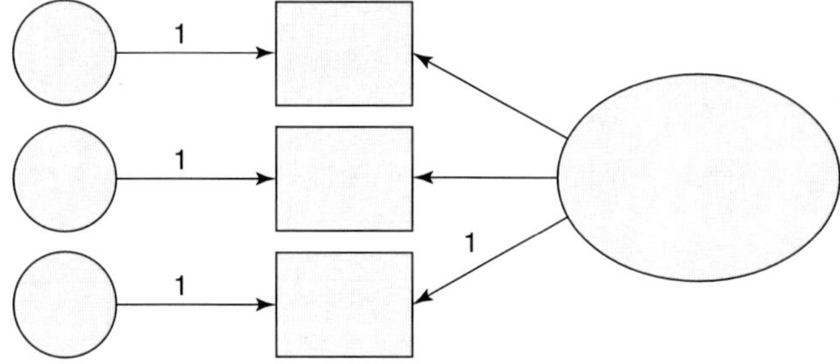

The oval in Figure 14.1 represents a latent construct. The rectangles represent observed variables, and the circles represent error terms. These shapes are *not* interchangeable. Any time you see an oval in a path model, it will refer to a latent construct. A rectangle will always refer to an observed variable. Also note that the arrows in this model are unidirectional. This means that the latent variable accounts for the change in all of our observed variables. That is why the arrows point *from* the latent variable to the observed variables. Remember, we are arguing that there is some underlying characteristic that we measured that explains the communality of our measured variables. This is also why our error terms point to our measured variables. We assume that there is error in every measured variable that influences responses.

Now, examine the following model.

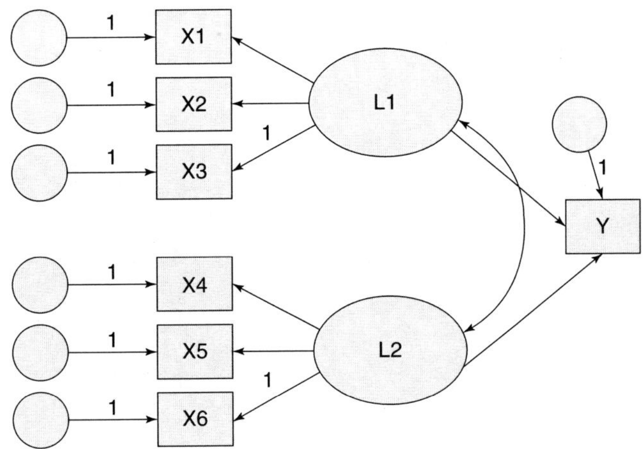

As you can see, we have added another latent construct and its observed variables, as well as an observed dependent variable. This model suggests that L1 and L2 both have a direct effect on Y. However, you will notice that there is a curved arrow that connects L1 and L2. This means that L1 and L2 are correlated with one another, or are nonrecursive. Single-pointed arrows will always represent recursive relationships, and double-pointed arrows will always represent nonrecursive relationships.

Building a Model in AMOS

Since AMOS is an add-on for SPSS, you have to work within SPSS to use the program. Open the data file that you want to work with. For this example, we will use GSS 2008. You will find the AMOS program under the Analyze tab. If AMOS is not an option on this tab, it means that the version of SPSS you are using doesn't include the add-on. It has to be purchased separately, so not every computer with SPSS will have AMOS.

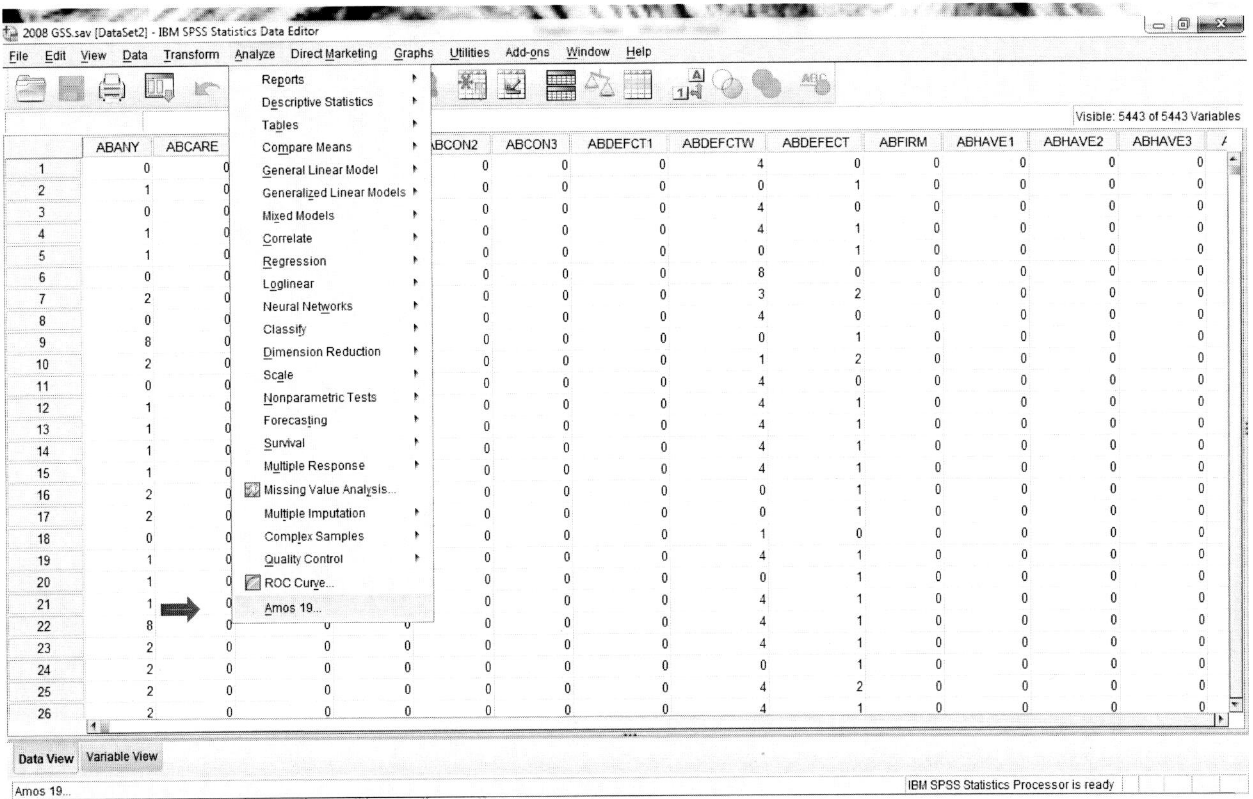

When you select AMOS, it will open the Analysis window.

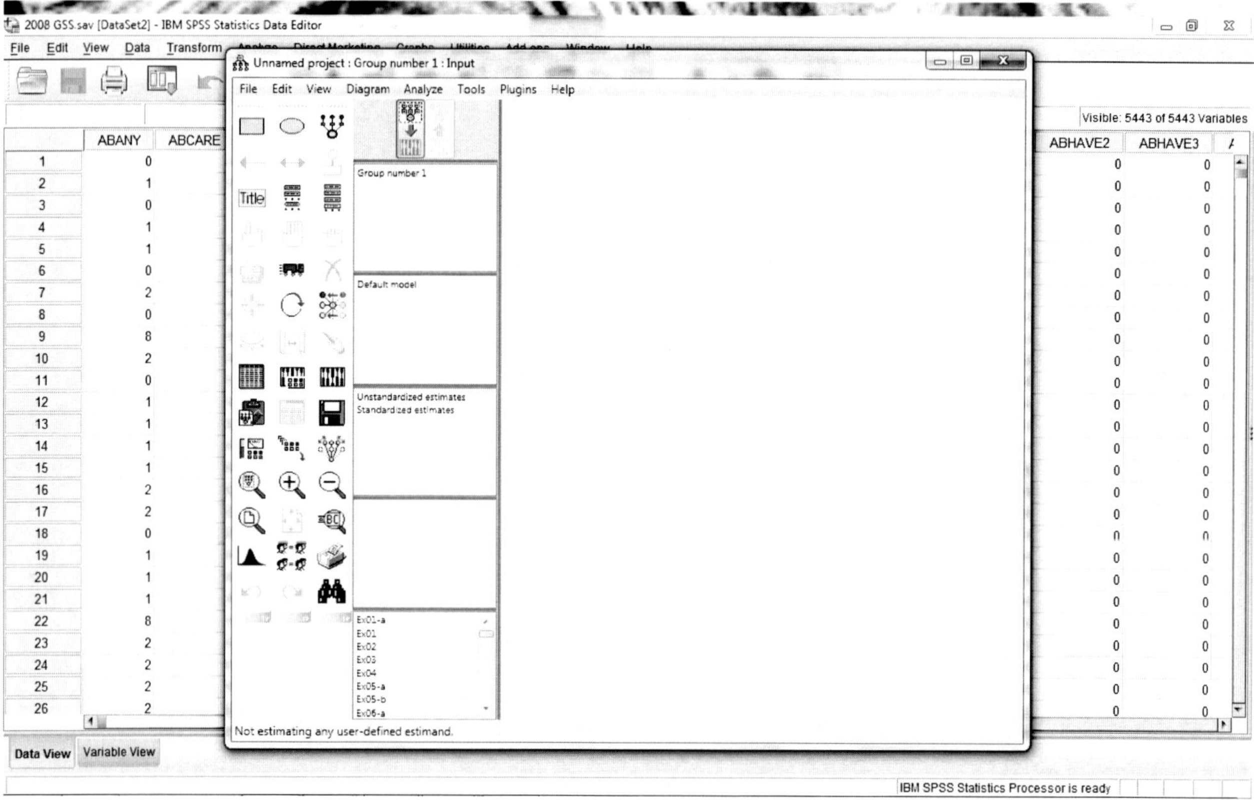

The top line on the menu gives you the options for drawing your variables. As you can see, you have the option of adding a latent or an observed variable to your model. The third option on the menu lets you draw the latent variable and add the observed variables. Draw the latent variable by dragging the cursor and click five times to add observed variables and their error terms.

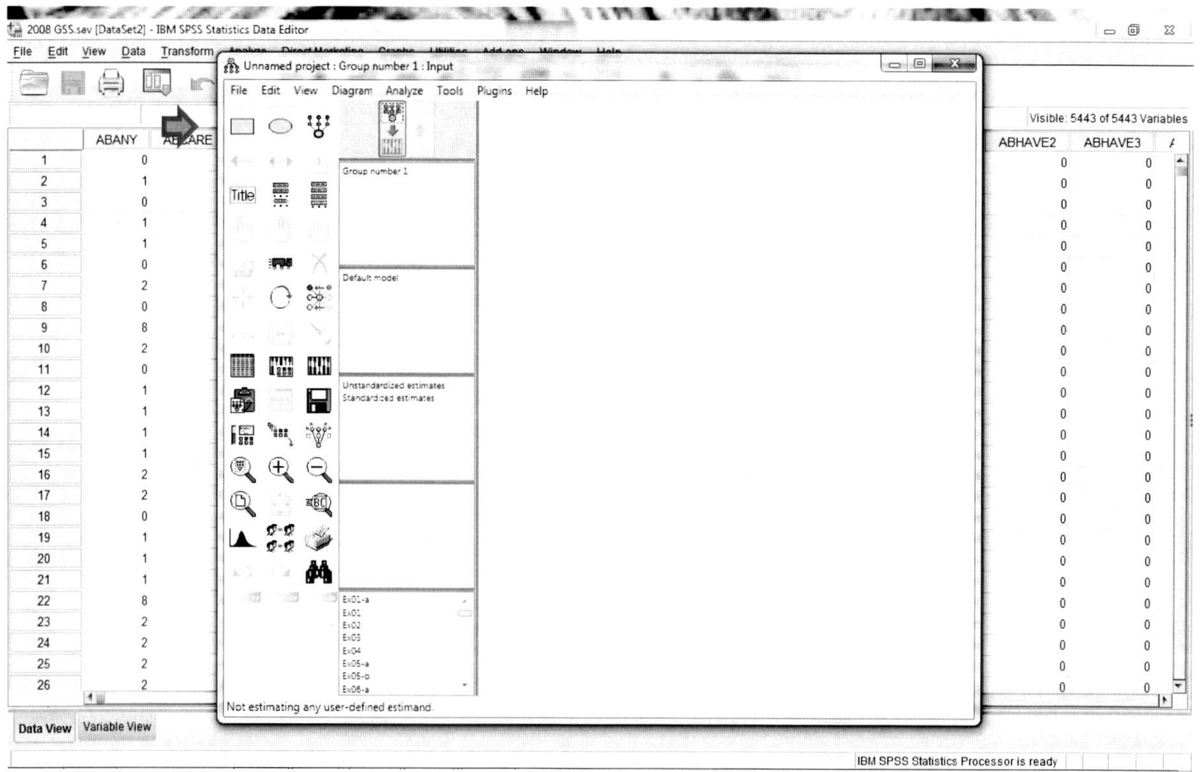

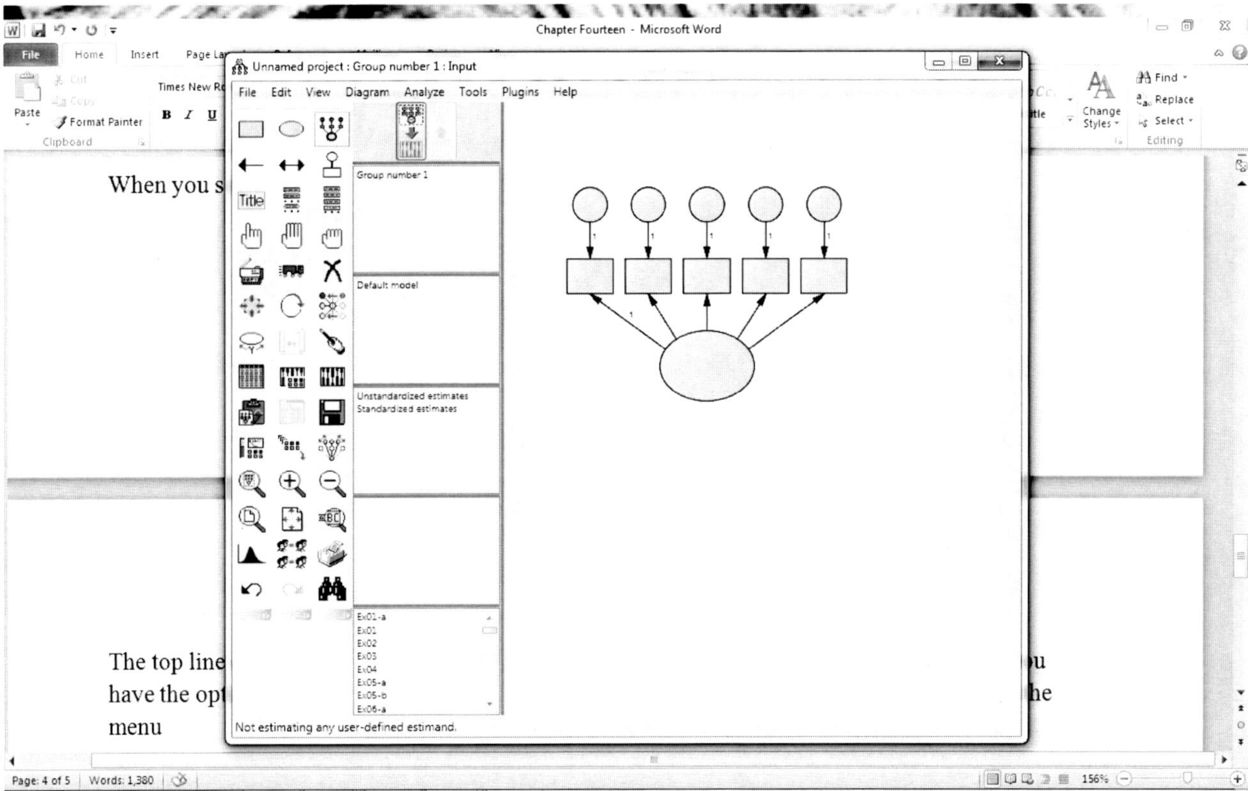

We will use the example here that we used for our factor analysis in Chapter 13. Now that we have drawn our model, we need to add our variables. We do this by choosing List Variables in Dataset from the menu.

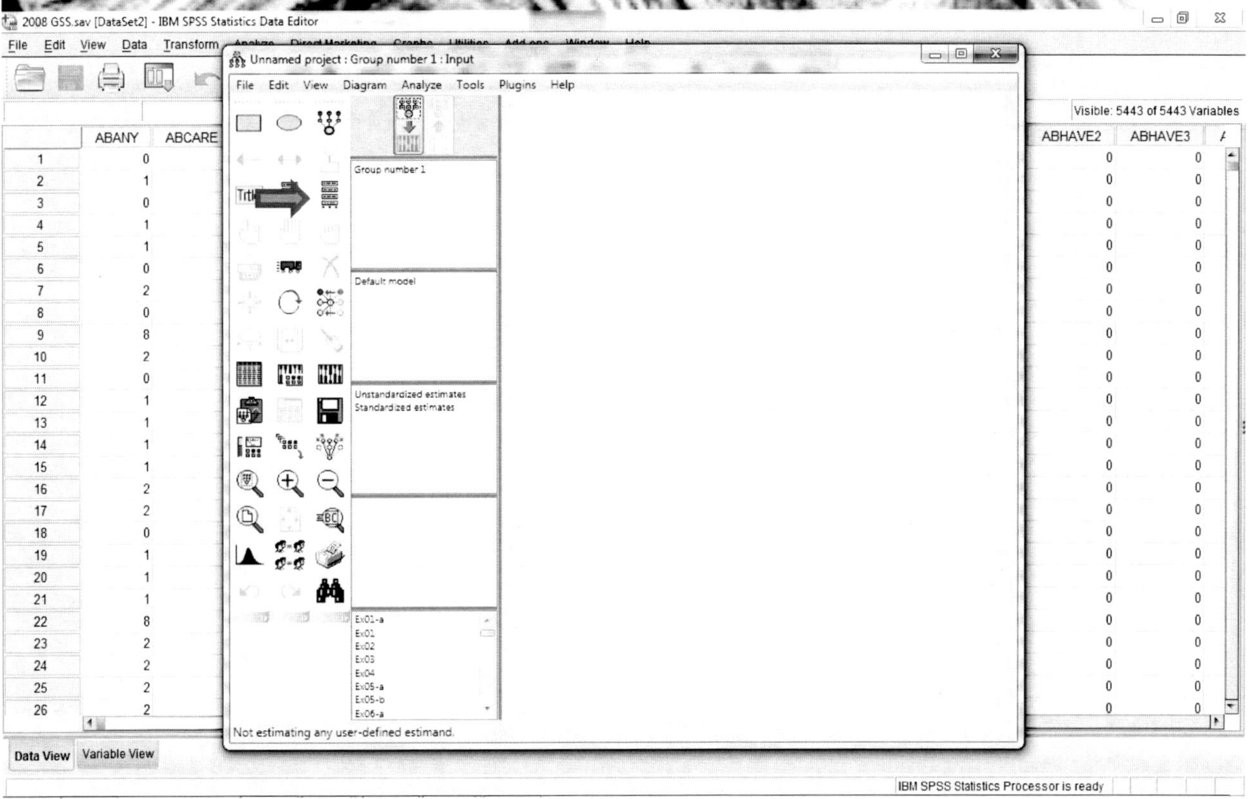

Selecting this option will bring up the list of variables in the open data set.

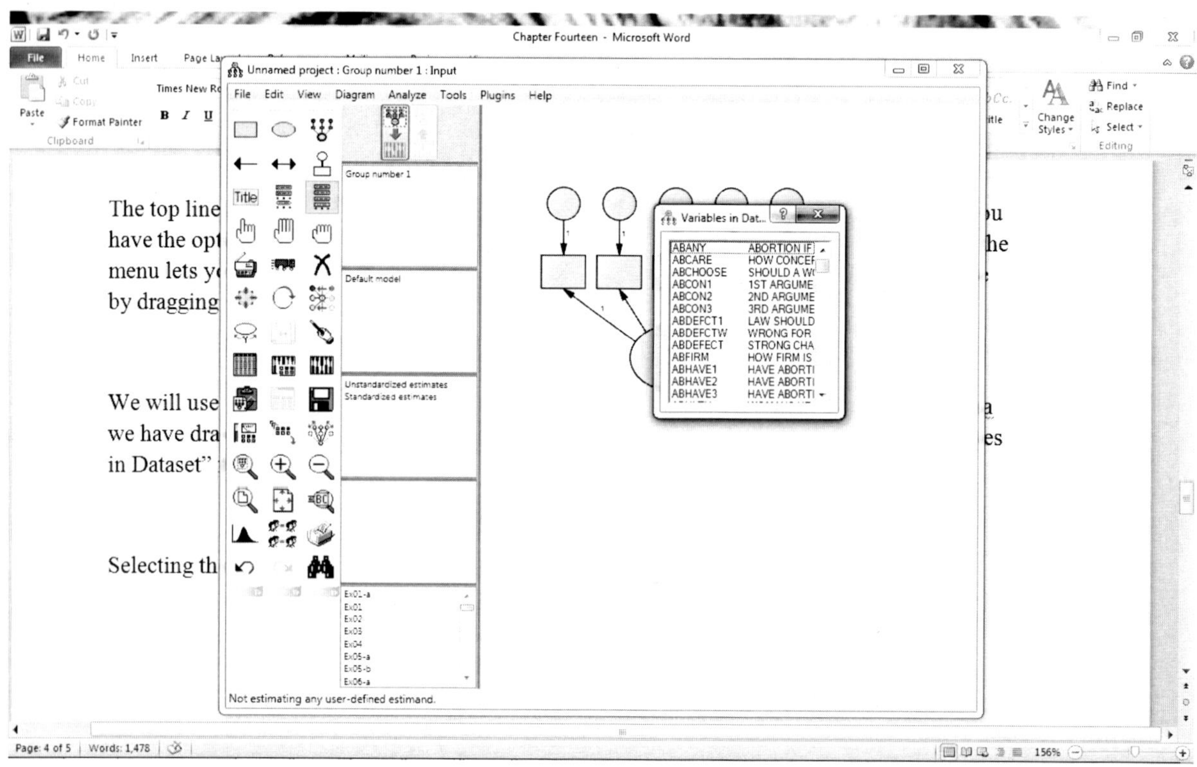

We add these variables to our model by dragging and dropping them. Remember, the latent variable and error terms are not measured, so we will only add variables to the rectangles in our model. Add ABHLTH, ABRAPE, ABNOMORE, ABPOOR and ABDEFECT to the observed variables. You will notice that AMOS uses the variable labels when you add them and not the variable names, which are not legible.

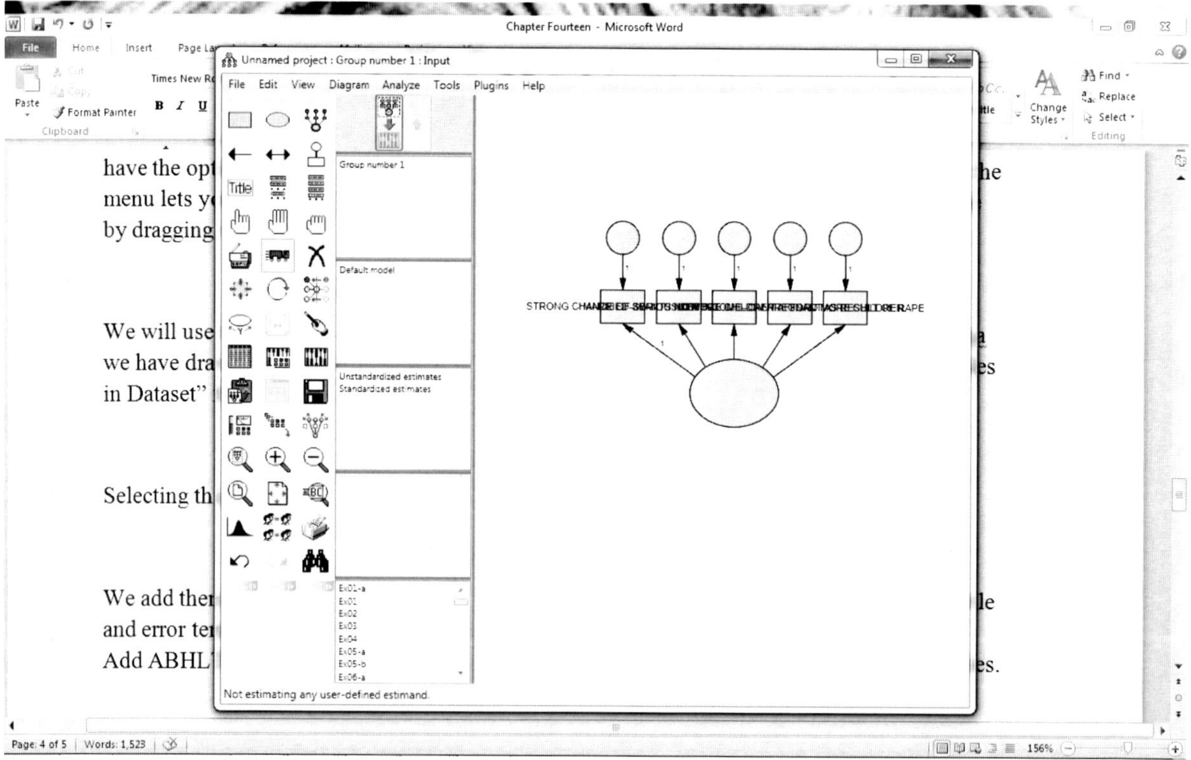

This means that we have to edit our model. Double click on the first observed variable to bring up the Edit menu. Change the variable label to the variable name by typing in the new information. Do this for each variable. When you have completed this, name your error terms. (Use E1–E5) and then add ABBEL to your latent variable. AMOS requires unique names for each variable in the equation before it will calculate your model.

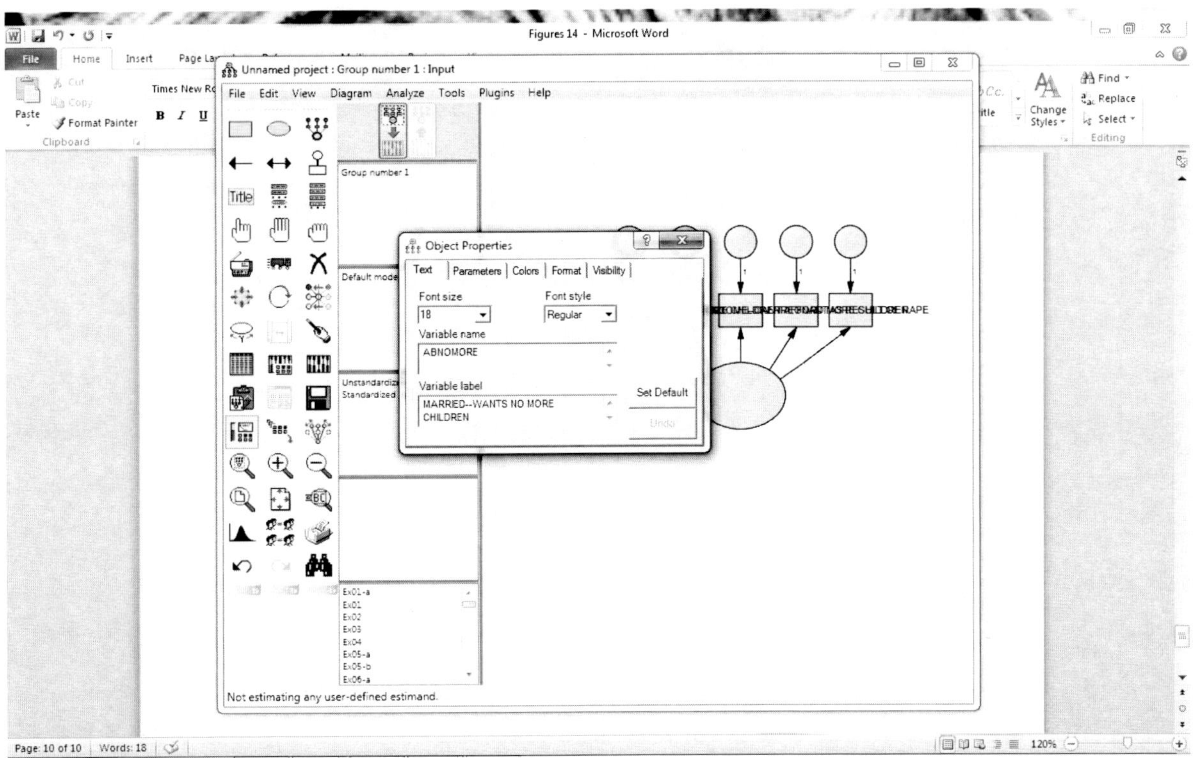

After you have named your variables and reformatted the labels, click on the Rotate Indicators tab on the menu until your model has the latent variable on the right.

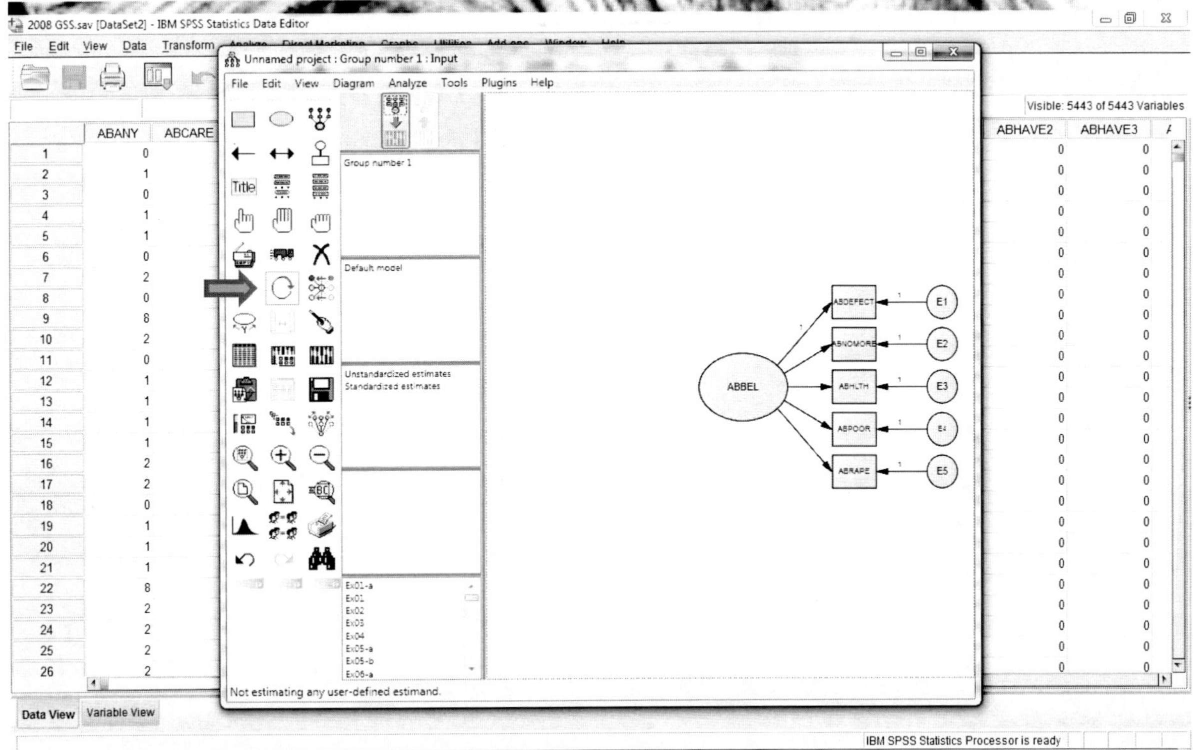

Now you can add the predictor variables. We will use gender and education as variables for this model. Add two observed variables to the model and create recursive relationships between both observed variables and the latent variable. Remember to drag and drop the variables from the variable list, to add error terms to the new variables and the latent variables, and to name your variables. Use E6–E8 for the new error terms.

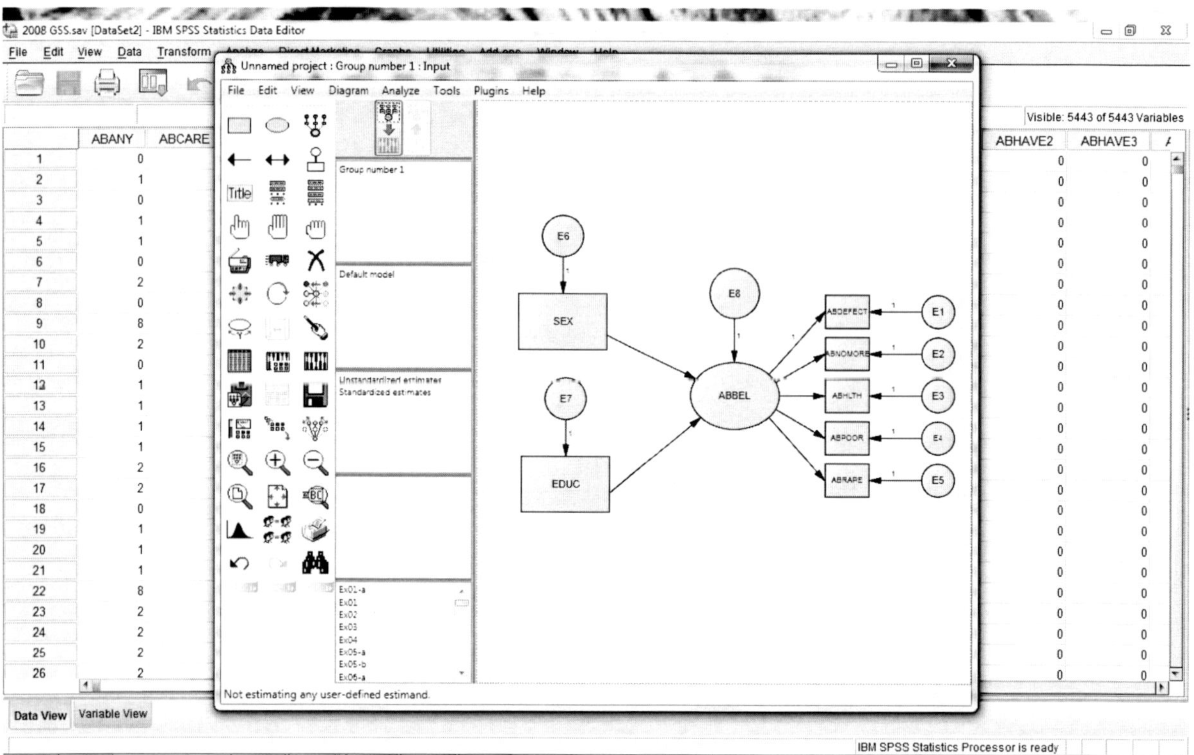

Since there are missing data in our file, you need to check Estimate Means and Intercepts before you can run the model. Select View and Analysis Properties on the menu and place a checkmark in the appropriate box.

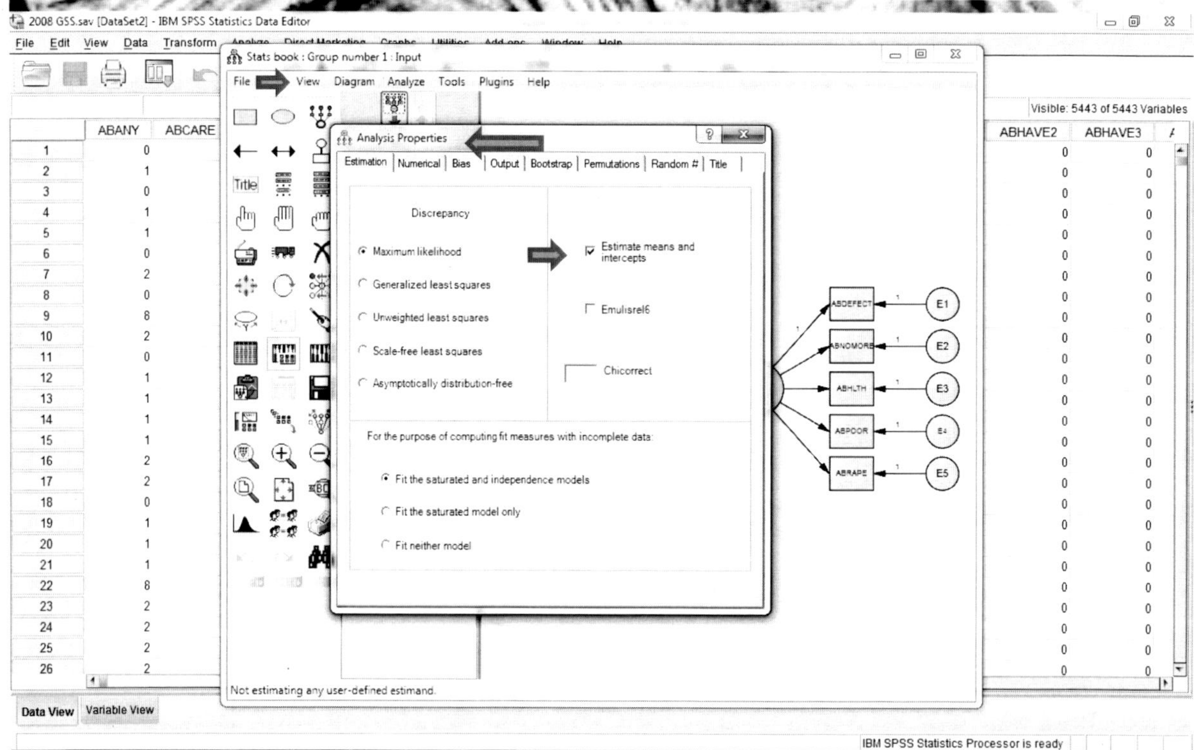

Next, choose calculate estimates from the Analyze menu. If there are errors in our model, a pop-up window will explain them. When the analysis is complete, it will give you preliminary results in the window on the left of the model. This will include whether the model has a significant χ^2 and the degrees of freedom associated with the χ^2. The results of the model are listed on a separate tab at the top of the window, as shown in **Figure 14.13**.

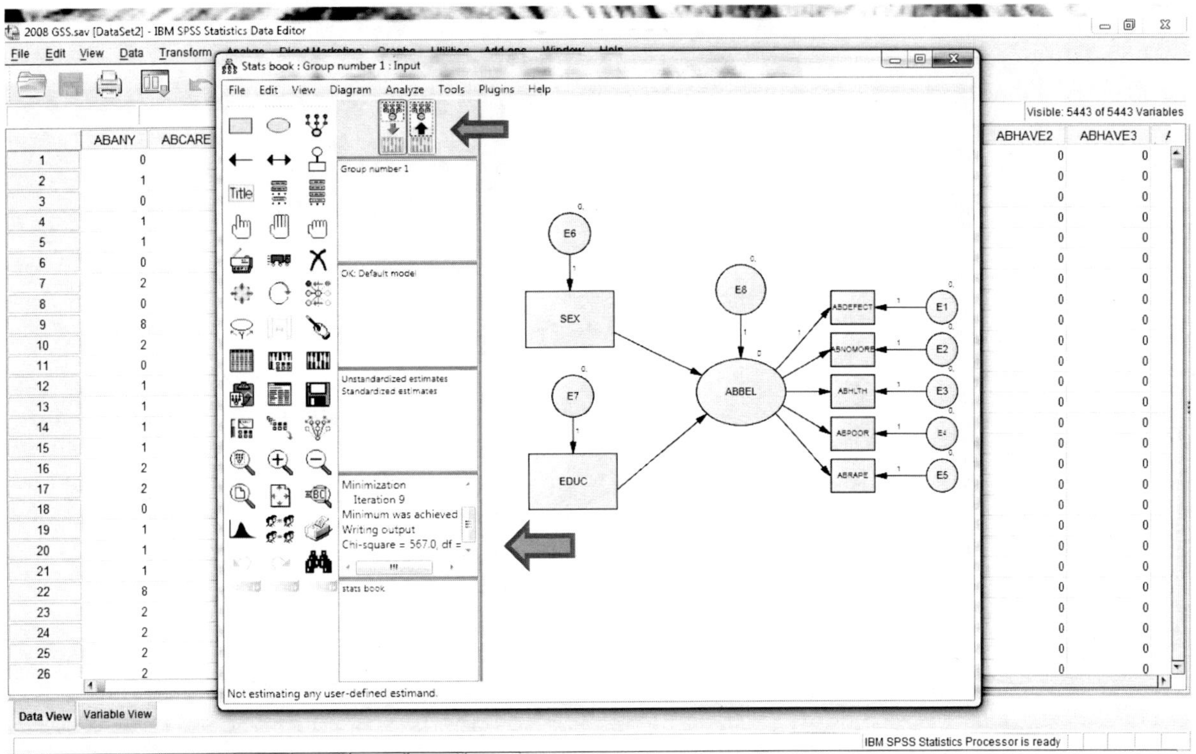

Select this tab and notice the change to the model.

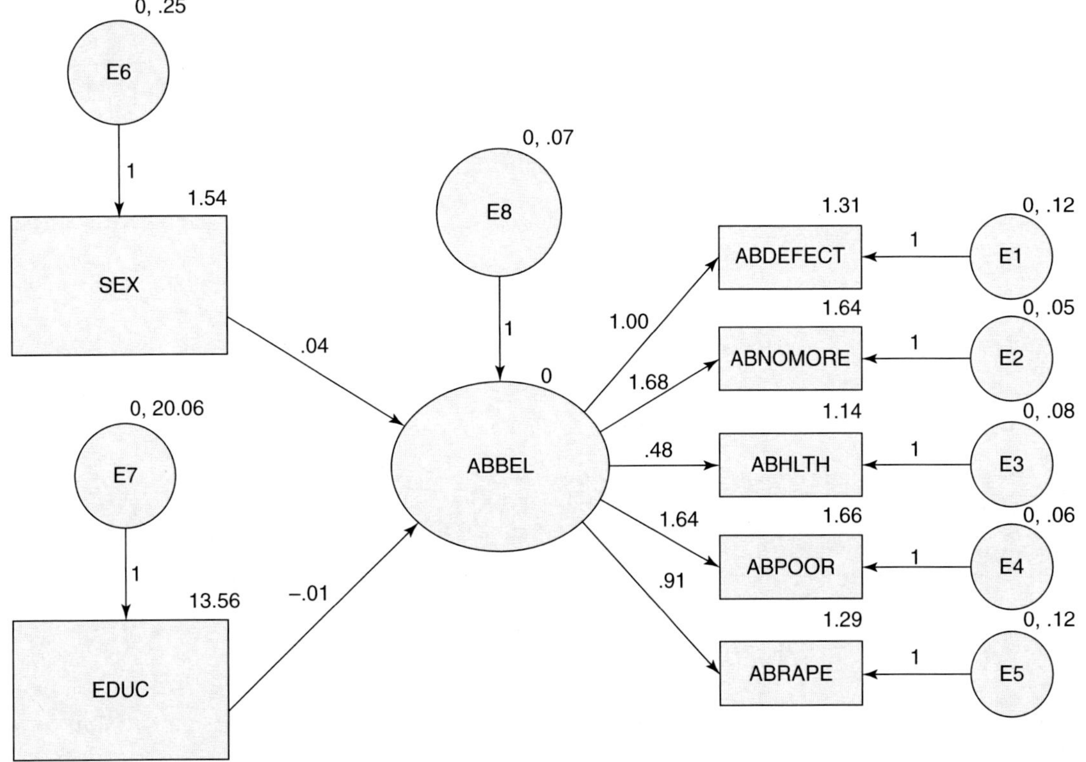

Interpreting the Output

The first thing that we interpret is the χ^2. As with factor analysis, the χ^2 is our model fit statistic. We want to have a small, nonsignificant χ^2. If we select Text Output from the View menu, we will find a full discussion of our results. Notes for Model provides the following table.

Table 14.1

Result (Default model)
Minimum was achieved
Chi-square = 567.035
Degrees of freedom = 14
Probability level = .000

As we can see, our χ^2 has a probability of .000, which means that it is significant. If we were using this analysis in research, this might send us back to the drawing board. Our independent variables are not good predictors for our dependent variable. There are ways to manipulate the analysis in SPSS when our fit indices are not adequate. However, this relies more on our data than on the theory that drives our model, so we will not discuss them here. Also, remember that the χ^2 is dependent on sample size. Thus, this is not always the best indicator of our model fit, so we need to look at a few more indices.

The next section to view is the Model Fit section. There are many statistics on this table, but we will only concern ourselves with three. The statistics that you interpret from this table are primarily dependent on two things. First, What are you interested in arguing? and second, Which discipline do you belong to? Sociologists typically use the Tucker Lewis Index (TLI), the Comparative Fit Index (CFI) and the Root Mean Square Error of Approximation (RMSEA). The TLI and CFI are listed under baseline comparisons, as shown in **Table 14.2.** We want these to be $\geq .95$. Since our TLI is .612 and our CFI is .806, you can see that we have further evidence that this model is not significant and a poor fit for our data.

Table 14.2 Baseline Comparisons

Model	NFI Delta1	RFI rho1	IFI Delta2	TLI rho2	CFI
Default model	.803	.606	.807	.612	.806
Saturated model	1.000		1.000		1.000
Independence model	.000	.000	.000	.000	.000

The RMSEA is found under its own list, shown in **Table 14.3.** We want this to be $\leq .05$. Since ours is .140, we clearly have not created a good model for our data.

Table 14.3 RMSEA

Model	RMSEA	LO 90	HI 90	PCLOSE
Default model	.140	.130	.150	.000
Independence model	.224	.217	.231	.000

At this point, you would go back to your theoretical understanding of the data and build a new model. This may include rethinking your latent variables and/or considering adding mediating variables or different independent variables.

We can test our assumptions by examining our estimates, found on the model and in the Model Estimates text output. Again, this table provides more data than we will use. The first table in this section provides our regression coefficients. Note that the estimate for Sex and ABBEL is .041. If you look back at the model, you will notice that the number .04 is listed there as well. This is the unstandardized estimate. The SE (.15) is the standard error and the CR is the critical ratio (.041/.15). The CR is the estimate/standard error, and we interpret it like a Z statistic. Finally, the p value is our probability that this CR would occur if the null hypothesis is tested. If we have a nonsignificant CR, this would indicate that this variable could be removed from our model. The decision to remove it would be based on several things. First, is the nonsignificance theoretically unexpected? If all the literature argues that X is a significant predictor and you devise a model that demonstrates that it is not, then you would want to rethink this. Assuming there is not a problem with the data and that you have conducted the analysis correctly, a nonsignificant finding in this example would be theoretically groundbreaking. The second reason to retain a nonsignificant variable is sample size. The CR is dependent upon a robust sample. When your sample size is small, you are far less likely to get significant p values.

Our estimate for EDUC is $-.009$. While both SEX and EDUC have significant estimates, they account for less than 5% of the variability in ABBEL, as shown in **Table 14.4**. This is why our fit parameters are not significant. **Table 14.4** also provides regression weights for our variables that create our latent variable. This provides similar results to those in Chapter 13, with ABHLTH and ABRAPE accounting for less change in our latent variable. AMOS 19 does not provide factor loadings. Rather, it provides weighted factor scores. These weighted scores take into account the amount of common variance that is explained by our latent construct.

Table 14. 4 *Regression Weights: (Group Number 1 - Default Model)*

			Estimate	SE	C.	p	Label
ABBEL	←	SEX	.041	.015	2.713	.007	
ABBEL	←	EDUC	−.009	.002	−4.982	***	
ABDEFECT	←	ABBEL	1.000				
ABNOMORE	←	ABBEL	1.681	.074	22.635	***	
ABHLTH	←	ABBEL	.482	.038	12.797	***	
ABPOOR	←	ABBEL	1.643	.073	22.559	***	
ABRAPE	←	ABBEL	.906	.054	16.775	***	

Also notice the error terms for our model. Some of them are very high. All of these indicators together mean that we need to consider a different model.

Reading the Model

The first thing you might think of when you examine the AMOS output is that there are a lot of numbers. What do they mean? Interpreting them is easier when you consider that SEM uses a series of regression models based on the paths that you created. Remember that we drew a unidirectional arrow between SEX and ABBEL. This means that our model predicts that the respondents' Sex will influence their overall pro-choice attitudes. When we look at our AMOS output, the first number we will examine is the one that rests over our line. This is the regression coefficient. As it is .04, we can say that the score of ABBEL increases .04 for an increase in Sex. Since females are 1, we can say that, in general, females answer slightly higher on the pro-choice indicators. When we look at our text output, we find that the relationship is significant, though extremely small. Thus, while there are sex differences in attitudes, it is not a strong predictor.

The next number to consider in the AMOS output is the number over the rectangle for the observed indicator. For SEX, it is 1.54, and for EDUC, it is 13.56. This number represents the variance for that variable. (See previous chapters for interpretation of variances.) Finally, each error term has an associated number. This number reflects the error associated for that specific variable when predicting the value of Y in our overall calculations. Higher levels of error mean less reliability in the coefficient as a predictor.

The last thing to interpret is the relationships between our observed variables and our latent variables. These numbers are also regression coefficients. They need to be explained. Remember, we are asking AMOS to calculate our latent variable (ABBEL) by measuring the communalities of these variables. AMOS still does this (and older versions of the program allow you to request the output of the factor analysis), but it does not give you the factor loadings. You can ask the program to give you factor weights, but they are less beneficial to understanding the data. The factor weights are used to calculate the regression coefficients in the model. A variable with a factor weight of .25 would have half as much influence in the model as one that has a factor weight of .5.

Most researchers don't discuss the regression coefficients for the indicators of the latent variables unless there is a problem with one of the indicators. With the newer version of AMOS, it might be more beneficial to test your model for the latent variables with factor analysis in SPSS prior to building them in AMOS. This is not necessary as long as you understand the concept of weighting, but we will not discuss weighting here.

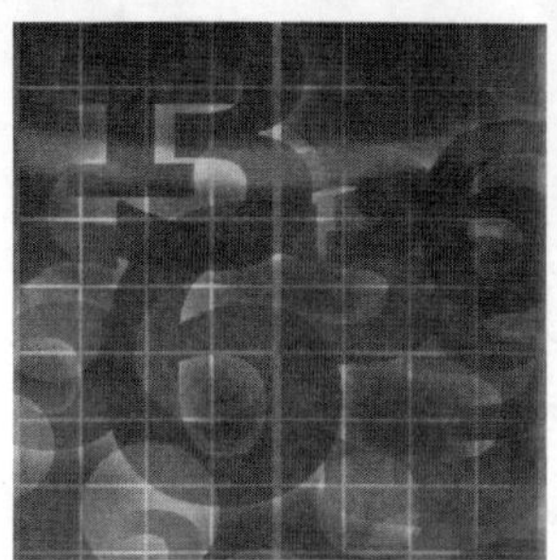

Exercises and Notes

Name: _________________________________   Date: _________________

1. _________________ is a variable that is measured during data collection while _________________ is a variable that is created using a data reduction method.

2. Gun ownership is dependent on a variety of social attitudes. This is an example of _________________ because we can predict it using other variables.

3. Age is an example of _________________ because it cannot be predicted using other variables.

4. A _________________ relationship exists when X has a direct effect on Y. A _________________ relationship exists when X has a spurious or suppressor effect on Y or when another variable suppresses the relationship between X and Y.

5. Variables that are found in the middle of a path analysis are known as _________________.

6. Consider the following path: $X_1 \rightarrow X_2 \rightarrow Y$. We can argue that X_2 has _________________ on Y because it contributes to the overall value of Y.

7. In the question 6, X_1 has _________________ because it contributes to the value of Y only in its relationship to X_2.

8. The following is an example of _________________ because the _________________.

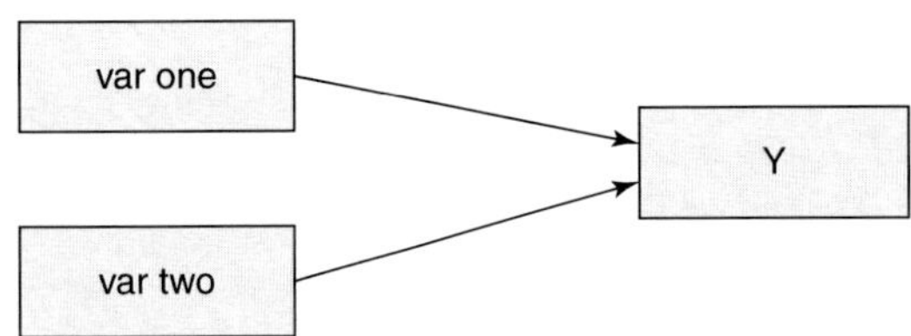

9. The following is an example of a _________________ because the _________________.

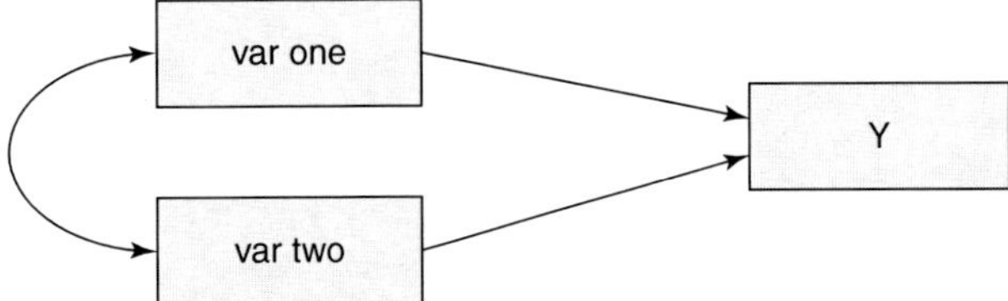

Use the following model to answer questions 10–14.

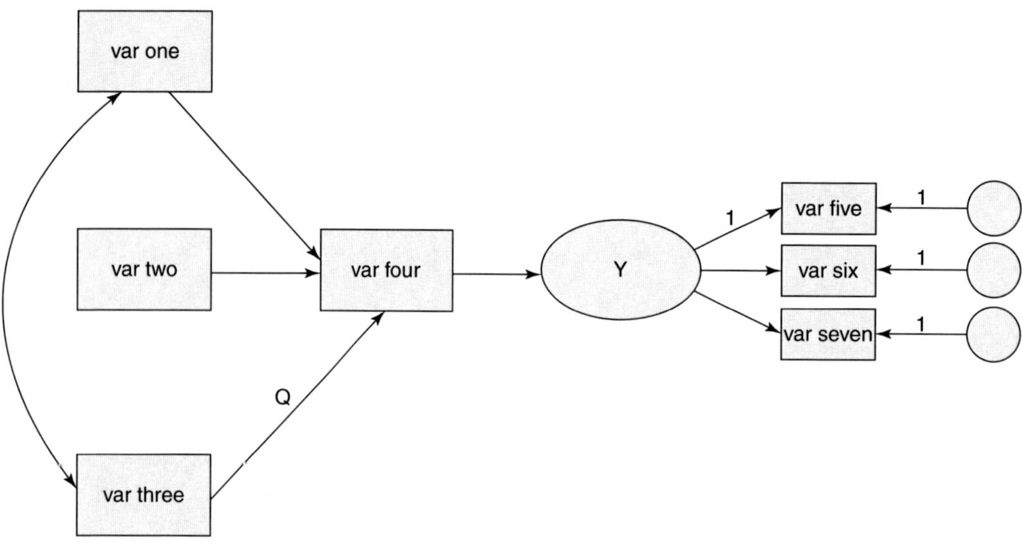

10. Which variable(s) is(are) latent in this path? How do you know?

11. Describe the role of variable four in this model.

12. Discuss the relationship between variables one and three.

13. Which variable(s) have an indirect effect on Y?

14. What does the letter Q on the model represent?

Name: _______________________________________ Date: ________________

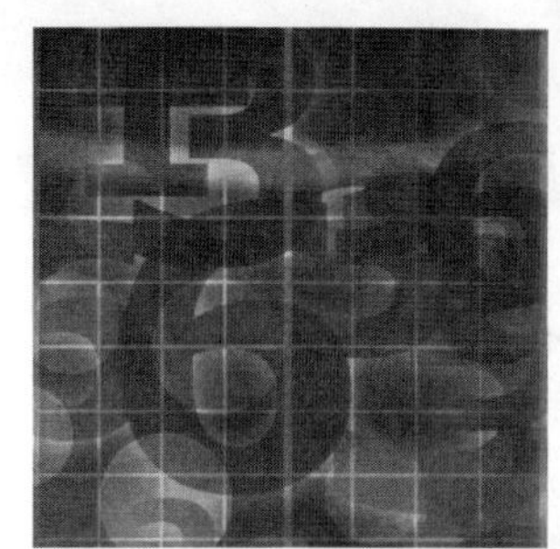

Learning Check

1. A data set has the following variables.

Sex
Age
Education
Income

How often do you pray?
How often do you attend church?
How important is religion in your life?
Were your parents married throughout your childhood?
Did you have dinner with your parents most nights?
Did your parents have consistent rules?
Did your parents make time for play?

Use these variables to construct a path model that explores the relationship between religiosity and family values. You must include at least one nonrecursive relationship, at least one latent variable, and one mediating variable. You must include both direct and indirect relationships. Submit your model with a short discussion of why you chose the design that you used.

Glossary

Addition rule the probability of multiple distinct outcomes is the sum of their individual probabilities.

Alternative (or research) hypothesis predicts a difference between two quantities or values.

Attitudinal variable measures only attitudes or beliefs.

Bar graph allows the comparison of how members of various groups differ on another variable. Along the horizontal (or X) axis of the bar graph are the categories of the variable, while the vertical (or Y) axis displays the frequency or percentage for each category of variables.

Causal relationships those relationships that have an effect on Y.

Central Limit Theorem With sufficient sample size, the sampling distribution of the means will be normal, even if the shape of the distribution in the population was not normal; the mean of the sample distribution equals the population mean; the standard deviation of the sampling distribution of means is smaller than the standard deviation in the population distribution.

Class interval each category of a group frequency distribution.

Coefficient of determination the r^2, which explains the amount of covariance between two variables.

Communality how much a latent variable accounts for the shared variability of responses in a data set.

Confidence interval a range of values within which the population may fall; used instead of using the single sample mean as an estimate of the population mean.

Confirmatory factor analysis (CFA) – used when a researcher already has a theoretical framework and hypothetical model that presuppose that there are latent variables at play in a set of data.

Constructs measures researchers develop when they believe that several variables are related or that several variables are neeed to understand the relationships between variables.

Continuous variables measure where values can fall anywhere on the number line.

Converse rule the probability of the event occurring; the probability of an event not occurring is 1.

Correlation coefficient a number used to determine if the independent variable and the dependent variable are related to one another.

Critical region the area, or areas, of the sampling distribution that indicates that one should reject the null hypothesis.

Cross-tabulation or cross-tab a cross-classification of one variable by at least one other variable.

Cumulative frequencies (cf) refer to the number of cases that have a given score *or less*.

Cumulative frequency (or percentage) polygon succeeding points are never lower and typically are higher than the previous point, and the highest point of a cumulative frequency is always the last point in the graph.

Data reduction reducing the number of variables in an analysis by combining related variables.

Degrees of freedom (df) measures of the unique information available when conducting a statistical test.

Dependent variable the variable that is expected to change, dependent on the values of the other variable(s) in an analysis.

Descriptive statistics - measures used to summarize the main characteristics of a data set.

Direct effect the amount of variation in Y that can be attributed to X.

Discrete variables measure where the values assigned are whole numbers.

Eigenvalue the total amount of variation across a sample that can be attributed to a component.

Endogenous variables those that are predicted by other variables.

Exhaustive having atleast one outcome or category for each case.

Exogenous variables those that are uses to predict other variables but are not predicted in the model.

Exploratory factor analysis (EFA) typically undertaken without a researcher having a theoretical framework or hypothetical test in mind.

Factor loadings individual measures of how much each variable contributes to the communality; interpreted like correlations.

Frequency distribution the number of cases that occur for each outcome of a variable.

Frequency polygon frequencies are displayed by points over the score values or the midpoint of each class interval, with each point connected with a straight line.

Histogram similar in appearance to a bar graph; in a bar graph, the bars depicting the different categories of the variable have space between each bar, while in a histogram, the bars touch each other.

Hypothesis a measurable and testable prediction of relationships between two or more variables.

Independent variable the variable that is believed will cause the change in the dependent variable.

Indirect effect the amount of variation that a variable causes through another variable.

Inferential statistics allow researchers to generalize to population by measuring properties of a sample and then deciding if those properties probably exist or do not exist in the population.

Interaction effect the combined effect of two or more variables on the dependent variable.

Interval data numbered data are consistent, and the distance between numbers is measurable.

Kaiser criterion a standard rule for using a cutoff of 1 for Eigenvalues because Eigenvalues < 1 means shared variability is not greater than variability for individual variables.

Latent variables -not observed but are inferred from the relationships that researchers believe exist in their data.

Lower confidence limit the lowest value of a confidence interval.

Margin of error the product of the Z score and the estimated standard error of the proportion.

Mean considered the most powerful measure of central tendency because it takes every case into account.

Median the middle point in an ordered raw-data array or frequency distribution.

Mediating variables those that mediate the relationship between X and Y.

Midpoint the middle score in a class interval, which provides the best single number to represent the class interval.

Mode the value that occurs most frequently in a distribution.

Multicollinearity two or more variables have a strong linear relationship, which indicates that they are measuring some common unknown characteristic of the sample.

Multiplication rule of probability the probability of a combination of independent outcomes is the product of their separate probabilities.

Mutually exclusive each case can be assigned to only one category.

Negative relationships the variables move in opposite directions; as one increases, the other decreases.

Nominal data -variables or constants that are complete and mutually exclusive, but the numbers are only a means of counting responses.

Normal curve the Bell curve; a theoretical probability distribution that is symmetrical and unimodal, having one peak.

Noncausal relationships those that have a spurious effect or a suppressor effect on Y.

Nonparametric tests tests used with data when researchers cannot assume a normal distribution.

Nonrecursive two or more variables are related or their error terms are related.

Null hypothesis predicts that there is no difference between two quantities or values.

Ordinal data data that can be rank ordered, using the $>$ and $<$ signs.

Ordinary least square (OLS) a method used to estimate the regression coefficients that minimizes the error sum of squares.

Orthogonal relationship X and Y are perpendicular to each other.

Outcome variable measures actual experiences rather than beliefs or attitudes.

Outlier a data point that differs significantly from the rest of the data in a set.

Parametric tests tests used with data whereby researchers can assume a normal distribution.

Percentage a proportion multiplied by 100.

Percentile rank the percentage of cases in a distribution that fall at or below a given score.

Pie chart the simplest graphical display that is useful for showing the frequencies or percentages of a nominal-level or ordinal-level variable; the categories are shown as part of a circle whose segments sum to 100 percent.

Population the entire group of individuals, objects, or events that the researcher is interested in.

Positive relationship means that the variables move in the same direction; as one increases, the other increases.

Principle component analysis an analysis used to find the location of the most shared variation and creates a slope.

Probability the likelihood that an outcome will occur.

Probability distribution a table that contains the possible categories for a variable and the likelihood that each category will occur.

Proportion the frequency of an outcome divided by the total number of cases.

Quantitative study uses large samples in order to test hypotheses that explain the characteristics of the group.

Qualitative study uses smaller samples and focuses on the quality of the experiences of the subjects.

Range the difference between the highest number and the lowest number in a data set.

Rate compares the actual number of events or occurrences to the potential frequency of events or occurrences (sometimes called the population at risk).

Ratio data 0 exists and actually has a real value, which means the absence of the value.

Ratio -directly compares the number of cases falling in one category with the number of cases falling into another category.

Recursive model all of the causal relationships are unidirectional, meaning that Xs are not correlated and neither are their error terms.

Relative frequency distribution the relative, as opposed to the absolute, number of times an outcome of a variable occurs in the data, thereby standardizing each frequency based on the total number of cases, *i.e,.* the N.

Research question is exploratory in nature; researchers ask a question when they don't know enough about the situation that they are interested in studying to make a prediction of a relationship.

Sample a subsct of the population.

Sampling distribution a theoretical probability distribution of all possible values for a particular statistic.

Sampling distribution of the means a theoretical probability distribution of sample means drawn from all possible samples of the same given size.

Sampling error a discrepancy between the population parameter and the sample statistics.

Standard deviation the average distance from the mean, or the square root of the variance.

Standard error of the mean ($\sigma_{\bar{x}}$) the standard deviation of the sampling distribution of means.

Strength the absolute value of the relationship between two variables.

***t* distribution** a family of curves, each based on the number of degrees of freedom.

Total association the noncausal effect plus the total effect.

Total effect the direct effect plus the indirect effect.

Type I error rejecting a true null hypothesis.

Type II error failing to reject a false null hypothesis.

Unit of analysis the nature of the sample *i.e.,* individuals, states, institutions.

Upper confidence limit the highest value of a confidence interval.

Variable a measure that varies.

Variance the average of the squared distance from the mean.

Varimax rotation the most commonly used rotation in factor analysis; allows researchers to maximize the variance in the data.

With replacement when an item is withdrawn, it is then replaced before another item is withdrawn.

Without replacement when an item is withdrawn, it is not replaced before another item is withdrawn.

Appendix A

Tables

Table A *Percentage of Area under the Normal Curve*

Column a gives the distance in standard deviation units from the mean (z). Column b represents the percentage of area between the mean and a given z. Column c represents the percentage at or beyond a given z.

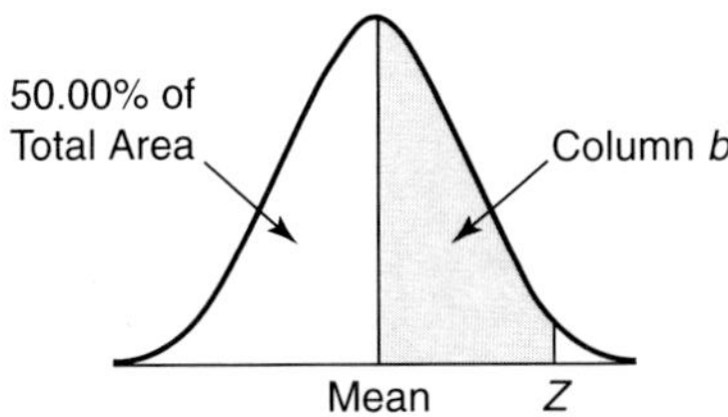

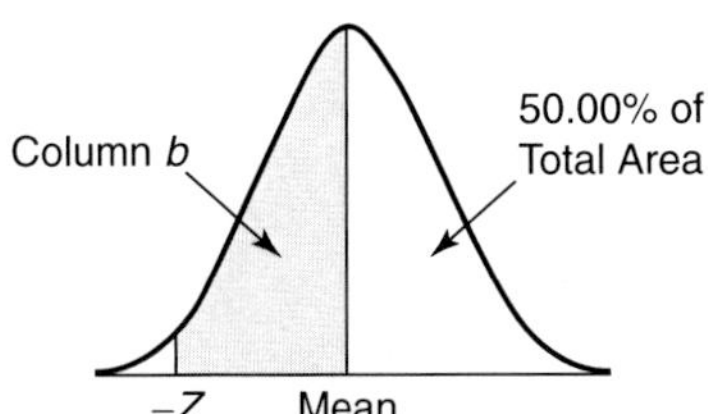

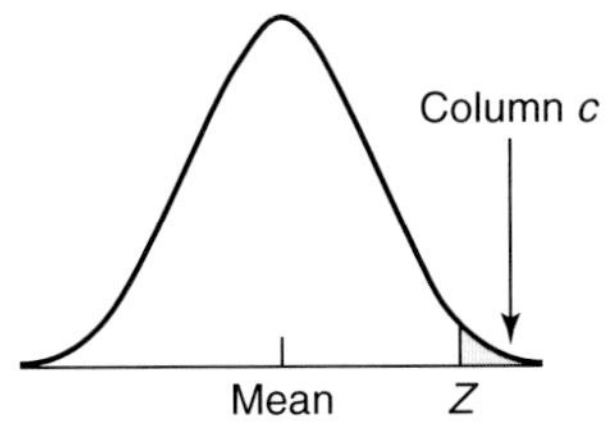

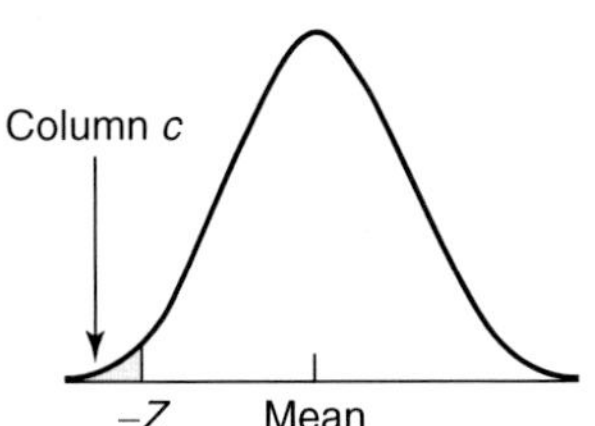

(*continued*)

Table A (*continued*)

(a) z	(b) Area between Mean and z	(c) Area beyond z	(a) z	(b) Area between Mean and z	(c) Area beyond z
.00	.00	50.00	.37	14.43	35.57
.01	.40	49.60	.38	14.80	35.20
.02	.80	49.20	.39	15.17	34.83
.03	1.20	48.80	.40	15.54	34.46
.04	1.60	48.40	.41	15.91	34.09
.05	1.99	48.01	.42	16.28	33.72
.06	2.39	47.61	.43	16.64	33.36
.07	2.79	47.21	.44	17.00	33.00
.08	3.19	46.81	.45	17.36	32.64
.09	3.59	46.41	.46	17.72	32.28
.10	3.98	46.02	.47	18.08	31.92
.11	4.38	45.62	.48	18.44	31.56
.12	4.78	45.22	.49	18.79	31.21
.13	5.17	44.83	.50	19.15	30.85
.14	5.57	44.43	.51	19.50	30.50
.15	5.96	44.04	.52	19.85	30.15
.16	6.36	43.64	.53	20.19	29.81
.17	6.75	43.25	.54	20.54	29.46
.18	7.14	42.86	.55	20.88	29.12
.19	7.53	42.47	.56	21.23	28.77
.20	7.93	42.07	.57	21.57	28.43
.21	8.32	41.68	.58	21.90	28.10
.22	8.71	41.29	.59	22.24	27.76
.23	9.10	40,90	.60	22.57	27.43
.24	9.48	40.52	.61	22.91	27.09
.25	9.87	40.13	.62	23.24	26.76
.26	10.26	39.74	.63	23.57	26.43
.27	10.64	39.36	.64	23.89	26.11
.28	11.03	38.97	.65	24.22	25.78
.29	11.41	38.59	.66	24.54	25.46
.30	11.79	38.21	.67	24.86	25.14
.31	12.17	37.83	.68	25.17	24.83
.32	12.55	37.45	.69	25.49	24.51
.33	12.93	37.07	.70	25.80	24.20
.34	13.31	36.69	.71	26.11	23.89
.35	13.68	36.32	.72	26.42	23.58
.36	14.06	35.94	.73	26.73	23.27

Table A (*continued*)

(a) z	(b) Area between Mean and z	(c) Area beyond z	(a) z	(b) Area between Mean and z	(c) Area beyond z
.74	27.04	22.96	1.11	36.65	13.35
.75	27.34	22.66	1.12	36.86	13.14
.76	27.64	22.36	1.13	37.08	12.92
.77	27.94	22.06	1.14	37.29	12.71
.78	28.23	21.77	1.15	37.49	12.51
.79	28.52	21.48	1.16	37.70	12.30
.80	28.81	21.19	1.17	37.90	12.10
.81	29.10	20.90	1.18	38.10	11.90
.82	29.39	20.61	1.19	38.30	11.70
.83	29.67	20.33	1.20	38.49	11.51
.84	29.95	20.05	1.21	38.69	11.31
.85	30.23	19.77	1.22	38.88	11.12
.86	30.51	19.49	1.23	39.07	10.93
.87	30.78	19.22	1.24	39.25	10.75
.88	31.06	18.94	1.25	39.44	10.56
.89	31.33	18.67	1.26	39.62	10.38
.90	31.59	18.41	1.27	39.80	10.20
.91	31.86	18.14	1.28	39.97	10.03
.92	32.12	17.88	1.29	40.15	9.85
.93	32.38	17.62	1.30	40.32	9.68
.94	32.64	17.36	1.31	40.49	9.51
.95	32.89	17.11	1.32	40.66	9.34
.96	33.15	16.85	1.33	40.82	9.18
.97	33.40	16.60	1.34	40.99	9.01
.98	33.65	16.35	1.35	41.15	8.85
.99	33.89	16.11	1.36	41.31	8.69
1.00	34.13	15.87	1.37	41.47	8.53
1.01	34.38	15.62	1.38	41.62	8.38
1.02	34.61	15.39	1.39	41.77	8.23
1.03	34.85	15.15	1.40	41.92	8.08
1.04	35.08	14.92	1.41	42.07	7.93
1.05	35.31	14.69	1.42	42.22	7.78
1.06	35.54	14.46	1.43	42.36	7.64
1.07	35.77	14.23	1.44	42.51	7.49
1.08	35.99	14.01	1.45	42.65	7.35
1.09	36.21	13.79	1.46	42.79	7.21
1.10	36.43	13.57	1.47	42.92	7.08

(*continued*)

Table A *(continued)*

(a) z	(b) Area between Mean and z	(c) Area beyond z	(a) z	(b) Area between Mean and z	(c) Area beyond z
1.48	43.06	6.94	1.85	46.78	3.22
1.49	43.19	6.81	1.86	46.86	3.14
1.50	43.32	6.68	1.87	46.93	3.07
1.51	43.45	6.55	1.88	46.99	3.01
1.52	43.57	6.43	1.89	47.06	2.94
1.53	43.70	6.30	1.90	47.13	2.87
1.54	43.82	6.18	1.91	47.19	2.81
1.55	43.94	6.06	1.92	47.26	2.74
1.56	44.06	5.94	1.93	47.32	2.68
1.57	44.18	5.82	1.94	47.38	2.62
1.58	44.29	5.71	1.95	47.44	2.56
1.59	44.41	5.59	1.96	47.50	2.50
1.60	44.52	5.48	1.97	47.56	2.44
1.61	44.63	5.37	1.98	47.61	2.39
1.62	44.74	5.26	1.99	47.67	2.33
1.63	44.84	5.16	2.00	47.72	2.28
1.64	44.95	5.05	2.01	47.78	2.22
1.65	45.05	4.95	2.02	47.83	2.17
1.66	45.15	4.85	2.03	47.88	2.12
1.67	45.25	4.75	2.04	47.93	2.07
1.68	45.35	4.65	2.05	47.98	2.02
1.69	45.45	4.55	2.06	48.03	1.97
1.70	45.54	4.46	2.07	48.08	1.92
1.71	45.64	4.36	2.08	48.12	1.88
1.72	45.73	4.27	2.09	48.17	1.83
1.73	45.82	4.18	2.10	48.21	1.79
1.74	45.91	4.09	2.11	48.26	1.74
1.75	45.99	4.01	2.12	48.30	1.70
1.76	46.08	3.92	2.13	48.34	1.66
1.77	46.16	3.84	2.14	48.38	1.62
1.78	46.25	3.75	2.15	48.42	1.58
1.79	46.33	3.67	2.16	48.46	1.54
1.80	46.41	3.59	2.17	48.50	1.50
1.81	46.49	3.51	2.18	48.54	1.46
1.82	46.56	3.44	2.19	48.57	1.43
1.83	46.64	3.36	2.20	48.61	1.39
1.84	46.71	3.29	2.21	48.64	1.36

Table A *(continued)*

(a) z	(b) Area between Mean and z	(c) Area beyond z	(a) z	(b) Area between Mean and z	(c) Area beyond z
2.22	48.68	1.32	2.59	49.52	.48
2.23	48.71	1.29	2.60	49.53	.47
2.24	48.75	1.25	2.61	49.55	.45
2.25	48.78	1.22	2.62	49.56	.44
2.26	48.81	1.19	2.63	49.57	.43
2.27	48.84	1.16	2.64	49.59	.41
2.28	48.87	1.13	2.64	49.59	.41
2.29	48.90	1.10	2.64	49.59	.41
2.30	48.93	1.07	2.65	49.60	.40
2.31	48.96	1.04	2.66	49.61	.39
2.32	48.98	1.02	2.67	49.62	.38
2.33	49.01	.99	2.68	49.63	.37
2.34	49.04	.96	2.69	49.64	.36
2.35	49.06	.94	2.70	49.65	.35
2.36	49.09	.91	2.71	49.66	.34
2.37	49.11	.89	2.72	49.67	.33
2.38	49.13	.87	2.73	49.68	.32
2.39	49.16	.84	2.74	49.69	.31
2.40	49.18	.82	2.75	49.70	.30
2.41	49.20	.80	2.76	49.71	.29
2.42	49.22	.78	2.77	49.72	.28
2.43	49.25	.75	2.78	49.73	.27
2.44	49.27	.73	2.79	49.74	.26
2.45	49.29	.71	2.80	49.74	.26
2.46	49.31	.69	2.81	49.75	.25
2.47	49.32	.68	2.82	49.76	.24
2.48	49.34	.66	2.83	49.77	.23
2.49	49.36	.64	2.84	49.77	.23
2.50	49.38	.62	2.85	49.78	.22
2.51	49.40	.60	2.86	49.79	.21
2.52	49.41	.59	2.87	49.79	.21
2.53	49.43	.57	2.88	49.80	.20
2.54	49.45	.55	2.89	49.81	.19
2.55	49.46	.54	2.90	49.81	.19
2.56	49.48	.52	2.91	49.82	.18
2.57	49.49	.51	2.92	49.82	.18
2.58	49.51	.49	2.93	49.83	.17

(continued)

Table A *(continued)*

(a) z	(b) Area between Mean and z	(c) Area beyond z	(a) z	(b) Area between Mean and z	(c) Area beyond z
2.94	49.84	.16	3.15	49.92	.08
2.95	49.84	.16	3.16	49.92	.08
2.96	49.85	.15	3.17	49.92	.08
2.97	49.85	.15	3.18	49.93	.07
2.98	49.86	.14	3.19	49.93	.07
2.99	49.86	.14	3.20	49.93	.07
3.00	49.87	.13	3.21	49.93	.07
3.01	49.87	.13	3.22	49.94	.06
3.02	49.87	.13	3.23	49.94	.06
3.03	49.88	.12	3.24	49.94	.06
3.04	49.88	.12	3.25	49.94	.06
3.05	49.89	.11	3.30	49.95	.05
3.06	49.89	.11	3.35	49.96	.04
3.07	49.89	.11	3.40	49.97	.03
3.08	49:90	.10	3.45	49.97	.03
3.09	49.90	.10	3.50	49.98	.02
3.10	49.90	.10	3.60	49.98	.02
3.11	49.91	.09	3.70	49.99	.01
3.12	49.91	.09	3.80	49.99	.01
3.13	49.91	.09	3.90	49.995	.005
3.14	49.92	.08	4.00	49.997	.003

Table B *Random Numbers*

Row	1	2	3	4	5	6	7	8	9	10	11	12	13	14	15	16	17	18	19	20
1	9	8	9	6	9	9	0	9	6	3	2	3	3	8	6	8	4	4	2	0
2	3	5	6	1	7	4	1	3	2	6	8	6	0	4	7	5	2	0	3	8
3	4	0	6	1	6	9	6	1	5	9	5	4	5	4	8	6	7	4	0	2
4	6	5	6	3	1	6	8	6	7	2	0	7	2	3	2	1	5	0	9	4
5	2	4	9	7	9	1	0	3	9	6	7	4	1	5	4	9	6	9	8	0
6	7	6	1	2	7	5	6	9	4	8	4	2	8	5	2	4	1	8	0	5
7	8	2	1	3	4	7	4	6	3	0	7	5	0	9	2	9	0	6	1	1
8	6	9	5	6	5	6	0	9	0	7	7	1	4	1	8	3	1	9	3	5
9	7	2	1	9	9	8	0	1	6	1	6	2	3	6	9	5	5	8	4	2
10	2	9	0	7	3	0	8	9	6	3	3	8	5	5	6	5	2	0	9	1
11	9	3	5	4	5	7	4	0	3	0	1	0	4	3	3	9	5	3	2	2
12	9	7	5	7	9	4	8	6	8	7	6	1	6	8	2	5	5	5	3	8
13	4	1	7	8	6	8	1	0	5	8	8	6	1	6	8	2	9	0	4	9
14	5	0	8	3	3	4	5	4	4	2	5	3	0	4	9	6	1	2	3	2
15	3	5	0	2	9	4	1	0	0	3	9	0	5	8	6	0	9	9	6	2
16	0	3	8	2	3	5	1	0	1	0	6	8	5	2	4	8	0	3	8	6
17	1	7	2	9	1	2	7	8	4	7	0	3	3	1	5	8	2	7	3	9
18	5	0	5	7	9	5	8	7	8	9	3	5	3	4	4	6	1	1	3	3
19	7	7	3	3	5	3	6	1	3	2	8	5	4	1	4	8	3	9	0	4
20	1	0	9	1	3	8	2	5	3	0	3	8	0	9	3	3	0	4	5	1
21	1	3	8	5	1	8	5	9	4	1	9	3	9	3	6	5	9	8	4	9
22	8	6	4	7	8	7	5	9	4	1	9	3	9	3	6	5	9	8	4	6
23	0	6	9	6	5	1	0	3	2	6	7	7	4	9	6	0	3	4	0	7
24	7	6	7	4	7	0	8	3	8	7	3	2	5	1	2	4	2	9	7	8
25	3	2	3	8	1	3	1	8	7	4	5	9	0	0	2	4	1	2	1	7
26	9	2	1	6	4	2	3	8	7	6	2	6	2	6	4	8	1	0	1	0
27	3	7	4	2	2	8	1	7	8	0	6	0	0	0	3	2	2	9	7	4
28	0	7	8	0	8	5	1	5	2	6	5	8	7	5	3	0	5	9	6	2
29	7	4	2	3	3	2	6	0	0	6	5	2	2	3	6	3	9	0	4	0
30	1	8	2	7	5	9	5	3	6	5	2	9	9	1	1	7	3	4	3	0
31	4	3	1	8	7	0	6	0	8	6	5	0	1	0	4	0	6	1	5	5
32	8	5	8	0	6	1	4	1	2	0	4	4	1	4	7	6	3	5	1	0
33	4	5	8	5	0	4	5	8	3	9	2	8	7	8	9	0	8	4	3	5
34	5	0	2	5	4	9	2	2	1	1	0	0	5	4	8	7	6	4	0	1
35	0	8	1	7	0	6	3	3	4	7	6	2	6	8	9	3	4	1	4	0
36	2	5	9	3	4	6	0	7	5	2	0	0	9	6	0	8	2	2	5	6
37	2	1	3	1	3	7	8	9	8	4	9	3	8	0	2	2	1	8	1	6
38	3	8	8	6	8	5	1	3	3	4	6	7	2	6	3	4	8	6	7	9
39	0	9	9	8	5	9	8	4	4	2	2	1	1	0	1	7	6	1	3	1
40	2	2	3	5	3	9	7	4	4	2	1	4	0	5	8	2	3	0	8	5

(continued)

Table B (*continued*)

									Column Number											
21	22	23	24	25	26	27	28	29	30	31	32	33	34	35	36	37	38	39	40	Row
9	7	1	1	9	1	2	7	3	5	1	8	4	0	4	1	0	6	0	3	1
3	7	7	9	1	4	9	9	5	9	2	0	1	6	1	2	6	6	7	0	2
5	6	3	7	8	3	3	8	4	3	9	3	9	0	0	9	8	3	5	2	3
7	0	8	6	6	5	9	6	2	7	3	5	9	0	1	8	0	9	6	9	4
9	8	7	3	5	6	8	8	1	2	0	2	3	2	6	4	3	1	9	7	5
1	8	8	4	7	0	1	7	6	8	2	1	6	3	2	1	8	1	8	3	6
3	7	8	6	9	5	4	1	7	3	8	7	1	5	6	5	6	4	3	6	7
9	0	1	5	2	8	6	5	5	7	8	1	8	7	1	2	4	0	4	1	8
2	5	5	2	1	8	6	9	8	9	8	0	5	8	9	9	4	1	3	4	9
3	4	2	8	5	0	7	9	8	4	3	5	8	0	9	4	6	6	0	5	10
6	8	6	6	4	7	1	5	1	6	4	6	7	6	0	8	7	3	5	2	11
6	0	1	4	2	9	8	6	8	0	7	6	5	1	9	1	3	7	0	3	12
5	7	0	9	8	7	6	9	0	6	5	4	0	3	6	5	6	3	5	0	13
2	3	4	7	8	0	2	0	8	0	3	4	9	2	5	7	7	8	6	4	14
4	6	1	0	5	0	6	1	4	9	4	7	3	9	1	7	6	4	5	8	15
3	4	8	1	6	9	5	6	2	0	4	6	1	6	8	1	9	9	1	1	16
0	5	1	3	6	1	9	5	4	1	2	5	4	2	9	5	6	2	4	0	17
6	7	0	3	5	3	7	4	1	7	5	4	8	3	7	4	8	5	7	2	18
3	6	6	3	6	3	0	0	9	4	2	2	5	1	8	9	5	1	9	7	19
0	6	9	0	2	7	3	9	8	4	0	6	9	8	2	3	2	8	0	4	20
1	3	5	7	9	6	2	4	3	4	6	4	9	1	3	1	7	5	2	2	21
4	2	2	2	1	4	5	2	2	8	3	2	1	2	6	6	0	1	8	9	22
2	6	9	0	7	5	3	2	5	6	2	7	6	3	8	1	4	1	5	1	23
2	8	2	4	4	4	2	9	1	9	8	3	4	4	1	0	4	6	9	6	24
3	1	4	3	0	4	7	1	3	7	4	8	6	7	3	2	6	6	2	0	25
6	4	5	8	3	1	4	8	1	8	3	1	6	4	3	0	2	8	7	3	26
2	2	8	3	2	1	9	3	0	1	7	5	9	0	9	1	2	5	8	2	27
9	8	7	2	0	6	4	0	2	7	1	3	1	6	8	7	0	9	2	5	28
8	0	5	6	8	2	4	3	6	1	3	5	2	3	5	9	8	6	2	1	29
1	7	6	1	5	7	9	0	3	5	3	4	2	4	8	5	6	4	0	6	30
1	9	8	5	2	4	5	1	7	5	3	2	4	6	7	9	9	6	7	2	31
3	6	6	3	7	8	6	9	7	2	8	9	0	7	2	9	4	0	8	6	32
0	0	0	2	0	8	9	0	1	0	6	2	0	4	6	9	6	5	4	9	33
9	4	4	2	6	4	2	4	1	0	2	7	9	6	8	7	5	6	9	3	34
0	5	3	8	3	2	7	5	0	4	7	6	4	6	3	0	4	7	5	3	35
2	6	2	0	6	0	1	4	8	9	6	5	9	7	3	6	7	6	5	4	36
3	9	0	3	5	0	9	1	2	0	5	9	7	3	2	5	9	3	0	2	37
7	3	3	5	4	0	6	4	9	4	7	9	1	4	3	9	7	7	1	8	38
8	6	2	9	4	2	9	7	0	3	8	9	5	7	0	6	9	7	2	5	39
9	4	5	8	6	2	3	0	6	2	9	8	6	3	0	4	1	0	7	6	40

Table C *Critical Values of t*

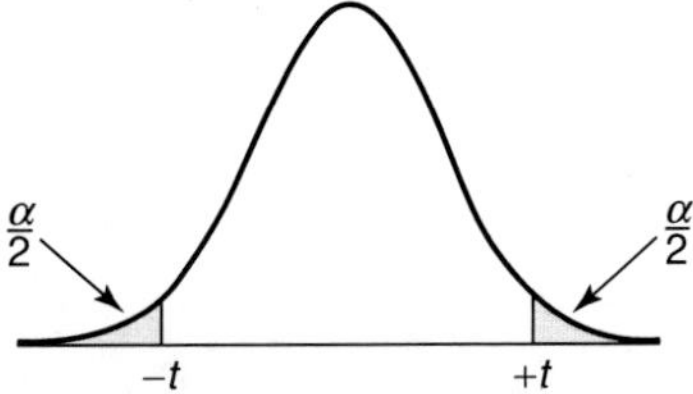

For any given df, the table shows the values of t corresponding to various levels of probability. Obtained t is significant at a given level if it is *larger than* the value shown in the table.

df	*Level of Significance (α)*					
	.20	.10	.05	.02	.01	.001
1	3.078	6.314	12.706	31.821	63.657	636.619
2	1.886	2.920	4.303	6.965	9.925	31.598
3	1.638	2.353	3.182	4.541	5.841	12.941
4	1.533	2.132	2.776	3.747	4.604	8.610
5	1.476	2.015	2.571	3.365	4.032	6.859
6	1.440	1.943	2.447	3.143	3.707	5.959
7	1.415	1.895	2.365	2.998	3.499	5.405
8	1.397	1.860	2.306	2.896	3.355	5.041
9	1.383	1.833	2.262	2.821	3.250	4.781
10	1.372	1.812	2.228	2.764	3.169	4.587
11	1.363	1.796	2.201	2.718	3.106	4.437
12	1.356	1.782	2.179	2.681	3.055	4.318
13	1.350	1.771	2.160	2.650	3.012	4.221
14	1.345	1.761	2.145	2.624	2.977	4.140
15	1.341	1.753	2.131	2.602	2.947	4.073
16	1.337	1.746	2.120	2.583	2.921	4.015
17	1.333	1.740	2.110	2.567	2.898	3.965
18	1.330	1.734	2.101	2.552	2.878	3.922
19	1.328	1.729	2.093	2.539	2.861	3.883
20	1.325	1.725	2.086	2.528	2.845	3.850
21	1.323	1.721	2.080	2.518	2.831	3.819
22	1.321	1.717	2.074	2.508	2.819	3.792
23	1.319	1.714	2.069	2.500	2.807	3.767
24	1.318	1.711	2.064	2.492	2.797	3.745
25	1.316	1.708	2.060	2.485	2.787	3.725
26	1.315	1.706	2.056	2.479	2.779	3.707
27	1.314	1.703	2.052	2.473	2.771	3.690
28	1.313	1.701	2.048	2.467	2.763	3.674
29	1.311	1.699	2.045	2.462	2.756	3.659
30	1.310	1.697	2.042	2.457	2.750	3.646
40	1.303	1.684	2.021	2.423	2.704	3.551
60	1.296	1.671	2.000	2.390	2.660	3.460
120	1.289	1.658	1.980	2.358	2.617	3.373
∞	1.282	1.645	1.960	2.326	2.576	3.291

Note: The bottom row (df = ∞) also equals critical values for z.

Table D Critical Values of F at the .05 and .01 Significance Levels

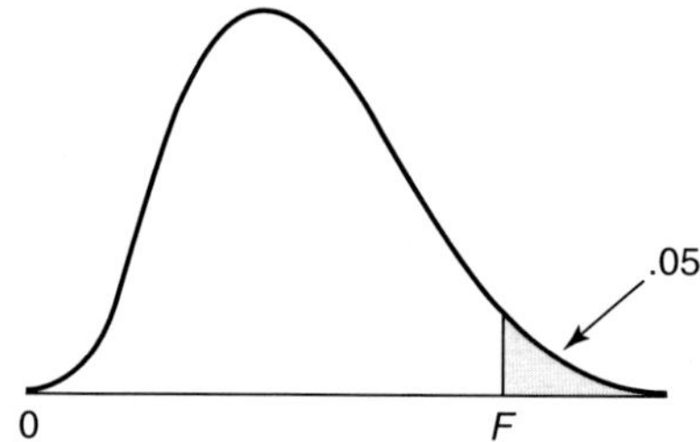

df for the Numerator α = .05

df for the Denominator	1	2	3	4	5	6	8	12
1	161.4	199.5	215.7	224.6	230.2	234.0	238.9	243.9
2	18.51	19.00	19.16	19.25	19.30	19.33	19.37	19.41
3	10.13	9.55	9.28	9.12	9.01	8.94	8.84	8.74
4	7.71	6.94	6.59	6.39	6.26	6.16	6.04	5.91
5	6.61	5.79	5.41	5.19	5.05	4.95	4.82	4.68
6	5.99	5.14	4.76	4.53	4.39	4.28	4.15	4.00
7	5.59	4.74	4.35	4.12	3.97	3.87	3.73	3.57
8	5.32	4.46	4.07	3.84	3.69	3.58	3.44	3.28
9	5.12	4.26	3.86	3.63	3.48	3.37	3.23	3.07
10	4.96	4.10	3.71	3.48	3.33	3.22	3.07	2.91
11	4.84	3.98	3.59	3.36	3.20	3.09	2.95	2.79
12	4.75	3.88	3.49	3.26	3.11	3.00	2.85	2.69
13	4.67	3.80	3.41	3.18	3.02	2.92	2.77	2.60
14	4.60	3.74	3.34	3.11	2.96	2.85	2.70	2.53
15	4.54	3.68	3.29	3.06	2.90	2.79	2.64	2.48
16	4.49	3.63	3.24	3.01	2.85	2.74	2.59	2.42
17	4.45	3.59	3.20	2.96	2.81	2.70	2.55	2.38
18	4.41	3.55	3.16	2.93	2.77	2.66	2.51	2.34
19	4.38	3.52	3.13	2.90	2.74	2.63	2.48	2.31
20	4.35	3.49	3.10	2.87	2.71	2.60	2.45	2.28
21	4.32	3.47	3.07	2.84	2.68	2.57	2.42	2.25
22	4.30	3.44	3.05	2.82	2.66	2.55	2.40	2.23
23	4.28	3.42	3.03	2.80	2.64	2.53	2.38	2.20
24	4.26	3.40	3.01	2.78	2.62	2.51	2.36	2.18
25	4.24	3.38	2.99	2.76	2.60	2.49	2.34	2.16
26	4.22	3.37	2.98	2.74	2.59	2.47	2.32	2.15
27	4.21	3.35	2.96	2.73	2.57	2.46	2.30	2.13
28	4.20	3.34	2.95	2.71	2.56	2.44	2.29	2.12
29	4.18	3.33	2.93	2.70	2.54	2.43	2.28	2.10
30	4.17	3.32	2.92	2.69	2.53	2.42	2.27	2.09
40	4.08	3.23	2.84	2.61	2.45	2.34	2.18	2.00
60	4.00	3.15	2.76	2.52	2.37	2.25	2.10	1.92
120	3.92	3.07	2.68	2.45	2.29	2.17	2.02	1.83
∞	3.84	2.99	2.60	2.37	2.21	2.09	1.94	1.75

Table D *(continued)*

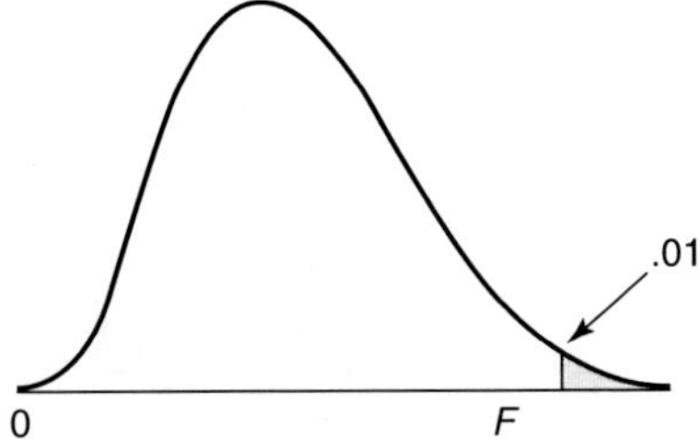

df for the Numerator $\alpha = .01$

df for the Denominator	1	2	3	4	5	6	8	12
1	4052	4999	5403	5625	5764	5859	5981	6106
2	98.49	99.01	99.17	99.25	99.30	99.33	99.36	99.42
3	34.12	30.81	29.46	28.71	28.24	27.91	27.49	27.05
4	21.20	18.00	16.69	15.98	15.52	15.21	14.80	14.37
5	16.26	13.27	12.06	11.39	10.97	10.67	10.27	9.89
6	13.74	10.92	9.78	9.15	8.75	8.47	8.10	7.72
7	12.25	9.55	8.45	7.85	7.46	7.19	6.84	6.47
8	11.26	8.65	7.59	7.01	6.63	6.37	6.03	5.67
9	10.56	8.02	6.99	6.42	6.06	5.80	5.47	5.11
10	10.04	7.56	6.55	5.99	5.64	5.39	5.06	4.71
11	9.65	7.20	6.22	5.67	5.32	5.07	4.74	4.40
12	9.33	6.93	5.95	5.41	5.06	4.82	4.50	4.16
13	9.07	6.70	5.74	5.20	4.86	4.62	4.30	3.96
14	8.86	6.51	5.56	5.03	4.69	4.46	4.14	3.80
15	8.68	6.36	5.42	4.89	4.56	4.32	4.00	3.67
16	8.53	6.23	5.29	4.77	4.44	4.20	3.89	3.55
17	8.40	6.11	5.18	4.67	4.34	4.10	3.79	3.45
18	8.28	6.01	5.09	4.58	4.25	4.01	3.71	3.37
19	8.18	5.93	5.01	4.50	4.17	3.94	3.63	3.30
20	8.10	5.85	4.94	4.43	4.10	3.87	3.56	3.23
21	8.02	5.78	4.87	4.37	4.04	3.81	3.51	3.17
22	7.94	5.72	4.82	4.31	3.99	3.76	3.45	3.12
23	7.88	5.66	4.76	4.26	3.94	3.71	3.41	3.07
24	7.82	5.61	4.72	4.22	3.90	3.67	3.36	3.03
25	7.77	5.57	4.68	4.18	3.86	3.63	3.32	2.99
26	7.72	5.53	4.64	4.14	3.82	3.59	3.29	2.96
27	7.68	5.49	4.60	4.11	3.78	3.56	3.26	2.93
28	7.64	5.45	4.57	4.07	3.75	3.53	3.23	2.90
29	7.60	5.42	4.54	4.04	3.73	3.50	3.20	2.87
30	7.56	5.39	4.51	4.02	3.70	3.47	3.17	2.84
40	7.31	5.18	4.31	3.83	3.51	3.29	2.99	2.66
60	7.08	4.98	4.13	3.65	3.34	3.12	2.82	2.50
120	6.85	4.79	3.95	3.48	3.17	2.96	2.66	2.34
∞	6.64	4.60	3.78	3.32	3.02	2.80	2.51	2.18

Table E   *Critical Values of Chi-Square at the .05 and .01 Levels of Significance* (α)

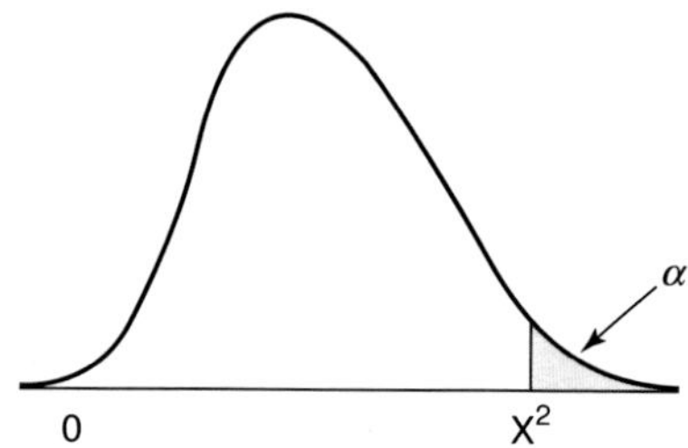

df	.05	.01	df	.05	.01
1	3.841	6.635	16	26.296	32.000
2	5.991	9.210	17	27.587	33.409
3	7.815	11.345	18	28.869	34.805
4	9.488	13.277	19	30.144	36.191
5	11.070	15.086	20	31.410	37.566
6	12.592	16.812	21	32.671	38.932
7	14.067	18.475	22	33.924	40.289
8	15.507	20.090	23	35.172	41.638
9	16.919	21.666	24	36.415	42.980
10	18.307	23.209	25	37.652	44.314
11	19.675	24.725	26	38.885	45.642
12	21.026	26.217	27	40.113	46.963
13	22.362	27.688	28	41.337	48.278
14	23.685	29.141	29	42.557	49.588
15	24.996	30.578	30	43.773	50.892

Table F *Critical Values of r at the .05 and .01 Levels of Significance (α)*

df	α .05	α .01	df	α .05	α .01
1	.99692	.999877	16	.4683	.5897
2	.95000	.990000	17	.4555	.5751
3	.8783	.95873	18	.4438	.5614
4	.8114	.91720	19	.4329	.5487
5	.7545	.8745	20	.4227	.5368
6	.7067	.8343	25	.3809	.4869
7	.6664	.7977	30	.3494	.4487
8	.6319	.7646	35	.3246	.4182
9	.6021	.7348	40	.3044	.3932
10	.5760	.7079	45	.2875	.3721
11	.5529	.6835	50	.2732	.3541
12	.5324	.6614	60	.2500	.3248
13	.5139	.6411	70	.2319	.3017
14	.4973	.6226	80	.2172	.2830
15	.4821	.6055	90	.2050	.2673

Table G *Critical Values of r_s at the .05 and .01 Levels of Significance (α)*

N	α .05	α .01	N	α .05	α .01
5	1.000	—	16	.506	.665
6	.886	1.000	18	.475	.625
7	.786	.929	20	.450	.591
8	.738	.881	22	.428	.562
9	.683	.833	24	.409	.537
10	.648	.794	26	.392	.515
12	.591	.777	28	.377	.496
14	.544	.714	30	.364	.478